The Ne[obscured by label]
of American
Foreign Policy

CONTRIBUTORS

W. Lance Bennett
I. M. Destler
Thomas W. Graham
Donald L. Hafner
Benjamin I. Page
Thomas Risse-Kappen
Bert A. Rockman
David W. Rohde
William Schneider
Robert Scigliano
Robert Y. Shapiro
John T. Tierney

The New Politics of American Foreign Policy

David A. Deese
BOSTON COLLEGE

ST. MARTIN'S PRESS
NEW YORK

Executive editor: Don Reisman
Managing editor: Patricia Mansfield-Phelan
Project editor: Alda Trabucchi
Production supervisor: Alan Fischer
Graphics: Academy Artworks
Cover design: Sheree Goodman
Cover photo: Copyright © Comstock Stock Photography

Library of Congress Catalog Card Number: 92-62743

Manufactured in the United States of America.
8 7 6 5 4
f e d c b a

For information, write:
St. Martin's Press, Inc.
175 Fifth Avenue
New York, NY 10010

ISBN: 0-312-09133-8 (paperback)
0-312-10267-4 (cloth)

Preface

Over the past ten to fifteen years, research has revealed striking new data and trends in the political forces underlying American foreign policy. *The New Politics of American Foreign Policy* synthesizes and presents these findings for the first time. Its purpose is to delineate the political changes transforming the American foreign policy process since the 1960s, which accelerated as opposition to the Vietnam War grew. The volume presents for students, experts, and general readers the specific ways in which domestic and global politics shape American foreign policy.

The book is the integrated work of thirteen leading scholars in American foreign policy, with many decades of experience as university teachers, who combined their efforts over a three-year period. Through intensive interaction in the American Political Science Association annual meetings and particularly the Thomas P. O'Neill Symposium at Boston College in 1992, the questions addressed by each chapter were focused and structured to encompass the broad scope of American foreign policy. The goal was to formulate a book with the coherency of a single-authored volume, but significantly more up to date and in depth. The multiple subfields of political science, history, and the social sciences generally, that are intrinsic to the U.S. foreign policy process no longer lend themselves to coverage by one or two authors. Understanding the American foreign policy process requires the collective knowledge of experts on public opinion, interest groups, the media, Congress, the bureaucracy, the president, comparative foreign policy, and international politics.

While fully recognizing the crucial role of the president, top officials, and the foreign policy bureaucracy, *The New Politics of American Foreign Policy* highlights the newer and fundamental roles of voters, interest groups, social and political movements, American and foreign experts, the media, and Congress. The organizing theme is that new issues, more actors, and a wider range of interests do not necessarily create a less effective foreign policy process. On the contrary, American foreign policy could be strengthened and improved by a more consensus-oriented process that is driven by "deliberative politics."

This book is intended for undergraduate and graduate courses in United States foreign policy, but it also addresses government officials, scholars, and general readers. It resists the tendency to "talk-down" to undergraduates. It is presented at a level that requires attention and careful reading, but it should be fully accessible to the college junior or to the sophomore with some background in history or the social sciences.

The entire project would not have been possible without the generous support of Boston College, particularly the Thomas P. O'Neill Chair in American Politics, and the intensive efforts of Kate Lynch and the staff of the Department of Political Science. I also gratefully acknowledge the invaluable advice and assistance of Donald Hafner, John Tierney, Mac Destler, William Schneider, and Pat Jacobs in discussing ideas and designing the conferences and volume. The book also reflects the advice and suggestions of government officials and experts across the country, as well as several university professors who participated in St. Martin's rigorous review process: John Garofano, University of Southern California; Scott Gartner, University of California at Davis; Richard Melanson, National War College; Brian Ripley, University of Pittsburgh; D. Michael Shafer, Rutgers University at New Brunswick; and David Tarr, University of Wisconsin at Madison. Our O'Neill Symposium in 1992 benefited in particular from the reviews and participation of Robert Art, Richard Eichenberg, Ellen Frost, Ronald H. Hinckley, Ellen Hume, Andrew Moravcsik, Brigette Nacos, Mark Peterson, Paul Peterson, Francis E. Rourke, and Barbara Sinclair. Executive Editor Don Reisman at St. Martin's provided skillful guidance throughout the project. Finally, I thank Pat, Heather, and Brian for their support and encouragement.

David A. Deese

Contents

Introduction: From Foreign Policy to "Politics as Usual"

In his valedictory speech delivered at Texas A&M University on December 15, 1992, President George Bush expressed concern that the United States might be tempted to abandon its world leadership role. "From some quarters, we hear voices sounding the retreat," the president declared. "We've carried the burden too long, they say, and the disappearance of the Soviet challenge means that America can withdraw from international responsibilities." Mindful of preserving his legacy, President Bush warned that "the alternative to American leadership is not more security for our citizens but less, not the flourishing of American principles but their isolation."[1]

Bush knew that foreign policy was one of the reasons he had just been defeated. It was not that the voters believed Bush's foreign policy had been a failure. They did not. Strong majorities gave Bush high marks for his accomplishments in foreign affairs, particularly the Persian Gulf War. But the same polls showed widespread resentment over the amount of attention Bush gave foreign affairs. The rap on Bush was that he cared too little about problems at home and too much about problems in the rest of the world. Instead of saving Bush's career, the victory over Iraq may have contributed to his demise. It provided a prime example of misplaced priorities.

"Our choice as a people is simple," President Bush advised. "We can either shape our times or we can let the times shape us. And shape us they will, at a price frightening to contemplate—morally, economically and strategically."[2]

Exactly where did this threat of withdrawal from the world come from? It did not come from Bill Clinton, the Democrat who had just defeated Bush. Clinton was just as much an internationalist as Bush. In 1991, both Governor Clinton and his running mate, Senator Albert Gore, Jr. (D-Tenn.) had broken with the majority of Democrats in Congress and supported going to war in the Persian Gulf. Clinton also broke with organized labor and endorsed the free trade agreement with Mexico.

During the 1992 campaign, Clinton tried to minimize his foreign policy differences with Bush. Clinton knew that foreign policy had been a problem for Democrats ever since the end of the Vietnam War. It was in the Democrats' interest for the voters to believe that there was no longer any real difference between the two parties on foreign policy—that the Democrats were just as tough as the Republicans, and maybe tougher. At one point, Clinton even moved to the right of Bush, urging the United States to "take the lead" in seeking United Nations authorization for air strikes in Bosnia. The idea was to neutralize the Democratic party's long-standing disadvantage on foreign policy. Clinton told reporters covering the campaign, "I just don't think that either side should play a lot of politics with this country's foreign policy. We ought to be working to develop a bipartisan foreign policy."[3]

It was not the Democrats who threatened to retreat from the world. It was the voters. What worried President Bush was that public pressure, applied equally to both parties, would shift the nation's agenda away from world leadership. With the end of the cold war, it is reasonable to expect Americans to have less interest in world affairs. After all, the nation has been through fifty years of struggle and sacrifice to save the world from the threat of totalitarianism—from Pearl Harbor in 1941 through the collapse of Soviet communism in 1991. The United States is now the world's only remaining military superpower. It faces sharply reduced military threats from abroad, but it has terrible problems at home. As David A. Deese explains in Chapter 1, some of these problems, including the threat of serious economic competition from abroad, are inseparable from foreign policy, even though many Americans would like to think they are not.

There is a more subtle and profound reason why Americans may turn away from the world, however. It is not simply a lack of interest in foreign affairs. It is also the fact that foreign policy has become far more political. Instead of being driven by a clear-cut, overriding sense of national interest, foreign policy is now becoming "politics as usual," as one of the authors of this volume puts it. That means Americans may no longer see foreign policy as a cause, an expression of great national purpose. Instead, they may see it as another area of complex contending interests, domestic as well as foreign. If that happens, Americans may not just lose interest in foreign policy. They may actively rebel against it, as they threatened to do in 1992.

Americans have been turning against "politics as usual" in record numbers. They have a taste for anti-Washington candidates like Jimmy Carter, Ronald Reagan, and Bill Clinton. Americans have voted for term limits and campaign reform whenever possible. They despise the political process and gravitate toward leaders who promise to rise above politics. Running as a pure outsider in 1992, Ross Perot managed to do better than any third-party presidential candidate has done in eighty years. Anger at politics as usual is fueling the revolt against Washington. If foreign policy turns into "politics as usual," then the United States may also end up with a revolt against world leadership—exactly as President Bush warned.

There are two reasons why foreign policy is turning into "politics as usual." One has to do with external changes in the world environment. The other has to do with changes in the domestic policy process.

The cold war created stability in the world. Many conflicts were contained because of their potential for spreading to additional countries or escalating to superpower confrontation. With the end of the cold war, the world could become less stable, less predictable, and more violent. In any case, it is becoming more difficult for the United States, or any other country, to control events. Instead of using force, the United States will have to rely more on bargaining and persuasion—that is to say, "politics"—in order to get other countries to do what it wants them to do.

Instead of one overriding east-west conflict, the new world order will be based on "mixed interests." Countries will find themselves allied on some issues and opposed on others. The United States and Russia have a mutual interest in preserving the military balance in Europe. The United States and western Europe want to prevent Japan from becoming too powerful economically. Russia and western Europe are worried about resurgent German military and economic power. The United States, Russia, and western Europe have a joint interest in ending the civil war in what used to be Yugoslavia. The United States, Russia, and Japan want to keep the European market open now that internal European trade barriers are coming down. The United States, western Europe, and Japan want to ensure the free flow of oil from the Persian Gulf. The United States and Japan share an interest in promoting the economic development of China. The United States favors the rapid development of Pacific rim export economies in South Korea, Taiwan, Singapore, and Hong Kong. Japan does not.

In other words, the world is moving toward complex competition in which there are fewer permanent enemies and fewer permanent allies. Americans found it easy to commit themselves to a world where there was a powerful, threatening enemy, a Hitler, a Stalin, or an imperial Japan. Now Americans are being called upon to adjust to a world of shifting allegiances. That world requires a subtle, complex, and sometimes devious style of diplomacy. It also requires a more political foreign policy, one that is based on a calculation of interests more than values.

Moreover, those interests are not simply national. The foreign policy environment is no longer the arena of contending national interests. It is increasingly dominated by international interdependence and transnational ties. As Deese observes in Chapter 1, those forces have "triggered the growth of domestic interests and groups increasingly at odds with traditional foreign policy goals." In a more complex world, foreign policy inevitably becomes more political.

It is not only the world that has changed. It is also the foreign policy process in the United States. This book focuses on the second set of reasons why foreign policy has become more political, namely, changes in the domestic policy environment. The overall picture is one of an erosion of consensus, a fragmentation of responsibility and a competition for influence. The result, according to Deese, is "a more political and less manageable foreign policy process."

Thus, John T. Tierney (Chapter 5) describes how "the politics of American foreign policy has become more like U.S. domestic policy—fractious and ideologically riven—and it yields policy outcomes riddled with multiple and contrary objectives." In Chapter 3, Bert A. Rockman shows how the environment of foreign policy making today has become more politicized.

The "always fragile" separation between domestic and foreign policy has become "increasingly permeable." Americans are seeing more conflicts between organized groups. As a result, Rockman argues, the role of the president as exclusive shaper of foreign policy has been reduced, although it still varies considerably from issue to issue.

In Chapter 2, Donald L. Hafner argues that, to be effective on a foreign policy issue, the president now has to assert dominance even within the executive branch, "which means besting his political competitors in gaining the attention and cooperation of the bureaucracy." In other words, Hafner writes, the president has to campaign "for the hearts and minds of his own subordinates."

This assessment would seem to conflict with Robert Scigliano's argument in Chapter 7 that "a new understanding of the president's war power has emerged since World War II: that the president may engage the country in war without the consent of Congress." Scigliano explains how, in the case of the Persian Gulf, that principle was broadened by the president to include collective action against an aggressor. And yet, Scigliano argues, even the best attempts of presidents to dominate in the use of force have not removed a fundamental role for Congress. At a minimum, Congress has reserved for itself the power to respond authoritatively to presidential decisions to use force. It has done so by debating and voting on whether to authorize the entry of American forces into military conflicts overseas. In the case of the Persian Gulf War of 1991, before the air war began both houses of Congress voted to approve the U.S. use of force as part of a United Nations action.

So, the doctrine that "the president knows best" does not always apply. The voters and the Congress are always ready to second-guess a president's use of force. Every decision to use force requires a careful political calculation. President Bush, for instance, made the calculation that the Persian Gulf War would be politically sustainable only if it stood in sharp contrast to the Vietnam War. Vietnam was a political war. Politicians placed strict limits on what could and could not be done. Bush made sure that the Persian Gulf War was a military war. The military was given a mandate to win a total victory as quickly and efficiently as possible. It did.

"By God," President Bush said the day Kuwait was liberated, "we've kicked the Vietnam war syndrome once and for all."[4] Sure enough, the Persian Gulf War ended with marine helicopters landing on the roof of the U.S. embassy in Kuwait—a dramatic contrast with the image from 1975 of U.S. helicopters evacuating the U.S. embassy in Saigon. What Americans learned in Vietnam was that the United States does not like to fight limited wars for political advantage. It prefers to fight total wars for total victory. The Persian Gulf experience did not contradict that lesson. It confirmed it.

Bush did not try to micromanage the war from the White House, and he did not let politics interfere with military decisions. Until the end. Gen. Norman Schwarzkopf argued that the president's decision to end the war after one hundred hours was essentially a political, not a military, decision.

He told the interviewer David Frost that Bush's decision to stop fighting "is one of those that historians are going to second-guess forever—why didn't we go for one more day versus why did we stop when we did, when we had them completely routed."[5]

The answer, of course, is politics. The White House was eager to pull American troops out as quickly as possible to protect the president's standing in the polls. The administration also wanted to preserve the territorial integrity of Iraq, even if the only way to do so was to leave a murderous dictator in power. And it wanted to avoid responsibility for what happened in Iraq, even if that meant abandoning the Kurdish and Shiite minorities to the tender mercies of Saddam Hussein.

So while the president's prerogative of using force has been enhanced, that decision may be more politically sensitive now than at any time in the past. The war-making environment, like the policy-making environment, has become much more intensely politicized. In his comparative essay (Chapter 11), Thomas Risse-Kappen finds American foreign policy much more sharply constrained by public opinion than either French or German foreign policy. He writes, "During the Persian Gulf crisis and war in which the president exerted strong leadership, George Bush's choices were nevertheless heavily influenced by the need to ensure public support."

In Chapter 9, Thomas W. Graham describes the "pervasive but now discredited elitist paradigm" relating public opinion and foreign policy, "that public opinion is volatile or moody, unstructured and poorly informed, changed through a top-down process and not particularly significant to decision making." The "emerging new paradigm" holds that public opinion is now a stable and independent actor in the process. Robert Y. Shapiro and Benjamin I. Page (Chapter 10) offer evidence of a "substantial degree of opinion stability" on foreign policy issues. They argue that public opinion responds to events "in ways that are regular, predictable, and generally sensible." As a result, they maintain, "The traditional view of little effect of opinion on foreign policy has to change."

The activation of public opinion is one reason for the more politicized environment of foreign policy making. There are others. According to David W. Rohde (Chapter 4), conflict over foreign policy has grown increasingly partisan over the years as a result of shifts in the mass electorate. Each political party has become more homogeneous ideologically (Democrats more liberal, Republicans more conservative), and the differences between the two parties in Congress has grown wider. That shift helps to explain why Congress has become increasingly more assertive in foreign and defense policy.

W. Lance Bennett argues in Chapter 8 that the press has also become a more active player. So "foreign policy has taken on a public relations, or media diplomacy, dimension of substantial proportions." Further complicating the process, according to I. M. Destler (Chapter 6), is the deep division within the government between two areas of foreign policy making—the security complex and the economic complex. According to Destler, decisions are often made in one sphere without examining repercussions in the other. The effort to isolate the security sphere from the more overtly political economic sphere has failed. Instead, security decisions have become embroiled in political controversy because of their unforeseen economic implications.

Case after case reveals a more fragmented, less autonomous, and more politicized foreign policy process. One might think the consequence would be more public involvement. But the opposite may be true. The more foreign policy looks like "politics as usual," the more likely Americans are to get turned off.

The 1992 presidential campaign was a good example of that turn-off. Foreign policy was conspicuously absent from the campaign. The public was not interested in foreign affairs, and that put both establishment candidates, George Bush and Bill Clinton, on the defensive. During the primaries, Bush and Clinton were both challenged by protest candidates. Jerry Brown and Pat Buchanan positioned themselves as antiestablishment populists. Isolationism is a theme with powerful populist appeal, and both Brown and Buchanan exploited it.

A 1991 National Opinion Research Center poll showed a 3 to 1 margin in favor of reducing foreign aid. More than 80 percent of respondents agreed that "we shouldn't think so much in international terms but concentrate more on our own national problems." True, a majority rejected the view that "the U.S. should mind its own business internationally and let other countries get along as best they can on their own." But a striking 44 percent endorsed this outright isolationist position.

George Bush was an ardent internationalist. That seemed to be one of his few hard-core convictions. Nevertheless, while Bush condemned Buchanan's isolationism, he never tried to engage Buchanan on the issues. Instead, he ran an ad in the Michigan primary campaign attacking Buchanan for driving an imported car.

Bill Clinton, too, is an internationalist. During the campaign, he called for a "strategy of American engagement" abroad. But he was careful to defend aid to Russia in terms of U.S. self-interest. Clinton said it would "save us billions in lower defense costs forever" and "increase trade opportunities dramatically." He also tried to sound tough on trade. Clinton promised that he would not approve a free trade agreement with Mexico without "the elevation of labor and environmental standards on the other side of the Rio Grande."[6]

As noted earlier, Clinton tried to minimize his foreign policy differences with Bush—and keep foreign policy out of the campaign. There were only shades of difference between Bush's internationalism and Clinton's internationalism. What they had in common was a sense of defensiveness, as both tried to contend with the pervasive mood of isolationism.

Bush seemed to favor a hegemonic role for the United States, along the lines of the Persian Gulf War. Bush said the United States would be the ultimate guarantor of world order and stability. The U.S. aim, according to a Bush administration document, would be "to discourage [other countries] from challenging our leadership or seeking to overturn the established political and economic order."[7] We'll supply the muscle; they'll supply the money. It was a vision of hegemony on the cheap.

Clinton's internationalism was more multilateral. He envisioned a "United Nations rapid-deployment force" to deal with threats to the peace. He was also more concerned with economic development, environmental protection, and human rights. Clinton's ideas for a "Democracy Corps" and "America Houses" involved very little money, however. It was a vision of democracy on the cheap.

Now that he is president, Clinton inherits two alternative models of U.S. military intervention. One is the Vietnam syndrome, in which every intervention looks like a potential quagmire ("Don't go in—it will be another Vietnam"). The other is the Persian Gulf syndrome, in which the United States prevails through a show of invincible force ("Win quickly and get out"). Vietnam was the formative experience of Clinton's youth. The Persian Gulf was the defining experience of the Bush administration. Whenever an issue of intervention comes up, Clinton will have to choose a syndrome.

The Bush administration did that in Somalia (another Persian Gulf) and in Bosnia (another Vietnam). What criterion did Bush use to distinguish the two cases? It was not national interest. The United States had a much stronger interest in Bosnia than in Somalia. Nor was it morality. "Ethnic cleansing," a policy of deliberate genocide, is at least as horrifying as starvation.

It was not, or not simply, public support. The intervention in Somalia certainly drew strong public support. In a December 1992 CNN–USA Today–Gallup poll, three-quarters of Americans said they approved of the decision to send U.S. armed forces to Somalia "as part of a United Nations effort to deliver relief supplies there." But 57 percent also felt that "U.S. armed forces should go into Bosnia as part of a United Nations effort to deliver relief supplies there" as well.

The Bush administration defined the difference in military terms: Somalia was simple, while Bosnia was complicated. As Secretary of State Lawrence S. Eagleburger put it, "In the case of what was Yugoslavia, it ought to be clear to everyone that the use of force as a means of bringing that war to an end would require far more in the way of troops and far more in the way of commitment."[8] Apparently, U.S. policy was to take on the easy interventions and say, "Remember the Gulf War." The nation would shrug off the difficult interventions and say, "Beware of another Vietnam."

The essential difference between Vietnam and the Persian Gulf, however, was political, not military. Once force is used and American lives are at stake, the American public's view is, "Keep politics out of it." In Vietnam, the country fought a proxy war to defend the South Vietnamese government, thereby deeply involving itself in that country's political conflicts. In the Persian Gulf, President Bush defined the U.S. mission as strictly military. The country fought to reverse a brutal act of aggression and uphold the rule of law (and, at the same time, protect the world's oil supplies). Bush insisted on keeping the war apolitical, even to the extent of ruling out an overthrow of Saddam Hussein.

The American public is sensitive to this distinction. By 2 to 1 in the December 1992 poll, Americans felt that the role of U.S. troops in Somalia should be "limited to delivering relief supplies there," a technical, military objective, rather than "attempt to bring a permanent end to the fighting," a political goal. In a letter to congressional leaders, President Bush promised that "U.S. armed forces will remain in Somalia only as long as necessary to establish a secure environment for humanitarian relief operations." Suppose the United States left and the country reverted to tribal warfare? It would no longer be a U.S. problem. Just as Saddam Hussein's brutality is no longer a U.S. problem as long as he complies with U.N. resolutions.

"I have two choices," Bill Clinton said in describing his approach to the New World Disorder. "We can either focus on these problems, come up with a decent policy and aggressively pursue it, or wait for it to explode."[9] The new president may discover that preemptive diplomacy is difficult because Americans do not want to get involved in other countries' problems. Once the situation explodes, however, narrowly aimed military intervention may be easier to sell—as long as Americans believe the United States can solve the problem and get out.

If foreign policy has been politicized, as this book argues, then it will be subject to many of the same populist pressures as domestic policy. And populist pressure in foreign policy is usually isolationist. In 1991, for example, congressional Democrats took two critical foreign policy votes. Both of them tended to reinforce the party's isolationist image. In January 1991, a majority of Democrats in both houses of Congress voted against giving President Bush authority to use force in the Persian Gulf. In May, most Democrats opposed giving the president fast-track authority to negotiate a North American free trade agreement.

President Clinton has to make the case for internationalism not only to the country but also to his own political party. "History's lesson is clear," President Bush said in his valedictory. "When a war-weary America withdrew from the international stage following World War I, the world spawned militarism, fascism and aggression unchecked. . . . But in answering the call to lead after World War II, we built from the principles of democracy and the rule of law a new community of free nations, a community whose strength, perseverance, patience and unity of purpose contained Soviet totalitarianism and kept the peace."[10] The issue the United States now faces, after victory in the Persian Gulf and victory in the cold war, is whether foreign policy will be an arena of high principle or become "politics as usual" with all the unfortunate consequences. Can President Clinton and Congress establish the consistency and strength of purpose President Bush was talking about as the political pressures become stronger and more debilitating?

William Schneider

Notes

1. *New York Times,* December 16, 1992, p. 1 and p. 25.
2. *New York Times,* December 16, 1992, p. 1 and p. 25.
3. *New York Times,* July 29, 1992, p. 1.
4. *New York Times,* February 28, 1991, p. 1 and p. 12.
5. Interview with David Frost: Norman Schwarzkopf—WETA (local public television station); March 1991; transcript p. 25–26.
6. *New York Times,* March 24, 1992.
7. *New York Times,* March 8, 1992, p. 1 and p. 14.
8. *Washington Post,* December 17, 1992, p. 1.
9. *Washington Post,* December 21, 1992, p. 1.
10. *New York Times,* December 16, 1992, p. 25.

Part I / The Global Environment

1 / The Hazards of Interdependence: World Politics in the American Foreign Policy Process

DAVID A. DEESE

Americans in the early decades of the cold war defined for themselves a role and image of global "leadership" that encouraged the belief that it was the United States which shaped international politics, not international politics which shaped the United States. There were occasional dissenters, such as J. William Fulbright, who worried that the global burdens of the nation were compelling fundamental changes in the Constitution's arrangement of checks and balances among the public, Congress, and the president. In the main, however, the view of the nation was echoed in the assertive confidence of its presidents that Americans were the masters of their own and the globe's fate: "With a good conscience our only sure reward, with history the final judge of our deeds, let us go forth to lead the land we love, asking His blessing and His help, but knowing that here on earth, God's work must truly be our own."[1]

Even with eroding American self-confidence in the wake of the Vietnam War, the oil shocks of the 1970s, and the rise of economic competition from Asia and Europe, Americans still have not fully grasped the reality that "the world out there" has not just disrupted lifestyles but is in fact gradually transforming the character of the nation. While it is more often observed by the 1990s that the U.S. economy is becoming part of a global marketplace, it is little appreciated that the American political system is also being forcefully shaped and molded by global forces.

The making of American foreign policy, according to the common wisdom, is a strategic, rational process, free of the heavy pressures and politics of domestic issues. As the president discovers a threat "out there" to U.S. ideals or interests, he gathers facts, formulates and weighs options, and deliberately selects an appropriate solution serving the national interest. Once the command is given for orderly policy execution, he is free to address the next challenge.

This view bears little resemblance, however, to the real world of foreign policy making, particularly during the post-Vietnam era. American presidents must at once address several audiences, both inside and outside of government, and inside and outside of the United States. While providing leadership for the foreign policy bureaucracy and Congress, they must also work closely with other heads of state, international organizations (both public and private), the media, and a wide range of persistent pressure groups. Equally important, their policies will often succeed or fail according to their ability to both follow and lead public opinion at home and even abroad. Indeed, increasingly in the 1990s presidential success in foreign policy will at times hinge on skillfully accepting co-leadership with foreign leaders, global organizations such as the United Nations, and even entrepreneurs in Congress.

The purpose of this chapter is to lay out the ways in which American politics, and foreign policy in particular, has been affected by its international environment. Somewhat like the period from the nation's founding to the early 1800s, the post-Vietnam era has witnessed deep, inevitable international impacts on U.S. domestic and foreign policy. But unlike the earliest years of the Republic, over the past two decades the nation's external and internal environments have become tightly enmeshed.

This important change is partly explained by significant but isolated international events. Crises and wars can trigger introspection and deliberate change in American political institutions and processes. Indeed, the Vietnam War and external economic shocks in the 1970s were critical catalysts for political change. However, to fully understand change in the American political system one must examine the unprecedented levels of international interdependence and transnational ties beginning in the late 1960s. These less visible, more routine external forces ushered in a new era in American politics, defined by a more political and less manageable foreign policy process, with significantly different opportunities for, and balances of, political influence.

This chapter analyzes the impact of international interdependence on American politics and policy making, first in historical perspective, and then through three brief case studies: (1) U.S. immigration in the 1970s and 1980s; (2) the U.S. movement for a "freeze" on nuclear weapons in the early 1980s; and (3) the politics of the U.S. dollar and exchange rate policy in the mid-1980s. The analysis and cases point to several key changes in American politics and policy making. In the comparative perspective of Chapter 11, the U.S. liberal democratic political system is uniquely endowed with a decentralized institutional structure and a strong, direct role of coalitions in the policy process, as well as the fragmentation of society along regional, ideological, and ethnic lines. These characteristics allow, even encourage, strong transnational linkages.

Thus the government finds itself at the nexus between domestic and international forces. "Central decision-makers strive to reconcile domestic and international imperatives simultaneously."[2] Inward and outward flows of ideas, people, money, goods, and services are generally monitored or regulated rather than controlled by the government. Yet, contrary to conventional wisdom, increasing private and public links to foreign individuals and firms, transnational institutions, and foreign governments are not diminishing the role or importance of the government.[3]

The greater the international penetration of U.S. society, economy, and politics, the greater the demands by different sets of opinion leaders, groups, and even regions for the government to mitigate or offset perceived harmful effects on values or interests or to encourage foreign trade and investment and avoid protectionist policies. The tension created by these opposing pressures is visible in those regions and sectors most involved in or affected not only by foreign trade, such as textiles, footwear, automobiles, and steel, but also by capital flows, immigration, and cross-national political movements. As a result, the 1980s witnessed a surge in the creation of international offices in traditionally domestic departments and agencies of the federal government. At the same time, international linkages have reduced domestic autonomy, or the effectiveness of the government's traditional public policy tools[4]—macroeconomic and otherwise—blurred the lines between domestic and foreign policy issues, and decreased foreign policy autonomy, or the ability to make policies mainly for foreign purposes rather than in response to domestic needs or demands. The result since the 1970s is a set of public policy challenges that are broader in scope and more complex than perhaps any period in American history.

Finally, these international pressures appear to have triggered increased attention, debate, and participation on related foreign and domestic issues among groups, elites, and at least segments of the public that follow foreign affairs. They have also contributed to significant changes in beliefs, coalitions, and institutions in the politics of American foreign policy, such as the increase in partisanship in Congress explained in Chapter 4. Those who gain from interdependence have not always been as well represented and organized as those who are most threatened by it. Yet the losers, for example the Midwest region in the 1980s, labor in the traditional manufacturing industries, and some of the least well-off in large cities, have generally not gained effective assistance from the U.S. public sector. Those industries most affected by foreign imports have usually won effective economic protection only after years of efforts and have sustained it for only a few years.[5]

Thus by the 1990s it makes even less sense than in the 1960s, for most purposes, to speak of the U.S. "national interest." Individuals and groups in different socioeconomic classes, economic sectors, or geographic regions are affected differently by international forces. Their interests, policy preferences, and opportunities for political participation will vary widely. It is difficult to predict whether the opponents or proponents of greater U.S. interdependence will prevail. Indeed it is likely that the outcome will vary across different issue or problem areas.

Therefore, when foreign policy issues are salient and there is no consensus, particularly among the highest officials, policy coherence or effectiveness is unlikely, unless the president has clear goals and a sustained interest, as explained in Chapters 2, 3, and 12. This relationship appears, in turn, to increase the level of critical judgments made by attentive publics as well as by opinion and coalition leaders about the government's performance in important policy areas such as international monetary and financial relations, trade, energy, immigration, and nuclear arms control. In sum, the issues have become more complex, posing greater challenges to the political system, and the foreign policy process may be more fragmented and less able to cope with the complexity.

POLITICAL AND HISTORICAL BACKGROUND TO THE MID-1960s

Over the period from 1789 to about 1820 the United States struggled for economic survival in an international environment controlled by Britain. Americans were almost completely at the mercy of the British not only for providing critical capital and manufactured imports but also for purchasing most U.S. raw material exports. Continued access to British markets was absolutely essential to Americans. Yet "by 1860, the United States was far less dependent on foreign trade and investment than it had been earlier," because it had developed a large and increasingly important national economy.[6] The basic shift from an agricultural to an industrial economy lessened the impact of international economic flows on domestic and foreign policy.

From the 1870s into the 1930s the nation made the transition to a position of international economic and financial leadership. Around the turn of the century Presidents William McKinley and Theodore Roosevelt established the real beginnings of an American foreign policy designed for leadership in the Western Hemisphere and beyond. Prior to World War I, when European economic interdependence was relatively high and Britain's dominant role in the international system was clearly in decline, the United States remained protectionist. Although it actively promoted the export of its goods beginning in the late 1880s, the process of lowering U.S. tariffs did not begin until 1913. From 1916 to 1920, and to a lesser extent in the 1920s, immigration into the United States was very high and U.S. trade dependence grew to relatively high levels, as its firms assumed a central role in providing manufactured goods to world markets. Increasingly the international economy depended not only on Britain but also the United States, as it emerged as a central power in establishing the liberal international political economic system of the 1920s.[7] Yet American domestic politics generally continued to dominate foreign policy concerns during the 1920s and 1930s.

By definition a great power has a high degree of flexibility and maneuvering space in not only setting but also shifting its strategic posture and policies. As the world's leading economic power by World War I and military power by the end of World War II, the United States was relatively immune to external influence and able to organize itself internally and determine foreign policy with relative independence of its international environment.

World War II triggered significant changes in American political institutions. In the mid- to late 1940s, for example, there were major reforms in

Congress, including the creation of the Joint Economic Committee and, in the executive branch, the establishment of the national security bureaucracy and the Council of Economic Advisers. These institutional changes were deliberate efforts by the government to respond to the international and domestic challenges of public sector management confronted during the Depression and war. The initiative for change in this period originated from inside the government.

By the 1950s the unprecedented international economic, financial, political, and military commitments made after World War II began to enmesh the United States in its international environment. U.S.- and British-led international regimes in trade, money, and finance were put into operation gradually during the most threatening period of cold war tension from the late 1940s to the mid-1950s, although the Bretton Woods international monetary system was not fully operational until 1958. Fundamental expansion of the U.S. president's role responded, in part, to international forces, particularly perceived threats.[8] There was serious domestic dissension in some periods over foreign policy, particularly during the McCarthy era in the early to mid-1950s, but relatively less direct effect on a continual basis of the international environment on U.S. domestic politics, as compared to the 1970s. The American government, led by a strong president, still generally managed the process of political debate and reform, which was centered within the institutions—the White House, the bureaucracy, Congress, and the political parties.

THE TRANSITIONAL ERA—"OUTSIDE-IN": THE DOMESTIC EFFECTS OF INTERDEPENDENCE

By the 1970s, however, with the erosion of U.S. international economic and financial predominance, international forces began to alter basic elements of American political ideas, institutions, and coalitions, including U.S. foreign policy process, to a degree not experienced since the early 1800s. Two sets of international forces are central: the international state system and the international political economy. First, the international state system, through both singular events and long-term structural shifts, affects American politics. The Vietnam War was the single most important international event leading to change in American politics. Beginning in the late 1960s, disillusionment with U.S. policies in the war contributed to basic changes in public and elite belief systems, interest group activity, and institutional roles.[9]

Singular "external shocks may force into view dissatisfaction with the choices offered by the existing political structure."[10] Often when the public opposes a policy, a reversal of the policy is enough to dissipate public discontent. But when an unpopular policy is sustained by policy makers long after the collapse of public support, as was the case with the Vietnam War, public opposition turns into skepticism of the policy makers who contrived the policy and more widely of the very political institutions that fostered both the policy makers and the policy. The change in American political institutions was centered in Congress. A large number of liberal democrats gained seats in the elections of 1968, 1972, and 1974, as a result, in part, of their opposition to the war. Many House members most opposed to the war were also central actors in the movement for House reform in the early 1970s. At the same time there was rapid turnover of committee chairs in the House. As explained in Chapters 4 and 5, the largest leadership and structural changes in congressional committees since the early to mid-1950s were enacted during the early to mid-1970s in the House and throughout the 1970s in the Senate.[11]

Simultaneously, the gradual, parallel process of international social, economic, and political interdependence and transnational relations enmeshed U.S. citizens and officials at all levels in a "web" of links across economies and societies, particularly among the western industrialized nations. These links exposed societies and economies to even greater domestic effects from global shocks such as the energy crises of the 1970s. While these nations grew at an average of about 5 percent per year during the 1960s, the volume of international trade grew at about 8.5 percent. Trade was accelerated by expanding international markets and emphasis on global business strategies, strong economic growth, trade liberalization, lower transportation costs, and technological improvements in many areas such as communications and marketing. An unprecedented proportion of U.S. firms engaged in international trade, and a large and growing share of all U.S. trade was conducted among the subsidiaries and branches of U.S. multinational corporations. The American economy became more dependent on the performance of foreign economies and companies than at any other time since the early 1800s. As illustrated in Table 1-1, by the 1990s U.S.-traded goods had become major components of total economic activity. Furthermore, the rate of increase in exports and imports as a percentage of total economic activity accelerated in the late 1960s and remained very high, despite a pause in the early 1980s and the problems posed by an overvalued dollar through the mid-1980s. Finally, the growth of exports "accounted for 25 percent of the growth in private industry jobs in the United States between 1986 and 1990."[12]

Despite the temptation to focus heavily on trade, even larger flows and rates of increase in transactions can be tracked in financial affairs. Table 1-2 documents the dramatically heightened role of foreign currencies and Special Drawing Rights (SDRs) in the official reserves held by the U.S. government. Tables 1-3 and 1-4 highlight the increasing stock of foreign assets in the United States and U.S. assets abroad. Both indicators grew sharply from low levels after World War II, but the rate of growth of foreign investment in U.S. markets has advanced much more rapidly since the late 1960s. The total stocks of foreign assets in the United States and U.S. assets abroad were

Table 1-1 / Exports and Imports and Total as % of U.S. GNP (constant 1982 dollars)

Year	*1935*	*1940*	*1945*	*1950*	*1955*	*1960*	*1965*	*1970*	*1975*	*1980*	*1985*	*1989*
Exports	4.7	5.2	2.6	4.9	5.2	5.8	6.3	7.4	9.6	12.2	10.1	14.4
Imports	5.7	4.1	4.0	4.6	5.2	6.1	6.5	8.6	8.9	10.4	13.0	15.7
Total	10.4	9.3	6.6	9.5	10.4	11.9	12.8	16.0	18.5	22.6	23.1	30.1

Source: U.S. Bureau of the Census, *Statistical Abstract of the United States, 1991.* (111th Ed.) Washington, D.C. 1991.

Table 1-2 / U.S. Reserve Assets: Foreign Currencies and SDRs as % of Total (billions of dollars)

Year	*Percentage*
1970	10
1975	15
1980	47
1985	47
1990	76

Source: U.S. Bureau of the Census, *Statistical Abstract of the United States, 1992.* (112th Ed.) Washington, D.C. 1992.

at about the same level by 1985, whereas only four years later the former exceeded the latter by about 50 percent. The share of foreign direct investment in total foreign investment flows into the United States has also been growing sharply. From 1985 to 1990 foreign direct investment in the United States grew at an average annual rate of 18 percent.[13] This growth is important because of the potential or perceived social, economic, and political effects of the foreign ownership or management of U.S. assets.

Most dramatic of all the changes is the absolute size and rate of growth of foreign exchange and global banking markets beginning in the 1970s in the wake of gradual easing of exchange controls by the industrialized countries and the growth of global securities markets in the 1980s.[14] By the late 1980s daily transactions in worldwide foreign currency markets alone were estimated to be in the range of $400–$500 billion. Persistent movements into and out of the U.S. dollar have affected not only domestic but also foreign policy autonomy, as shown in the case developed later in this chapter of the overvalued U.S. dollar in the mid-1980s, the associated trade deficit, and the severe pressures for protection from import-competing and export industries. Highly mobile international capital is likely to further integrate U.S. and global markets and accentuate the divergence of policy positions held by domestic sectors from those of international investors and traders.

Table 1-3 / U.S. Assets Abroad (trillions of dollars)

Year	*Amount*	*Increase (%)*
1950	32	
1960	86	
1965	120	40
1970	165	38
1975	295	79
1980	922	213
1985	1,174	27
(1990)	(1,764)	(64)

Source: OECD (Organization for Economic Cooperation and Development) *Economic Survey;* U.S. Bureau of the Census, *Statistical Abstract of The United States, 1991 and 1992.*

Table 1-4 / Foreign Assets in the United States (trillions of dollars)

Year	*Amount*	*Increase (%)*
1950	18	
1960	41	
1965	59	44
1970	107	81
1975	221	107
1980	542	145
1985	1,110	105
(1990)	(2,176)	(96)

Source: OECD *Economic Survey;* U.S. Bureau of the Census, *Statistical Abstract of the United States, 1991 and 1992.*

A useful social indicator of interdependence, demonstrated in Table 1-5, is the number of overseas telephone calls placed from the United States. Since the first overseas call was placed in 1927, the number increased slowly until World War II. Thereafter the number doubled roughly every five years until the late 1960s, when the rate of increase began to surge. Federal Communications Commission data on international minutes of U.S. telephone calls show, for example, that business use grew at 25–30 percent annually from 1989 to 1991. Experts expect continued high annual growth rates in this market well into the 1990s, while U.S. domestic markets grow more slowly.

Table 1-5 / Overseas Telephone Calls Placed from the United States (tens of thousands)

Year	*Number*	*Increase (%)*
1930	3	
1935	3	
1940	7	133
1945	40	47
1950	100	150
1955	170	70
1960	370	118
1965	800	116
1970	2,300	188
1975	6,200	170
1980	20,000	223
1985	41,200	106
(1990)	(120,070)	(191)

Source: U.S. Bureau of the Census, *Statistical Abstract of the United States, 1992.*

A final indicator, which is inherently social, economic, and political, is immigrants admitted to the United States each decade, along with estimates of illegal immigration. While the data in Table 1-6 do not include illegal immigrants, they still illustrate that the levels and rates of immigration in the 1970s and 1980s were unusually high. Adding estimated annual averages of 250,000 for illegal immigrants and 645,000 for legal immigrants from 1981 through 1989 provides a rough average for total immigration of about 900,000 per year. A Census Bureau report released in December 1992 calculated that almost 5 million people immigrated to the United States between 1985 and 1990.[15] This is a staggering number of people to be employed and integrated into even a very large society and economy. But it is also a channel through which beliefs and interests from, or against, their countries of origin are introduced into American politics, complicating U.S. foreign policy making. These pressures can be seen in Cuban Americans strongly opposed to normalizing U.S. relations with Cuba, the demonstrations by immigrant Haitians against military rule in Port-au-Prince, or the conflicting pressures of Haitians urging the government to stop turning Haitian boat people back on the high seas, set against the clamor of Cuban Americans in Florida arguing that their community cannot endure yet another wave of Caribbean refugees.

These elevated levels of social and economic interdependence also became domestic political issues as a result of structural changes across the economies of the industrialized nations. First, during the 1980s both Japan

Table 1-6 / Immigrants Admitted to the United States (thousands)

Period	*Total*	*Rate per 1,000 U.S. Residents*
1901–10	8,795	10.4
1911–20	5,736	5.7
1921–30	4,107	3.5
1931–40	528	.4
1941–50	1,035	.7
1951–60	2,515	1.5
1961–70	3,322	1.7
1971–80	4,493	2.1
1981–90	(7,338)	(3.1)

Source: *U.S. Bureau of the Census, Statistical Abstract of the United States, 1992.*

and the newly industrialized countries captured an increasing share of manufactured goods exports markets to the industrialized nations. This shift placed severe pressure on some U.S. and European import-competing industries. Second, some of the important cross-national differences among the industrialized nations, for example in manufacturing processes and labor costs, were reduced. This shift required firms to work harder at retaining the competitive edge with new technologies and global production strategies. Their comparative advantage, already changing at a faster pace, was affected by relatively smaller changes at home.[16] National, regional, and local economies were struggling to become ever more specialized. Increased entry into, or expansion of, the same industries across countries led by the 1970s to a severe global problem of surplus production capacities in several large industrial sectors such as steel, chemicals, and petroleum refining.

Third, where labor in the industrialized nations is less able to move between occupations and regions, exports from East Asia can be highly threatening because they compete with labor-intensive industries in areas with high unemployment. The American Midwest had difficult times in the 1980s across industries such as cars, tractors, consumer electronics, and machine tools. A fourth change is the increasing impact of global recession, crises, and instabilities, when hardship at home makes adjustment to international change more difficult. This impact includes, for example, the misalignment of exchange rates in the mid-1980s that caused serious damage to U.S. exports, such as machinery and agricultural products and import-competing industries. In the 1970s energy crises became highly visible and damaging in their economic and social impacts, which were caused by sharp, rapid price increases for crude petroleum and its products. The natural result of these changes, both alone and in combination, was more pressure from threatened industries, sectors, and regions for assistance and protection by the national governments.

In this increasingly specialized and complex international division of labor the United States was confronted by a "crisis of international competition,"[17] particularly relative to Japan. This crisis increasingly mobilized U.S. groups affected by competition from imports and surplus production capacity. At times defensive, even protectionist policies in sectors such as apparel, textiles, footwear, steel, and consumer electronics appeared necessary as relief valves for legitimate domestic reactions or as components of intentional trade policies, while at other times they seemed contrary to longer-term competitive adjustment. By the early 1980s "unemployment was concentrated in those industries with the highest levels of foreign competition."[18] Table 1-7 demonstrates that the United States was responsible for most of the overall increase in "hardcore" import barriers in the industrialized nations from 1981 to 1986. Despite the clear increase in U.S. trade protectionism from the mid-1970s, the overall response of the U.S. government was not definitively protectionist, due to the continuing U.S. presidential preference for liberal trade policies, the high and increasing stake of some U.S. firms in the international economy, and the cross-pressures developed from complex and shifting political coalitions in U.S. trade policy.

Increasingly, international forces and foreign governments acted directly on domestic society rather than, or as well as, through the government. Prominent pathways for this direct influence include advanced information and telecommunications technologies and the media's real-time presentation of events abroad, transnational political movements, immigrants, external economic shocks, and foreign agents and actors in the United States. Increased interdependence and pressure from international socioeconomic and political-economic forces decreased U.S. domestic and foreign policy autonomy. The U.S. government became more involved, but less effective, in managing both domestic and international forces. International events and threats, for example in oil markets or nuclear weapons and doctrine, caused higher visibility for key issues and sometimes shifts in public opinion or dissension among elites that could lead public policy in new directions.[19] Foreign governments with significant influence in the United States, particularly those of Israel, Japan, Taiwan, and South Korea, turned to indirect lobbying strategies to build coalitions, develop elite support, and mobilize attentive publics in order to multiply their diplomatic bargaining leverage.[20] Foreign governments have also successfully circumvented the government by hiring leading U.S. public relations firms to change their image among the media, coalitions, and attentive publics.

By the early 1970s the unprecedented degrees of international interdependence and transnational links contributed to strong internal tendencies toward political decentralization, pluralism, regionalism, and complication of public policy processes. As it had been from 1789 to 1820, the economy was increasingly dependent on foreign markets and decisions, but unlike during

Table 1-7 / Industrial Country Imports Subject to "Hard Core" Nontariff Barriers, 1981 and 1986 (percent)

Importer	*1981*	*1986*
European Community	32	36
Japan	51	51
United States	23	32
All industrial countries	32	37

Source: World Bank, *World Development Report, 1987* (New York: Oxford University Press, 1987), p. 142.

this period, international forces penetrated the society, economy, and political system at all levels and helped decentralize power at both the national and subnational levels.

As the effects of the international system and the world political economy on the United States grew, so, too, did individual and group demands for the government to soften threats, slow changes, and ease adjustment to social and economic dislocations. International influences and institutions do not displace as much as complicate and increase demands on the role of the government, even in the extraordinarily liberal, free market American case. Because regions of the country, industrial sectors, industries, even individual towns and firms were affected differently by globalization, they were divided in their preferences about how the public sector should respond. Such divisions create multiple, complex challenges and opportunities for government officials negotiating in "two-level games" of both domestic and international coalitions simultaneously.

Government activity and expenditure levels have increased in response, in part to the problems of economic decline and international competitiveness, particularly under higher levels of interdependence. Government spending in western industrial nations increased sharply from the 1970s to the 1980s.[21] In fact, these challenges not only stimulate new government activity but also, under some conditions, legitimate an expansion of roles into new socioeconomic and political-economic areas.[22]

At the same time, governments have not had an easy time managing the pressures of interdependence and transnationalism. For just as the challenges of governance have increased on several fronts, the resources available to governments have been seriously questioned. At least among the OECD nations, governments generally have been called upon to manage a wider range of increasingly difficult tasks with slower rates of growth in resources and more balanced budgets. International influences such as those from multinational corporations, western consumerism, and the revolution in information, telecommunications, and related media technologies can also complicate governance in at least two ways. First, they reduce the central role of the government as the gatekeeper between citizens and transnational or foreign interactions. Second, they serve as a mechanism for spreading the neoclassical or liberal beliefs in markets and capitalism that underlie the movement to reduce public sector roles in national economies.

CASE STUDIES

Three case studies were selected to illustrate and provide evidence for the main points of this chapter. Their purpose is to highlight some of the central structural trends in American politics caused in part by external forces since the early 1970s, as opposed to the 1950s and 1960s. While they represent a wide array of issue areas from the flow of people into the United States, to the transfer of ideas through transnational political movements in national security, and capital flows and the role of the dollar, they are not intended to represent a comprehensive set of cases. Their function is more to suggest trends and raise questions than to offer definitive proof.

U.S. Immigration in the 1970s and 1980s: "Holding a Finger in the Dike"

An important function for any government is regulating the composition and rate of people entering the country to work or live over an extended period of time. In particular, the U.S. government must try to find a proper balance between allowing newcomers in for humanitarian or economic reasons and preventing strain on the social fabric. This case will demonstrate, however, that by the 1970s the federal government appeared to be losing control over the flow of immigrants into the United States. This loss of control is likely to have important implications for future levels of U.S. immigration and for relations with Mexico and other countries in the hemisphere.

The United States was to a significant degree formed and reformed by waves of immigrants starting in the early seventeenth century. It appears to have experienced the largest and most varied flows of immigrants, including refugees, of any modern industrial nation. Net U.S. immigration since 1820, when systematic data collection began, has amounted to about 50 million people.

The era of mass migration, triggered by peace in Europe, started for the United States in 1815 and continued until the 1920s, when Congress passed highly restrictive legislation. As a result, the foreign-born U.S. population decreased sharply from about 15 percent in 1910 to about 5 percent in 1970.

In 1875 Congress began a long progression of restrictive immigration policies to exclude Chinese, contract labor, and other Asians, culminating in comprehensive legislation in 1917 that consolidated existing restrictions and banned persons of Asiatic ancestry. This legislation established the context for watershed legislation in 1921 enacting the basic national origins quota system that is still in place. It allowed a total of 350,000 immigrants per year, but was heavily tilted toward people from Germany, Britain, and Ireland and excluded tens of thousands of people on waiting lists who were born in Italy, Greece, Poland, and Portugal.

Presidents Harry Truman, Dwight Eisenhower, John Kennedy, and Lyndon Johnson each sought, but failed, to end the discriminatory national origins system for allocating annual immigration quotas, which excluded nonwhites from outside western Europe and the western hemisphere. Significant changes were, however, introduced by the Immigration and Nationality Act of 1965, implemented in 1968. It established preference for relatives of U.S. citizens and resident aliens, individuals with special skills, and refugees. Despite President Johnson's strong preference for allowing unlimited entry from the western hemisphere, Congress—pressured by various groups—insisted on an annual ceiling of 120,000 from the region. Thus the law reallocated, more than eased, the restrictions on immigration, but its emphasis on family members established an important and continuing principle.

The law was intended to limit annual immigration to a total of about 320,000, whereas actual levels from 1968 to the late 1970s averaged about 400,000. As shown in Table 1-6 earlier, legal immigration flows increased both in levels and rates in the 1970s and 1980s. Illegal immigration also increased over this period. In 1987 the federal agency responsible for executing U.S. immigration policy, the Immigration and Naturalization Service (INS), reported more than 1.2 million apprehensions for immigration violations. Based on this number, experts estimate that at least 200,000 immigrants arrived illegally in 1987. For 1988 and 1989 estimates of illegal immigration are closer to 300,000 per year. Despite increasing public and group pressure and government efforts to control immigration flows, including the Immigration Reform and Control Act of 1986 and the Immigration Reform Act of 1990, the levels and rates continue to escalate.

It appears that the level of flows to the United States "is related in important ways to the tension between legal and 'illegal' immigration."[23] The U.S. government has relatively little control over a huge pool of illegal immigrant workers. It is caught in the middle under pressure from strong groups, with conflicting interests, to (1) protect the rights of resident immigrants and minorities generally; (2) defend agricultural sector needs; and (3) regulate immigration flows. All three goals have existed since the late 1800s, but in the recent period the first and second appear to be dominant. There is heightened concern among U.S. publics and groups, including Latinos, about the economic impact of undocumented workers and their families, and this concern will likely continue to rise under the pressure of the emerging U.S.-Canadian-Mexican free trade agreement. Also of increasing concern is the fear of loss of autonomy and sovereignty, particularly by border states such as Texas, New Mexico, and California when, for example, they are obliged by the courts to provide school or hospital facilities on an equal basis to all within the community but lack the means or authority to prevent a crippling influx of immigrants. They naturally turn to Washington to lobby for assistance or restrictions to help address their problem, but their requests arouse objections from other groups and states with different needs.

The relative permeability of U.S. labor markets and society has both encouraged increasing immigrant (including refugee) flows and made it difficult for the government to monitor and regulate them, never mind enforce the exact legislated quotas. Government regulation is specifically thwarted by: (1) domestic forces—powerful agricultural interests that rely on seasonal labor and the individual and civil rights protections guaranteed in the United States; (2) international pressures—the high levels of U.S. interdependence and transnationalism and the differential U.S.-Mexican wages and employment opportunities; and (3) foreign policy—the acceptance of refugees for particular strategic or foreign policy, as well as humanitarian, purposes.

Particularly since the 1970s, most initiatives for change in U.S. immigration policy have begun in Congress under the pressure of contending interest groups.[24] Mexican Americans, civil liberties groups, labor organizations, growers of fruits and vegetables, and restrictionist and nativist groups have been the most active in the efforts to reform immigration law and policy. In addition to the growing importance of these groups, individual members of Congress have championed specific causes and legislation at key points in time. The courts have become actively involved in enforcing the rights of immigrants and since the 1970s have significantly reduced the federal executive and administrative powers to control aliens.

The U.S. Immigration and Naturalization Service has a narrow jurisdiction and is considered to be generally ineffective in controlling the entry of illegal workers. It has been further weakened by the intervention of elected officials, the increasing role of state governments and law, pressure from interest groups, and the new assertiveness of the courts.

Beginning in the 1970s, ever higher levels of immigration have prompted a wide range of interests to exert pressure on an executive and administrative system that has largely lost control of this policy area. Even within the government there appears to be an increasing diversity of interests among Congress, the federal judiciary, the INS, and state and local governments. With higher salience of both the domestic and international implications of immigration, additional agencies and committees engaged the issue. "Indeed, as one looks at both sending and receiving countries, it is striking to see how many actors have entered into the political struggle over immigration."[25]

A December 1992 Census Bureau report showed rapid increases over the past two decades in the number and proportion of foreign-born residents of the United States.[26]

Year	*Number*	*Percentage*
1970	9.6	4.7
1980	14.0	6.2
1990	19.8	7.9

This trend is most pronounced in California, with 21.7 percent foreign population (38.4 in Los Angeles), but also in New York with 15.9 percent, Florida with 12.9 (59.7 in Miami), and Texas with 9.0. By far the most important group of foreign-born residents, Mexican Americans, who account for 21.7 percent of the total, grew sharply from 2.2 million in 1980 to 4.3 million by 1990. As an ever larger percentage of the U.S. population comes from Hispanic roots and the percentage of immigrants in this group increases (32 percent in 1980; 40 percent in 1990), the electoral power of Mexican American and other Hispanic groups will keep growing and more local, state, and federal elected officials can be expected to help protect the rights of resident immigrants and their families. Particularly in Texas, California, and New Mexico, "Latino candidates in the 1970s and 1980s won elected offices at all levels of government at an unprecedented rate."[27] With greater fragmentation of interests and decentralization of power, the levels of interdependence and transnationalism in immigration will continue to increase and reduce the government's power to govern in this issue area.

The Commission on Agricultural Workers, a panel created by Congress in 1986 at the behest of the farm industry, observed in late 1992 that illegal immigration continues largely unchecked, particularly workers for fruit and vegetable farming. The 1986 law's sanctions against employers for hiring unauthorized workers have been undermined by a large black market in counterfeit documents. After five years of study the panel concluded that, with decreasing effectiveness over time, "the law has produced none of the major benefits that lawmakers expected, especially improvements in agricultural wages and working conditions."[28]

The increasing and less-controlled flows of immigrants into the United States demonstrate the depth of international interdependence and the fragmentation of national interests as well as the decentralization of political power in response to powerful interest groups. By protecting at least some of the rights of immigrants and minorities, U.S. society and law have reinforced the relative power of society and central coalitions over the government and paved the way for increased levels of interdependence and transnational linkages. The courts, state and local governments, and Hispanic populations and officials will increasingly make the key decisions defining the level and composition of immigration flows.

If the U.S.-Canadian-Mexican North American Free Trade Agreement (NAFTA) of 1992 is ratified and implemented during the 1990s, a new level of regional interdependence will challenge U.S. society, economy, and politics. For at least the first few years, NAFTA is likely to trigger even greater movement of Mexicans into the United States. New coalitions of domestic forces are likely to oppose increased immigration, but it is unclear how they would effectively reduce the inflow, both legal and illegal. The only apparent structural solution is significant improvement in Mexican wages and the standard of living, which would narrow the economic gap between the two nations and reduce migration to the United States over the long term.

One way for the U.S. government to address the issue is to follow the model of the European Community, which is establishing completely free movement of people among the member states in the context of the Single European Act of 1986 and the broader political-economic integration program of 1992. With higher levels of interdependence, the U.S. government can reassert itself by moving to higher levels of international policy coordination and formal, institutionalized arrangements such as NAFTA.

In any case, an important implication of the trends outlined above is that U.S. officials can be expected to carefully anticipate and take account of the likely political reactions of Mexican Americans and Mexico as they formulate policies toward immigrants and Mexico. For example, when the U.S. executive branch tries to clamp down on the flow of illegal immigrants from Mexico, in response to demands by Congress or border states or organized labor, it faces protests from the Mexican government, which allows its citizens to leave because emigration helps relieve severe unemployment problems in Mexico. Despite the domestic pressures, the U.S. government cannot ignore Mexican demands when it must have Mexico's cooperation on other urgent priorities such as controlling drug trafficking and importing crude petroleum from Mexico.

The U.S. Freeze Movement of the Early 1980s

An illuminating and important example of a transnational political movement, with major political effects in the United States, is the western European peace movement and U.S. freeze movement of the early 1980s. This case, as explained in part in Chapters 9 and 11, helps to show the "outside-in" political impacts in security as well as economic issue areas. While the exact chain of causality, particularly from Europe to the United States, is extremely difficult to establish, the sequence of events and transnational and transgovernmental influences back and forth between the United States and western Europe is quite clear.

An important baseline for comparison with this movement of the early 1980s is the case provided by nuclear weapons tests, radioactive fallout, and test ban negotiations from the late 1950s to the Limited (Nuclear) Test Ban Treaty of 1963. Both were periods with significant tension in U.S.-Soviet relations, U.S. public skepticism about the Soviet Union, and fears associated with nuclear weapons and the possibility of nuclear war. From the late 1950s there was consistent public concern about the effects of radioactive fallout on humans and activity by peace groups in support of negotiating a broad international agreement to halt the testing, particularly in the atmosphere, of nuclear weapons. As shown below, however, the public, interest group, congressional, and transnational pressures were considerably less important to U.S. nuclear policies and negotiations of the early 1960s than they were to those of the early to mid-1980s.

Sustained media and public attention to nuclear weapons tests began in several countries when the first full U.S. tests of the hydrogen bomb, at Bikini in 1954, caught Japanese fishermen on the ship *Lucky Dragon* and others at Bikini in the direct fallout. Significant public concern about the effects of fallout, organizational activity especially among concerned scientists, and pressure from other countries helped to get nuclear weapons test bans on the U.S., British, Soviet, and United Nations negotiating agenda. During the mid- to late 1950s the actions of U.S. and British scientists, science advisers, and transgovernmental coalitions, including the U.S. Department of State, were important to keeping nuclear test bans on the U.S. foreign policy agenda, swaying the debate among U.S. political elites, and exerting pressure for key compromises and policy shifts during the U.S.-British-Soviet negotiations.[29] Foreign influence, perhaps particularly transgovernmental, was significant in part because U.S. executive departments and agencies, as well as congressional committees, have been deeply divided over the merits of test bans over the entire era since World War II.

Nevertheless, under the constraints of fear and mistrust generated by the cold war and U.S. anticommunism, it took a crisis of unprecedented proportions—the Cuban missile crisis of October 1962—and the personal commitment and determined leadership of President John Kennedy and General Secretary Nikita Khrushchev to begin the détente process and forge the first test ban agreement in 1963. In the months immediately after the crisis Kennedy decided that the time had come for decisive action to stabilize the nuclear balance of terror, reduce the risk of war, and begin the process of halting the spread of nuclear weapons to additional nations. After favorable communications between Kennedy and Khrushchev, in June Kennedy delivered the speech at American University that established the basic political groundwork for détente, in both the United States and Soviet Union. After eight years in the negotiating process, the United States, Britain, and the Soviet Union signed the Limited Test Ban Treaty on August 5, less than ten months after the crisis, and it was signed into U.S. law on October 7. The treaty prohibited nuclear tests in the atmosphere while allowing continued testing underground. Public concern and the activities of peace groups opposing nuclear testing may have helped to encourage test ban supporters in the Senate and restrain the opponents, including the military, but nuclear fallout did not emerge as a principal factor in the Senate debate over ratification of the treaty.[30] Public opinion played at most an indirect role in national security issues, and peace groups simply did not attract widespread U.S. public support during this period.

The idea for a freeze on the deployment, and possibly the development, of U.S. and Soviet nuclear weapons originated in 1980 in the United States. Yet in the midst of the Carter-Reagan presidential campaign and the crisis involving U.S. hostages in Iran, in 1980 the nuclear freeze lacked salience for most Americans. By the late 1970s American public opinion, and particularly that of people who followed foreign affairs, was concerned about Soviet ventures in the third world, including the invasion of Afghanistan, as the two superpowers embarked on a new era of "dangerous relations." Despite Ronald Reagan's strong opposition to the freeze during the campaign, he was elected in part because the public believed that he would be tougher than Jimmy Carter in dealing with the Soviets.

As president, Reagan was indeed tough in pushing sharp increases in defense spending, accelerated nuclear weapons development, and strident rhetoric about not only containing but even rolling back Soviet influence in eastern Europe. But in pressing the European allies hard in 1981 on continuing the deployment of the new intermediate range nuclear forces (INF) in Europe, the president undermined European assurance in the reasonableness of their American ally and catalyzed the anti-INF movement in Europe. In one particularly vivid instance, the president remarked in a press conference that he thought a superpower nuclear war in Europe could indeed be kept "limited." The percentage of European populations considering the outbreak of world war to be probable "within the next ten years" peaked in 1981. With public fear of war and nuclear weapons and parliamentary opposition to the North Atlantic Treaty Organization (NATO) deployments of INF missiles in West Germany, Holland, and Belgium (and later in Italy and Britain), the western European peace movement grew rapidly.

Mass demonstrations began late in 1980 and continued at a high level from early 1981 into 1983. Reacting in part to the news of U.S.–western European disputes over nuclear weapons policies, U.S. media coverage of the western European demonstrations and peace movement began to be prominent by December 1980, increased throughout 1981, and continued until the winter of 1982. By the fall of 1981 the peace movement was featured on the front pages of magazines and newspapers and increasingly on national television evening news.

The U.S. peace movement was very small in 1981, while the western European movement burgeoned and received heavy coverage in the American media. There was relatively very little coverage of any activities by the U.S. movement until months after the most intensive coverage of western Europe began. But the U.S. media coverage of Europe fanned freeze sentiments in the United States, giving a sense of legitimacy and urgency to the minority who led and supported the U.S. movement. The effects on the U.S. movement and population were also transmitted through the Protestant and Catholic churches, which were prominent in the European movement, and the U.S. State Department, led by Secretary Alexander M. Haig, and President Reagan, who argued in 1981 that the European movement was financed in part from, and inspired by, Moscow and clearly intended to divide NATO and increase neutralist tendencies in western Europe.

As U.S. press coverage of the European movement grew, so, too, did the reaction from U.S. conservatives in government and the media, who attributed it to the Soviets. This interpretation, in turn, stimulated further debate and press coverage. In response to the European movement, U.S. and western European governmental fears that the NATO deployment would be stopped by European political parties and opposition, and pressure on the Reagan administration to negotiate with the Soviets from key European allies, Reagan announced his double zero proposal for canceling the NATO nuclear deployment if Moscow would dismantle its intermediate range forces based in the European theater.

The U.S. freeze movement began to gather nationwide grass-roots support and the ability to launch protests by 1982, while local governments across the country debated resolutions on a freeze. During the winter and early spring of 1982, led by a few central representatives and senators, Congress gradually began to debate the issue; Senators Edward Kennedy (D-Mass.) (who was focusing on a presidential campaign) and Mark Hatfield (R-Oreg.) introduced a freeze resolution in the Senate in March. "By the spring of 1982, the nuclear freeze had become a major foreign policy issue in Washington, D.C. and across the country."[31] Public opinion polls indicated roughly 3:1 approval of a "mutual, verifiable" freeze on the deployment of U.S. and Soviet nuclear weapons.

On May 9, President Reagan aimed again to recapture the initiative, this time from the U.S. freeze movement. In his Eureka College speech he formally launched the Strategic Arms Reduction Talks (START) proposal for reducing significantly the superpowers' strategic nuclear weapons forces. In June the U.S. freeze movement reached its peak with a large demonstration in New York City. A freeze resolution failed by a close vote of 202–204 in the House in August.

As public fears of war and support for defense spending receded in 1982, the freeze movements were drawn into, and taken over by, the European and U.S. political processes. By 1983 top officials in the Reagan administration were more moderate and pragmatic in their positions for U.S. strategic nuclear policy and negotiation strategy. In March 1983 the West German Green party gained seats in the parliament, based on the disarmament issue. In May the U.S. House passed a freeze resolution by a 278–149 vote, but the Senate resolution failed by a 40–58 vote in October. U.S. freeze activists and supporters moved to Washington, D.C., to become influential experts or form lobbying groups on arms control and defense issues.

Both European and U.S. movements passed through several clear stages of political activities from 1980 to 1984. First, they joined experts, political activists, and some politicians across several nations through meetings, conferences, and projects. Second, these transnational networks helped stimulate grass-roots support, and hundreds of local and regional organizations emerged, triggering local referenda initiatives. Finally, a few key politicians adopted the cause, triggering attacks by top officials, intense political debate within government and outside, and constant media coverage.

U.S. public opinion was affected by broad public, then legislative debate and critique of the most technical and secretive component of national security policy—nuclear weapons doctrine and policies. This debate, in turn, contributed to the rapid, sharp shift in U.S. mass public opinion in 1980–82 against increased defense spending and direct overseas use of military force, and for arms control.[32]

On policy outcomes, the movements failed to directly stop the NATO deployment or to enact a U.S. freeze. Yet long-term transnational links on nuclear weapons policies were established across elites and governments of several OECD nations. And the shifts in U.S. public and elite opinion appear to have been primary factors in Reagan's shift in nuclear arms control strategy, which had been not to negotiate, to an aggressive agenda for reducing both strategic and intermediate range weapons. The U.S. House resolution was passed in 1983, European elections reflected major shifts in political party positions toward the peace movement proposals starting in 1983, and nuclear weapons became a major issue in U.S. congressional campaigns in 1982 and elections in 1984.

The U.S. peace movement of the early 1980s demonstrates the central political influence of the global flow of ideas and transnational political movements on U.S. public opinion in the post-Vietnam era. It also illustrates how transnationalism and political movements can help shift and splinter elite opinions, which, in turn, have long-term effects on the media, public opinion and policy outcomes.

As explained in Chapter 8, public and elite attention may often be necessary to cause the media to cover an issue, but public attention and political parties in one country can help trigger media interest in another and gradually attract official and grass-roots participation in an issue area. The policy outcomes of such activity vary across differing national political structures—for example, public opinion and interest groups influence U.S. officials and politicians relatively easily—but the resulting policies are likely to be less enduring and institutionalized in a formal sense in the United States, as compared to western Europe. While the U.S. system "institutionalizes" new interest groups and individual congressional leaders, new political parties and party policy positions are institutionalized in Europe.[33] Finally, this case shows that public, interest group, congressional, and transnational pressures were considerably more important to U.S. nuclear policies and negotiations by the 1980s than they were in the early 1960s. This shift resulted from various factors, including both the restraining influence of the cold war and anticommunism on protest politics in the United States during the earlier period and the heightened role by the 1980s of both social and political movements generally in the United States, and foreign and transnational influences on American foreign policy.

The U.S. Dollar and Exchange Rate Policy in the Mid-1980s

This case must be traced back to the postwar Bretton Woods international monetary system of 1944, its collapse in the period around 1971, and the ensuing efforts to reestablish a workable system of market-driven, or floating, exchange rates. The original Bretton Woods system of 1944 relied heavily upon markets and national governments but included fixed exchange rates (in terms of gold), limited international management (by the International Monetary Fund [IMF]), and open monetary and trade relations. Its design did not anticipate the total collapse of the European economic system that confronted leaders by 1946–47. In place of the intended Bretton Woods system, the United States assumed leadership of international monetary relations, and the dollar, fixed at a value of 1/35 ounce of gold, provided both the confidence and liquidity required for the international economy.

This system of special U.S. aid and arrangements, particularly a large outflow of dollars, succeeded in speeding the recovery of European nations and Japan. It functioned well in the immediate postwar period because the U.S. economy was expansive and foreign governments were pleased to use dollars as a global currency as long as they could be reliably exchanged for American goods. As the United States habitually spent more overseas than it earned by selling its goods, however, during the 1950s this system created increasingly larger U.S. balance of payments deficits, and foreign governments found themselves with larger and larger surpluses of dollars. This situation, too, was workable for a time because the global economy was recovering from World War II and the expanding economies of western Europe and Japan needed more and more dollars as capital to invest and as a currency for global exchange. Ultimately, however, the willingness of foreigners to accept and hold dollars depended upon their confidence that the dollars could be redeemed by the United States for desired things—American goods or gold.

By 1960, the total dollars held by foreigners exceeded the total gold reserves of the United States, and the goods of other countries were becoming as desirable as American products. As confidence in the dollar fell, international cooperation and institutions, including the IMF, took on an increasingly important role, thus preventing collapse of the system. Through the IMF, initiatives by the Group of Ten finance ministers from the leading industrial nations, and U.S. policies to reduce the deficit and strengthen the dollar, the Bretton Woods system operated quite effectively during the 1960s. The unique, privileged U.S. position allowed the government to operate domestic and foreign policies largely independent of the balance of payments constraints on other nations. Monetary and fiscal policies did not have to be adjusted, or tightened, as long as there was confidence in the dollar as the primary source of liquidity worldwide.

Once again, however, events overtook these arrangements. From 1968 to 1971 a crisis grew as a result of a combination of U.S. abuse of the system from heavy spending and debt to finance the Vietnam War and the Great Society programs of the 1960s; serious inflation; the refusal of Japan, West Germany, and other nations to absorb more of the costs of adjustment; and major structural problems in the Bretton Woods system. Over this period, the American society and economy were for the first time seriously affected by the exchange rate and related international capital and trade flows. The dollar increased in value due to high U.S. inflation and interest rates as well as the efforts of other countries to keep their currencies from rising and making their exports expensive and less competitive. As a result, capital flowed out of the United States, exports dropped, and imports increased.

Yet there was relatively little activity by organized interest groups in this issue area due to closed decision making and the lack of visibility and public understanding of the crucial effects of the dollar's value. President Richard Nixon, consumed by the Vietnam War, relations with the Soviet Union and China, and other issues, largely ignored the problem until he was forced to act in 1971 as a result of massive selling of the dollar, dwindling gold stocks, the first U.S. trade deficit in the twentieth century, and rapidly increasing inflation and unemployment.

Despite increased U.S. monetary and financial interdependence with its major trade partners by 1971, foreign exchange and monetary policy was still conducted behind a shroud of secrecy by a small group of officials who were unwilling to adjust domestic policies to the demands of newly integrated international markets.[34] To cope with the crisis, the dollar was delinked from gold, thus ending formally the Bretton Woods system that had already broken down, and allowed to "float" or move according to demand and supply in foreign exchange markets. As the dollar dropped in value, a 10 percent surcharge was levied on imports, and restrictions were negotiated on key imports from Europe and Japan. Instead of providing international leadership, Nixon's policy was an aggressive unilateral attempt to blame other nations for the dollar's problems, force as much as possible of the burden of adjustment onto allies, and minimize the costs to the United States. The Nixon surprise was also aimed at spurring economic growth in time for the economy to help him win the presidential election of 1972.

What followed during the 1970s was a series of attempts at reestablishing minimal international procedures and rules, interspersed among international energy, economic, and financial crises, including in 1975 the worst recession of the postwar era. Markets and private actors assumed increasingly important roles in monetary, financial, and trade relations. Exchange rates fluctuated much more widely than in the 1950s and 1960s, and the dollar dropped in value again in 1973. The bargains of Bretton Woods and the experience of the 1950s, U.S. leadership and explicit rules for balancing domestic needs with international commitments, were lost. Every western industrial economy and society, finally including even the United States, became more interdependent—less manageable by national authorities and more driven by large transnational monetary and financial flows.

The early 1970s were a watershed both in the loss of U.S. leadership and control over international monetary, capital, and energy markets and the new vulnerability of the U.S. economy and society to external forces. The mid- and late 1970s demonstrated the dilemma of interdependence without Bretton Woods: the ever more visible domestic effects of exchange rates and energy and economic crises forced governments to act, but unilateral policies were increasingly ineffective. For example, when governments act to prop up the value of their currencies, even the richest seldom have more than the equivalent of tens of billions of dollars in their official reserves to use in an effort to manipulate international currency markets. In contrast, private firms and traders exchange hundreds of billions of dollars daily in these markets. So when Britain and Germany, for instance, attempted late in 1992 to avoid devaluation of the pound, they spent a few tens of billions of dollars in markets that ran in the hundreds of billions daily. Finally, as energy and other crises demanded increased capital on an unprecedented scale worldwide, public national and international lenders were replaced by private banks as the primary sources of liquidity,[35] and currencies such as the yen and deutsche mark became more important relative to the dollar.

The immediate background to the dollar politics case of the mid-1980s was further major changes underlying the unstable, nonsystem of floating exchange rates and international capital flows in the 1980s. Rapidly advancing technologies in information processing, computers, and telecommunications revolutionized international financial markets. Along with the deregulation and opening of national foreign exchange and financial markets, these advances allowed multinational corporations and large financial institutions to move ever larger volumes of funds across markets worldwide with increasing speed and efficiency. While U.S. interest groups, Congress, and the media closely monitored trade flows, financial transactions in globally integrated markets that were many times larger and highly significant in their effects on U.S. markets went largely unnoticed until the emergence of the third world debt crisis, led by Mexico in 1982, and the U.S. trade deficit "crisis" of the 1980s.

The first term of the Reagan administration was marked by a passionate free market ideology that dictated strict adherence to floating, market rates in foreign exchange and heavy reliance on tight monetary policy at home. This strategy might have worked, except that there was no attempt at providing international leadership or even monetary policy coordination, and fiscal policy (particularly defense spending) was not restrained. The result, in the midst of deepening U.S. economic recession, was an increasingly overvalued dollar and growing deficits in the federal budget and in trade flows. Foreign, particularly Japanese, funds flowed into U.S. markets in response to high interest rates and other factors, thus subsidizing U.S. government spending. If managed, this flow could have been a positive development, but unmanaged it created a vicious cycle of more federal government borrowing, high interest rates, an overvalued dollar, and an unprecedented net debtor position for the U.S. economy.

As other currencies, such as the yen and the deutsche mark, fell against the dollar, U.S. exports grew too expensive to be competitive and the sale of American goods slowed from 1981 through 1986, most noticeably in 1983 and 1984. Since foreign goods were relatively inexpensive, imports rose sharply during 1983 through 1986.[36] The result was a series of U.S. trade deficits that were massive and unprecedented, at least as reported in terms of traditional measures and as transmitted by the media.

Just as the U.S. national economy was recovering in 1983–84 from a long recession, surging imports displaced domestic manufacturing production and jobs and extended the recession in manufacturing, most noticeably in the Midwest, until 1987. Over this period the effects of floating exchange rates and the overvalued dollar gradually became more visible to industrial and labor groups and the public, but interest groups could not bring consistent, effective pressure to bear until 1985. Indeed, beginning only in the mid- to late 1980s did the media regularly track the domestic effects of the dollar.

Part of the difficulty in U.S. political mobilization over the dollar is its complex and differing effects across sectors and regions of the United States. An overvalued dollar damages manufacturing and agricultural industries producing for export, in the 1980s particularly in the Great Lakes region and the frost belt, but not as much in the South because southern products compete more with those from countries whose currencies are linked to the dollar. At the same time, an overvalued dollar benefits nontrade sectors such as health care and real estate, as well as U.S. investors and traders in financial services who want to buy foreign assets.[37]

An expensive dollar also discourages foreign investment in U.S. plants and encourages U.S.-based multinational corporations to invest in plants and facilities overseas. In the mid-1980s, this situation led to a major movement of U.S. jobs overseas and a loss of important production and jobs in already vulnerable sections of major U.S. cities and regions, such as the Midwest. Later in the 1980s, as the dollar declined sharply, foreign investment in U.S. industry benefited mainly the South with new plants and jobs.[38]

Dollar politics are further complicated by the effects on domestic monetary and fiscal policies and differing preferences about how to manage the relationship between monetary policy and the exchange rate. Owners of capital generally and international investors and traders, including multinational corporations, prefer stable exchange rates and more integrated global markets, whereas nontrade sectors of the U.S. economy want the U.S. government to retain as much control as possible over the domestic economy. Industry and labor-producing traded goods exert pressure for a low dollar and expansive monetary policy, as opposed to nontrade sectors and international investors, who will generally lobby for a higher dollar and tighter monetary policy.[39]

In the second Reagan term, with the surprise exchange of positions by Chief of Staff James Baker and Treasury Secretary Donald Regan, Baker assumed leadership from the Treasury Department of U.S. monetary and international financial policy in coordination with the Federal Reserve. He operated with a much more pragmatic set of beliefs than those that governed the first Reagan administration. Indeed, there was a near-complete transformation of U.S. international monetary and financial policies, in part because of powerful domestic political pressure for change.

The definitive case study of this issue documents that U.S. industry "mounted a major assault on the administration's international economic policy in 1985."[40] With the dollar still rising in value and clearly accelerating federal budget and trade deficits, as well as a seemingly aloof president who would not even acknowledge the damaging effects of the high dollar, a cross-cutting coalition grew, including business groups and leaders, expert analysts, journalists, congressional leaders, and cabinet members. By the spring and summer of 1985 not only this coalition but also a storm of pending protectionist trade initiatives from Congress and continuing pressure from trading partners was weighing heavily on the president to intervene to bring the dollar down. Even some of the administration's staunchest supporters from multinational corporations and cabinet members Malcolm Baldrige (Commerce), William Brock (U.S. trade representative), John Block (Agriculture), and George Shultz (State) actively supported multilateral intervention. In the end it was perhaps the forceful action of Congress, with bipartisan support, more than any other single pressure, that triggered a new U.S. policy. Propelled by what appears to have been total frustration with a stubborn, even hostile president on this issue, who in the midst of the crisis left the U.S. trade representative position vacant from March to May, Congress threatened to take control of trade and then monetary policy by every available means. This period thus established some of the strongest precedents for retaliating against U.S. trade partners with large surpluses and directly linking trade and monetary policy, even requiring foreign exchange intervention in response to large trade deficits.

In this situation, Baker led a successful effort by the Group of Five finance ministers and central bank governors to forge a multilateral response to the misaligned exchange rates. The Plaza Agreement of September 1985 focused on coordinated selling of the dollar, which accelerated its decline through early 1987, and looser monetary policy in Japan and Germany, which was intended to expand purchases of U.S. exports and reduce the U.S. trade deficit and protectionist pressures.

The events of 1985 marked both the first time since World War II that exchange rates became a central issue of American politics and a new, much more multilateral U.S. approach to international monetary and financial issues. In rapid succession there followed four important actions: a new presidential policy for trade, the Baker initiative in the international debt issue, multilateral agreements to coordinate macroeconomic policies, and in February 1987 the Louvre Accord to stabilize the dollar at its lower level and institute coordinated government intervention to maintain currencies within specified bands. After the Plaza Agreement, news announcements about the U.S. trade balance began to have a direct, noticeable effect on the dollar, with increases in the trade deficit driving the dollar down in value, and vice versa.[41] The Plaza Agreement turning point demonstrated that international cooperation was necessary to achieve even minimal domestic needs. With the globalization of markets, the actions of foreign governments, large banks and corporations, and international institutions became central to managing the U.S. economy and policy process.

This case highlights three primary sets of U.S. domestic political changes triggered by events and developments in international, and closely related national, money and finance. First, the underlying ideas and basic policy approach to international money management changed abruptly six times: from the original Bretton Woods system of 1944 to U.S. dominance in 1947; to IMF and multilateral management in 1958–60; to markets and national policies in the early 1970s; to multilateral management under Carter in 1977; to markets in 1981; and back to multilateral management from 1985. The dollar crisis of the mid-1980s paralleled that of the early 1970s in one important respect. The U.S. public policy response was driven by crisis and intended to benefit the United States, but overall it was "too little, too late" either to assist the adjustment of U.S. producers of traded goods or to stabilize exchange rates. Given the enormous volumes of private capital flows by the 1980s, U.S. macroeconomic policy began to work less through interest rates than through exchange rates. And as long as the United States or its major trade partners have significantly over- or undervalued currencies, exchange rates will not reflect the fundamental competitive positions of these leading nations. Finally, U.S. trade policy became tightly linked to monetary issues and capital flows in a way that could make it more difficult to achieve important goals in either area.

A second domestic change flowing from this case was more visibility and political salience for the level of the dollar and its relationship to the continuing trade deficit by the late 1980s. The U.S. and international media, as well as U.S. publics, appear to follow these issues more often. Third, unlike in 1971, in the mid-1980s the U.S. executive could no longer act on its own in international monetary policy, more or less regulate the domestic economy, and push much of the adjustment burden off onto trade partners. Although the interests of congressional delegations, geographic regions, and coalitions diverged in a world of highly mobile capital that distributed benefits and costs differently,[42] interest groups and Congress actively intervened on international monetary issues and cabinet officers and executive agencies fought over how best to respond. The process for making exchange rate policy was opened and broadened significantly to involve executive departments beyond the Treasury Department. Consultation with the major industrial nations and regular reports to the Congress twice per year, as institutionalized in the Trade and Competitiveness Act of 1988, grew directly from this case. These changes opened regular opportunities for Congress and interest groups to monitor and intervene in exchange rate policy making.[43]

This case shows that the dollar is central to the nation's international competitiveness, production and jobs related to exports, regions of the country most vulnerable to competition from imports, and the allocation of resources across the economy. It demonstrates the high costs—for example, more than 1 million jobs lost over the period of this case in industries producing traded goods—of sporadic, inconsistent exchange rate policy making. It highlights the increase in public awareness and political activity and the challenges to governance that grow from the ever-increasing interactions among the value of the dollar, trade policy, federal budget deficit reduction, and macroeconomic management. Finally, it clarifies and confirms the crucial continuing role of U.S. leadership in catalyzing international cooperation and mechanisms and the new, essential role of other key nations, such as Japan and Germany, to successful international policy coordination.

CONCLUSION

For most of the twentieth century, and particularly since World War II, scholarly research has emphasized the relative autonomy of the American society and government as well as the nature of the nation's role as a world leader. Instead of this "inside-out" orientation, the focus here is "outside-in." This approach ignores neither the important power of the government as initiator nor the importance of particular presidents, advisers, and officials, but it also assumes that generally there is no unified or single U.S. national interest or foreign policy and that domestic interests and goals now diverge more often from foreign ones. It highlights the unique nature beginning at least by the mid-1970s of (1) the linkages between the external environment and the U.S. society, economy, and government; and (2) the associated impacts, both direct and indirect, of international forces and foreign governments on U.S. domestic politics.

By early in the Nixon administration, 1969–70, the United States was engulfed in an economic crisis at home, collapsing international monetary and petroleum regimes, and an unwinnable war in Southeast Asia. Its international political, economic, and military power had declined in relative terms, as its industrial competitiveness lost ground to Japan and western European nations. By the early 1970s, the legitimacy of its global leadership and the purposes of its foreign policy were under attack both at home and abroad.

These deep challenges to the United States were in part the reflection of broad global changes—the hazards of interdependence. The major industrial, or OECD, nations were increasingly vulnerable to external events and global flows of money, finance, trade, and oil. This interdependence, accompanied by growing governmental and private links across societies and economies, was growing rapidly by the late 1960s and beginning to affect even American politics.

Interdependence had two primary effects on the United States. In the short term it undermined U.S. domestic policy autonomy, or the ability to make decisions and enact policies that work without cooperation, or at least coordination, with the major allies, particularly Japan, Germany, France, Britain, Canada, and Italy (the Group of Seven). Over the longer term, interdependence triggered the growth of domestic interests and groups increasingly at odds with traditional foreign policy goals, particularly the openness of U.S. borders. This divergence and dissension was also caused by international economic shocks, such as the energy crises of 1973–74 and 1979–80 and the worldwide recession of 1975, disagreement over basic ideas and purposes of U.S. foreign policy, and divided government for twenty of the twenty-four years from 1969 through 1992.

From the early 1970s American institutions, coalitions, policies, media, and public opinion were increasingly influenced by international interdependence and transnational ties as well as by the Vietnam War. Indeed, the combination of singular international political-economic shocks and the cumulative effects of interdependence have helped to create increasingly permeable, penetrated U.S. social and political structures. This situation, in turn, has contributed to the: (1) increased pluralism and regionalism of American politics, including particularly conflict and fragmentation in the foreign policy process and partisanship in Congress; (2) reduced ability to achieve primary goals in foreign, and closely related domestic, policy areas; and (3) sharply reduced autonomy of foreign policy making, or the ability to make policies mainly for foreign purposes rather than in response to domestic needs or demands.

International events, links, and pressure will be of increasing importance in the continual shaping of U.S. political institutions, groups, and ideas. As U.S. attentive publics and elites devote more attention to the effects on their security and welfare, they will participate more often and actively in public policy debates. Since interdependence and transnationalism create new alliances and issue linkages, as well as new interests among actors, the policy preferences of individuals and groups will diverge and broaden the spectrum of debate.

U.S. publics and coalitions have increasingly called upon the state and federal governments at the same time to protect against and remain open to the international influences on domestic society and economy. Despite the power of liberal political-economic ideology in the United States, therefore, the level of public sector activity has not declined.

The primarily domestic orientation of an increasing number of U.S. agencies in the foreign policy process complicates that process because domestic and foreign policy problems must be addressed simultaneously and trade-offs must be made between domestic and foreign policy priorities. Executive and administrative agencies at various governmental levels join transnational alliances and interests and, along with individuals, groups, and regions, find increasing divergence in policy preferences within their own units as well as between traditional U.S. foreign policy goals and domestic priorities. Political activity is decentralized and more pluralistic, with more private and public participants.

Finally, some policy problems cannot be solved by the U.S. government alone. With increased interdependence and less domestic policy autonomy, increased policy coordination with foreign governments is required to address even routine socioeconomic and macroeconomic needs at home. Broad global forces posed by advancing communications and information technologies, multinational corporations, capital flows, and consumer expectations further undermine the ability of governments to manage national problems effectively.

While governments struggle to cope with the impacts of interdependence, they are likely to be less successful in enacting public policies that achieve intended objectives. Their activity level is unlikely to decline, but their effectiveness and stature will decline unless they enact deeper levels of policy coordination and take radical steps to achieve various forms of economic and financial, or even political-economic, integration. For the United States, political decentralization and fragmentation hobbles domestic and foreign policy autonomy. Attempts to create new trade and energy agreements, and the Uruguay Round of General Agreement on Tariffs and Trade (GATT) talks in particular, have been challenged in part by the wide range of U.S. political actors and interests pressuring the government and affecting international diplomacy. For the foreseeable future, negotiations among domestic elites and groups must be expected to be at least as difficult and as important to successful diplomacy as negotiations with foreign governments. Thus, more consistent, patient, and creative political leadership from both the executive and Congress will be essential to solving foreign policy problems and energizing effective multilateral and global negotiations. Strong leadership by the U.S. executive branch is absolutely crucial, but so is a sharing of the burdens of leadership with leaders in Congress and the major industrial nations.

Notes

1. John F. Kennedy, Inaugural Address, Friday, January 20, 1961, *Inaugural Addresses of the Presidents of the United States from George Washington 1789 to George Bush 1989* (Washington, D.C.: U.S. Government Printing Office, 1989), p. 308.

I am grateful to Donald Hafner for his penetrating comments, and to Robert Art, Mac Destler, Jeffrey Frieden, Patricia Jacobs, Thomas Risse-Kappen, John Tierney, and the anonymous reviewers for their important suggestions.

2. Robert D. Putnam, "Diplomacy and Domestic Politics: The Logic of Two-Level Games," *International Organization* 42 (Summer 1988): 460.

3. David R. Cameron, "The Expansion of the Public Economy: A Comparative Analysis," *American Political Science Review* 72 (December 1978): 1243–61; Robert Gilpin, *The Political Economy of International Relations* (Princeton, N.J.: Princeton University Press, 1987); Peter J. Katzenstein, *Small States in World Markets* (Ithaca, N.Y.: Cornell University Press, 1985); Ronald Rogowski, *Commerce and Coalitions* (Princeton, N.J.: Princeton University Press, 1989), p. 89.

4. R. N. Cooper, *The Economics of Interdependence: Economic Policy in the Atlantic Community* (New York: McGraw Hill, 1968).

5. Vinod K. Aggarwal, Robert O. Keohane, and David B. Yoffie, "The Dynamics of Negotiated Protectionism," *American Political Science Review* 81 (June 1987): 345–66.

6. Kinley J. Brauer, "1821–1860: Economics and the Diplomacy of American Expansionism," in William H. Becker and Samuel F. Wells, Jr., eds., *Economics and World Power: An Assessment of American Diplomacy since 1789* (New York: Columbia University Press, 1984), p. 113.

7. David A. Lake, "International Economic Structures and American Foreign Economic Policy, 1887–1934," *World Politics* 35 (July 1983): 517–43. See also David A. Lake, *Power, Protection, and Free Trade* (Ithaca, N.Y.: Cornell University Press, 1988).

8. Randall Strahan, *New Ways and Means* (Chapel Hill, N.C.: University of North Carolina Press, 1990).

9. Ibid.; William Schneider, "The Old Politics and the New World Order," in Kenneth A. Oye, Robert J. Lieber, and Donald Rothchild, eds., *Eagle in a New World* (New York: Harper Collins, 1992), pp. 35–68.

10. Miles Kahler, *Decolonization in Britain and France* (Princeton, N.J.: Princeton University Press, 1984), p. 376.

11. Roger H. Davidson and Walter J. Oleszek, *Congress against Itself* (Bloomington, Ind.: Indiana University Press, 1977).

12. *Economic Report of the President: Transmitted to the Congress February 1992* (Washington, D.C.: Government Printing Office, 1992), p. 196.

13. Ibid., p. 205.

14. Richard O'Brien, *Global Financial Integration: The End of Geography* (New York: Council on Foreign Relations Press, 1992).

15. "Percentage of Foreigners In U.S. Rises Sharply," *New York Times*, December 20, 1992, p. 36. This tension appears, for example, to have been an important contributing cause of the Los Angeles riots of 1992.

16. Joan Spero, *The Politics of International Economic Relations*, 4th ed. (New York: St. Martin's Press, 1990), p. 74; see also Gilpin, *Political Economy of International Relations*.

17. Peter Alexis Gourevitch, *Politics in Hard Times: Comparative Responses to International Economic Crises* (Ithaca, N.Y.: Cornell University Press, 1986).

18. Spero, *Politics of International Economic Relations*, p. 75.

19. See chapter 10.

20. Chung-In Moon, "Complex Interdependence and Transnational Lobbying: South Korea in the United States," *International Studies Quarterly* 32 (1988): 67–89.

21. See, for example, United Nations, *National Accounts Statistics: Analysis of Main Aggregates, 1988–89* (New York: United Nations, 1991); Geoffrey Garrett and Peter Lange, "Political Responses to Interdependence: What's 'Left' for the Left?" *International Organization* 45 (Autumn 1991): 539–64. On government strategies for coping with protectionist pressures, see especially Rogowski, *Commerce and Coalitions*.

22. Peter Evans, "Transnational Linkages and the Economic Role of the State: An Analysis of Developing and Industrialized Nations in the Post–World War II Period," in Peter Evans et al., eds., *Bringing the State Back In* (New York: Cambridge University Press, 1985).

23. See generally the insightful work of James F. Hollifield, *Immigrants, Markets and States* (Cambridge, Mass.: Harvard University Press, 1992) on the U.S., comparative, and international political economy of immigration.

24. Ibid., p. 185.

25. Myron Weiner, "On International Migration and International Relations," *Population and Development Review* Vol. 11 (September 1985): 441–55.
26. *New York Times,* December 20, 1992, p. 36.
27. Harry Pachon and Louis DeSipio, "Latino Elected Officials in the 1990s," *PS: Political Science and Politics* Vol. 25 (June 1992): 212–17.
28. *New York Times,* October 22, 1992.
29. See Thomas Risse-Kappen, "Cooperation among Democracies: Norms, Transnational Relations, and the European Influence on U.S. Foreign Policy," November 25, 1991, unpublished.
30. See Ivo H. Daalder, "The Limited Test Ban Treaty," in Albert Carnesale and Richard N. Haas, eds., *Superpower Arms Control: Setting the Record Straight* (Cambridge, Mass.: Ballinger Publishing Company, 1987), pp. 9–39.
31. Douglas C. Waller, *Congress and the Nuclear Freeze: An Inside Look at the Politics of a Mass Movement* (Amherst, Mass.: University of Massachusetts Press, 1987), p. 68. See also David S. Meyer, *A Winter of Discontent: The Nuclear Freeze and American Politics,* (New York, N.Y.: Praeger, 1990), p. 268 and chapter 7; and Thomas R. Rochon, *Mobilizing for Peace: The Antinuclear Movements in Western Europe* (Princeton, N.J.: Princeton University Press, 1988).
32. See Waller, *Congress and the Nuclear Freeze,* chapter 9; Meyer, *A Winter of Discontent,* chapter 16; and Rochon, *Mobilizing for Peace,* chapter 9.
33. See Thomas Risse-Kappen, *Cooperation Among Democracies: Alliance Norms, Transnational Relations, and the European Influence on U.S. Foreign Policy* (forthcoming 1994.) and "Public Opinion, Domestic Structure, and Foreign Policy in Liberal Democracies," *World Politics* 43 (July 1991): 479–512.
34. I. M. Destler and C. Randall Henning, *Exchange Rate Policymaking in the United States* (Washington, D.C.: Institute for International Economics, 1989); Raymond Vernon and Deborah Spar, *Beyond Globalism: Remaking American Foreign Economic Policy* (New York: Free Press, 1989).
35. See Benjamin J. Cohen, *In Whose Interest? International Banking and American Foreign Policy* (New Haven: Yale University Press, 1986).
36. I. M. Destler, *American Trade Politics,* 2d ed. (Washington, D.C.: Institute for International Economics, 1992), chap. 5.
37. Ann R. Markusen and Virginia Carlson, "Deindustrialization in the American Midwest: Causes and Responses," in Lloyd Rodwin and Hidehiko Sazanami, eds., *Deindustrialization and Regional Economic Transformation: The Experience of the United States* (Boston: Unwin Hyman, 1989) pp. 48–49; and Jeffry Frieden, "National Economic Policies in a World of Global Finance," *International Organization* 45 (Autumn 1991): 425–51.
38. Markusen and Carlson, pp. 74–76.
39. Frieden, "National Economic Policies."
40. See Destler and Henning, *Exchange Rate Policymaking;* Vernon and Spar, *Beyond Globalism.*
41. Michael Klein, Bruce Mizrah, and Robert G. Murphy, "Managing the Dollar: Has the Plaza Agreement Mattered?" *Journal of Money, Credit, and Banking* Vol. 23 (November 1991): 742–51.
42. Frieden, "National Economic Policies."
43. Destler and Henning, *Exchange Rate Policymaking.*

Bibliography

Richard F. Bensel, *Sectionalism and American Political Development, 1880 to 1980* (Madison, Wis.: University of Wisconsin Press, 1984).

April Carter, *Peace Movements: International Protest and World Politics since 1945* (New York: Longman, 1992).

John E. Chubb and Paul E. Peterson, eds., *Can the Government Govern?* (Washington, D.C.: Brookings Institution, 1989).

Congressional Quarterly, *Trade: U.S. Policy since 1945* (Washington, D.C.: Congressional Quarterly, 1984).

Rodolfo O. de la Garza et al., "Ethnicity and Attitudes Toward Immigration Policy: The

Case of Mexicans, Puerto Ricans and Cubans in the United States." Paper Presented at the Annual Meeting of the American Political Science Association, Palmer House Hotel, Chicago, September 3–6, 1992.

Richard C. Eichenberg, *Public Opinion and National Security in Western Europe* (Ithaca, N.Y.: Cornell University Press, 1989).

Samuel P. Huntington, "Transnational Organizations in World Politics," *World Politics* 25(April 1973): 333–68.

Robert W. Jerome, *U.S. Senate Decision-Making: The Trade Agreements Act of 1979* (New York: Greenwood Press, 1990).

Robert O. Keohane and Joseph S. Nye, *Power and Interdependence,* 2d ed. (Glenview, Ill.: Scott, Foresman, 1989).

———*Transnational Relations and World Politics* (Cambridge, Mass.: Harvard University Press, 1972).

C. P. Kindleberger, "Mass Migration, Then and Now," *Foreign Affairs* July, 1965, pp. 647–58.

Stephen Krasner, *Defending the National Interest* (Princeton, N.J.: University Press, 1978).

Paul Krugman, *The Age of Diminished Expectations: U.S. Economic Policy in the 1990s* (Cambridge, Mass.: MIT Press, 1992).

Edward E. Leamer, *Sources of International Comparative Advantage* (Cambridge, Mass.: MIT Press, 1984).

Stefanie Lenway, *The Politics of U.S. International Trade* (Marshfield, Mass.: Pitman, 1985).

Jarol Manheim and Robert B. Albritton, "Changing National Images," *American Political Science Review* 78 (September 1984): 641–57.

Helen V. Milner, *Resisting Protectionism: Global Industries and the Politics of International Trade* (Princeton, N.J.: Princeton University Press, 1988).

———, "Trading Places: Industries for Free Trade," *World Politics* 40 (April 1988): 350–76.

E. C. Morse, *Modernization and the Transformation of International Relations* (New York: Free Press, 1976).

Robert Pear, "Immigration Bill: How 'Corpse' Came Back to Life," *New York Times,* October 13, 1986, p. A16.

Richard Rosecrance, *The Rise of the Trading State* (New York: Basic Books, 1986).

Theda Skocpol, *States and Social Revolutions: A Comparative Analysis of France, Russia, and China* (Cambridge: Cambridge University Press, 1979).

Jack Snyder, "International Leverage on Soviet Domestic Change," *World Politics* 42 (October 1989): 1–30.

Strobe Talbott, *Deadly Gambits* (New York: Alfred A. Knopf, 1985).

Goran Therborn, "Migration and Western Europe: The Old World Turning New," *Science,* September 4, 1987, pp. 1183–88.

Peter Trubowitz, "Sectionalism and American Foreign Policy: The Political Geography of Consensus and Conflict," *International Studies Quarterly* 36 (1992): 173–90.

John A. Vasquez, "Domestic Contention on Critical Foreign-Policy Issues," *International Organization* 39 (Autumn 1985): 643–66.

World Bank, *World Development Report, 1987, 1988, 1989, 1990* (New York: Oxford University Press, annual).

Part II / The Executive Branch

2 / Presidential Leadership and the Foreign Policy Bureaucracy

DONALD L. HAFNER

Jimmy Carter and I started work in Washington on the same day in January 1977. His desk was in the Oval Office; mine was at the Arms Control and Disarmament Agency, in the art deco section of what was once the "new" State Department building when President Franklin Roosevelt exiled his diplomatic staff to Foggy Bottom in the 1930s.

I was pretty busy in those first several weeks, learning what my new agency expected of me, and I'll confess I lost track of what Jimmy Carter was doing. I had been hired by the previous administration during its final days, no doubt in an effort to salt the agency with presumably like-minded souls before Gerald Ford surrendered the seals of office. I occupied a slot in the bureaucratic hierarchy at what we called the "gerbil level," an apt image of small animals scurrying around energetically on their small-animal business. When I wanted to brag, I would say that I was separated from the president by only three layers of bureaucracy; when I wanted to deny responsibility, I insisted I was nothing more than a working-group analyst. Mostly I was a gerbil. And at my desk some weeks later, I was expressing a gerbil's irritation over a copy of a paper I had done on an arms control issue, a copy that had been returned with a series of remarks handwritten in the margins, highlighting a fault in logic here and a grammatical error there—until all at once I realized that the corrections had been scrawled by James Earl Carter, Jr., thirty-ninth president of the United States.

I tell this story as a provocation to thought about the character of presidential leadership over the foreign policy bureaucracy. Jimmy Carter has been roundly criticized for precisely this kind of behavior—for his obsession with minutiae, for squandering that precious resource of presidential leadership, his attention, and for undercutting his own cabinet officers before their subordinates by intruding in such details.[1] The argument is pressed further by some: the bane of American foreign policy is the increasing tendency of presidents to inject themselves and their political passions into the bureaucracy, producing foreign policies that are idiosyncratic, inflamed with partisanship, unstable, and lacking in continuity.[2]

Down at gerbil level, I saw the matter somewhat differently at the time. I was involved in two arms control negotiations during my two years in Carter's administration. The first was with the Strategic Arms Limitation Talks (SALT) II, where I served as an analyst with both the National Security

Council's interagency SALT working group and the American delegation in Geneva. As a jaded academic, versed in the literature, I had been told to expect bureaucratic parochialism and sabotage.[3] And yet Jimmy Carter had been able to convey to us gerbils that SALT was important to him, that he wanted progress, that our efforts would enjoy his support. The effect was electric. We pulled together—State Department, Defense Department, the Joint Chiefs, the Central Intelligence Agency—we were soldiers of the president, and we conspired together (occasionally against our own superiors) to anticipate our president's wish and serve his cause.

Sometime later, my responsibilities shifted to the negotiations on anti-satellite weapons (ASAT) that Carter had initiated. There I watched an initial surge of bureaucratic cooperativeness vanish into petty interagency bickering and delay, as all of us felt the withdrawal of presidential support and attention. The position of chair of our interagency working group was transferred from one National Security Council (NSC) staff member to another. The Joint Chiefs of Staff expressed their displeasure with the president's ASAT policy by repeatedly changing the set of analysts they sent to the working group meetings, each team professing ignorance of the issues and demanding more time to "get up to speed." The Joint Chiefs played this trick with impunity, despite our appeals for a show of authority by our NSC chair. Another official from one of the intelligence agencies who disliked Carter's policy tried the tactic of insisting that NSC working group sessions be escalated to increasing levels of security clearance, hoping to expel from the room some of the more vigorous advocates of the president's program who lacked the required clearances. I later learned that indeed Carter had decided that ASAT was too much to demand of the Joint Chiefs until SALT was ratified—a presidential judgment on priorities that I would not have disputed, had I known of it at the time.[4]

The contrast in bureaucratic performance between these two policy realms is not explained by a difference in organizational structure or presidential management style, for these were unchanged. It lies in something more elemental, even antithetic to bureaucracy, something more akin to Max Weber's notion of charismatic leadership. When Jimmy Carter graced our enterprise by his attentive gestures, however small, we pulled together because we felt ourselves connected to the president's vision, called to break from the precedents of old leaders, to repudiate the past and be a revolutionary force, to make his mission our own mission.[5] My experience, in sum, stands in apparent tension with the judgment of those who argue that American foreign policy suffers from too much presidential charisma—that what it needs is less political passion, less drama, more calm, more civility, more consistency.[6]

If a president neglects the charismatic element of leadership that seeks to rouse the ranks of public servants and rally them to battle on behalf of his own vision, then he leaves himself with little more than bureaucratic tools to elicit bureaucratic action. More than thirty years ago, Richard Neustadt reinvigorated the study of the presidency with his reminder that the president is not assured of obedience merely because he issues commands, no matter what the Constitution or the organizational charts imply. A president is powerful only when he can persuade his subordinates that "what the White House wants of them is what they ought to do for their sake and on their authority."[7] Surprisingly, this notion of leadership—indeed the very word "leadership"—has all but vanished now, displaced in the writings of scholars and practitioners alike by an emphasis on the president as "manager." Yet a president content to lead by deftly constructing organizational charts and then "managing" no more leads the bureaucracy than a Swiss clockmaker leads the cuckoos in his clocks. And like the clockmaker, the president-as-manager will never extract anything more imaginative, adaptive, or responsive from his bureaucratic devices than what he has built into them. Moreover, the rising complexity of post–cold war foreign policy and the bureaucracy needed to confront it threaten to overwhelm the president-as-manager. If already "a President cannot hope to keep watch on all those whom he pays to keep watch, let alone take care of business," then clearly it is time to resurrect a more elemental notion of leadership, one that emulates the general on horseback and not his quartermaster.[8]

The character of presidential leadership must be linked to the challenges confronting U.S. foreign policy in coming decades, and if we cannot foresee these in detail, at least some facets are apparent. For one, the United States will be conducting foreign policy with considerably diminished resources, both military and nonmilitary. Second, to avert the waste of diminished resources and the loss of sunken investments already made in policy, the United States will have to maintain greater coherence in policy. Third, to achieve coherence, the United States will require coordination among policy spheres not traditionally thought of as "foreign policy," such as economic, environmental, health policies, and so on. Fourth, to enhance the leverage of its own diminished resources, the United States will have to cooperate increasingly with other nations and thus will need the flexibility of a "strategic" policy, that is, a policy that can adjust swiftly to the actions of other international players.[9] These challenges dictate that dominant control of U.S. foreign policy must reside with the president. Neither Congress nor the executive agencies are capable of producing coherent, strategic policy on their own, so the task must fall to the president. And if the president is not currently dominant in policy realms that must now be brought together into a coherent foreign policy, he must exert the kind of leadership that will allow him to prevail.

Presidents have found it quite challenging enough to cope with the complexities of bureaucratic politics among the agencies traditionally involved in foreign policy: the State Department, Defense Department, the Joint Chiefs of Staff, the Arms Control and Disarmament Agency (ACDA), and the Central Intelligence Agency (CIA). Now other policy spheres and their bureaucracies must be connected and coordinated. The academic wisdom is that all bureaucratic structures embody politics, that each policy agency and realm will have its own political culture, practices, and dominant players.[10] The emerging task for presidents will be to cope with the political mutations that arise when these disparate policy realms must be drawn together and leadership must be asserted to secure a coordinated foreign policy. Indeed, a major challenge confronts the president in drawing together the two policy realms—economics and national security—that have in the past been handled by rather separate bureaucratic fiefdoms (see Chapter 6).

In the National Security Act of 1947, Congress thought it was making an important step toward policy coordination when it established the National Security Council, "to advise the President with respect to the integration of domestic, foreign, and military policies relating to the national security." Congress was reacting to the perceived organizational disarray in U.S. policy during World War II. Since that disarray—seemingly contradictory and overlapping presidential instructions about which agency was to do what—was in fact President Franklin Roosevelt's deliberate political style for cutting some officials out of policy formulation and for ensuring that all controversial issues came to him for resolution, in a sense the National Security Council was Congress's effort to compel presidents to listen to the advisers Congress thought were important.

The effort was unsuccessful. By law, the formal NSC is composed of the president, the vice-president, the secretary of state, and the secretary of defense, with the director of the CIA, the chair of the Joint Chiefs of Staff, and other officials of the President's choice serving as advisers. But presidents will seek advice where they wish, and despite President Dwight Eisenhower's brief enthusiasm for elevating the formal NSC to a more prominent role in policy formation, it became progressively irrelevant. In its place, policy coordination shifted to the NSC staff, under the direction of the special assistant to the president for national security affairs, a position created by Eisenhower in 1953. Originally members of the NSC staff were merely administrative aides to the council. Under John Kennedy, they began to serve the president directly. Under Richard Nixon and Henry Kissinger, the staff swelled from fifteen to nearly one hundred and assumed the initiative in directing the foreign policy agencies in the formulation of policy. Despite some contraction in staff and responsibilities under Jimmy Carter and Ronald Reagan, the dominance of the NSC staff and the national security adviser persisted, even in the face of occasionally intense efforts by the secretaries of state and defense to reassert their preeminence.[11]

What also persists is the place of the traditional foreign policy agencies within the NSC system. By the 1992 presidential campaign, the importance of integrating such matters as economic and environmental policy into foreign policy had become uncontested wisdom. Yet when President Bill Clinton drew upon the NSC model and constructed a new NEC—a National Economic Council—to coordinate U.S. economic policy, he built it as a separate entity. If bureaucratic structure is indeed the embodiment of politics, foreign policy shall apparently remain foreign to domestic policy.

THE CASE OF EXPORT CONTROLS

As it happens, every president since Harry Truman has learned a little about what the future portends. For forty years presidents have had to cope with the mélange of politics thrust upon them at the intersection of economics, science and high technology, international trade, national security, and alliance policy, because throughout the cold war the United States has tried to keep military-related goods out of the hands of its enemies by imposing controls on the exports of western products, both unilaterally and in coordination with allies. The bureaucratic system that administers these export controls is staggeringly complex, and yet arguably it has produced a very consistent export control policy over the decades and has insulated itself from political intrusions by Congress and interest groups. In this respect, the case of export controls might serve as a model for presidents of how to link disparate policy fields and their bureaucracies to achieve a stable and coordinated foreign policy. On the other hand, the factors that have produced consistency in export regulation policy have threatened to insulate the bureaucracy from the president's control as well. When the president stands before his bureaucratic troops, mantled in all the formal braid of his office, and still cannot secure obedience, he is unlikely to succeed merely by issuing orders to rearrange the organizational chart. He needs a different concept of leadership.

In short, there are things to be learned from the politics of export control policy, enough to warrant a brief excursion into the bureaucratic complexities.

The legislative authority under which the executive branch regulates the export of goods and services has been modified from time to time by Congress. At present, the basic mandate lies in the Arms Export Control Act of 1976, which governs trade in military weapons and services, and in the Export Administration Act (EAA) of 1979 and its amendments in 1985, which regulate commercial products and services.[12] However, there has never been a time since World War II when the United States has not had export controls. In 1940, Congress gave the president authority to prohibit or constrain the commercial export of any goods (including technical data) under such regulations as he might prescribe for reasons of national security, foreign policy, or domestic shortage of supply. That authority was renewed for two-year intervals up through 1949, when the Export Control Act codified the control procedures that had evolved and extended them into the cold war. The United States subsequently linked its own export control policies with those of its allies through a series of agreements establishing the Coordinating Committee on Multilateral Export Controls, or COCOM, which now encompasses the United States, its North Atlantic Treaty Organization (NATO) allies, Australia, and Japan. Subsequent revisions of Export Regulations in 1969, 1979, and 1985 have preserved the core features of the original Export Control Act. In both its legislative mandate and structure, therefore, the export control apparatus of the executive branch has remained substantially unchanged as the United States has moved from hot war to cold war to post–cold war.

That bureaucratic apparatus is extraordinarily fragmented and complex, built on statutory provisions, presidential executive orders, and administrative agency directives. Primary authority for establishing the Commodity Control List (CCL) of regulated goods, and responsibility for granting export licenses for these goods, is vested in the Bureau of Export Administration in the Commerce Department—a peculiarity to begin with, since this arrangement places a national security function in a cabinet department that is not a statutory member of the president's National Security Council. However, the Commerce Department is instructed to seek information and recommendations from other agencies whose policies and operations have an important bearing on exports. One of these is the Department of Defense, which must concur with the Commerce Department on what items are to be included in, or removed from, the CCL and which enjoys a broad right to review and make recommendations on licenses for the export of controlled goods. The Defense Department is required to draw up its own Military Critical Technologies List (MCTL), but it must gain the Commerce Department's agreement to get MCTL items on the CCL. If the Commerce and Defense Departments fail to concur, the matter is referred to the president for resolution, an arrangement that in practice has meant that the dispute goes to a specialized group within the National Security Council.

The State Department's primary role is to conduct negotiations with COCOM members over additions and deletions of items from COCOM's control list, a role challenged in recent years by the Defense Department, whose own MCTL is invariably longer than the CCL or COCOM lists.[13] The State Department also implements and must formally approve all exports of arms, munitions, military services, and military technical data through its Center for Defense Trade, although the Defense Department is largely responsible for drawing up the list of controlled military items. The Department of Energy has acquired a joint responsibility with the Department of Commerce for control of nuclear exports, and it provides technical review of license applications in this field referred to it by the Commerce Department. Criminal investigation and enforcement of export regulations is shared by the Commerce Department and the Treasury Department's Customs Office. Other agencies with occasional roles include the Justice Department, Transportation Department, the National Aeronautics and Space Administration (NASA), ACDA, the Nuclear Regulatory Commission, and the intelligence agencies. Indeed, Congress by law has granted the president wide discretion in deciding which executive agencies to involve in export decisions, with one notable exception. In the Export Administration Act of 1979, Congress specified that "no authority under this Act may be delegated to, or exercised by any official of any department or agency the head of which is not appointed by the President, by and with the advice and consent of the Senate." Congress's intent clearly was to limit the role of the national security adviser and the NSC staff—that portion of the foreign policy bureaucracy with which the president has the closest personal links. To complete this bureaucratic mulligan stew, Congress in its great wisdom reposed congressional oversight responsibilities with the House Foreign Affairs Committee—an arrangement that makes intuitive sense—and with the Senate Committee on Banking, Housing, and Urban Affairs.

Although the complexity and fragmentation of bureaucratic responsibilities over export controls is extreme, it is probably no worse than what presidents will face in the future when they seek coherent foreign policy in other intersecting realms—for instance, in responding to the need for a global environmental policy that integrates health, technology, economic development, military security, and trade concerns.

In a bureaucratic apparatus as baroque as this, one could imagine that a president would face formidable challenges in securing coordination and responsiveness. One conceivable problem would be getting each element of the bureaucracy to pay attention to his wishes, rather than to the congressional committees or private interest groups that are the potential long-term constituencies for each agency. Engaged in each decision about the composition of the Commodity Control List, for instance, are the economic fortunes of every business whose commercial products get swept into a controlled category because some agency judges that the products might have a military-related use.[14] And at stake in each licensing decision is the well-being of a specific exporting firm and the employees and voters in whose district or state the firm resides. Although no more than a few hundred export licenses are denied each year, in 1989 American firms submitted 135,000 license requests, covering $118 billion worth of goods, or roughly 40 percent of total U.S. exports of manufactured goods. Administrative costs to firms in applying for licenses have been estimated at $500 million annually, and total costs to firms in administrative costs, lost sales, and so forth have been estimated at $9.3 billion annually.[15] Moreover, the lead agency in export controls is the Commerce Department, the executive agency most closely linked to the business community and the agency whose mandate is to promote overseas sales by American firms. In sum, strong incentives for bureaucratic capture or an alliance among agencies, congressional protectors, and interest groups would seem to be present here.

But that alliance does not appear to have happened.[16] Certainly businesses have applied pressure on the Commerce Department, both directly and indirectly through Congress. Repeatedly when Congress has renewed or revised export control authority, it has attempted to compel the executive to liberalize restrictions and speed up licensing review. Congress has even mandated that the Commerce Department establish advisory panels composed in part of experts from private industry, to review existing and prospective controls. Yet at each juncture, the president and the agencies have defended the principle of executive autonomy in national security and foreign policy matters. And where Congress has succeeded in incorporating some statutory prod or mandating some bureaucratic reform, the actual behavior of executive agencies has seldom altered. When instructed by Congress to speed up the license review process, for instance, the agencies largely ignored the mandatory deadlines and then got Congress to rescind them by threatening to meet deadlines simply by rejecting applications out of hand.[17] In this respect, the politics of controls on exports appear to be markedly different from the politics of imports, where congressional and interest group intrusions have traditionally been more successful.[18]

The reasons for executive branch success in resisting encroachment by Congress and interest groups are instructive. Some have to do with the institutional advantages enjoyed by the executive branch in confrontations with Congress: the executive agencies can command expertise in a highly technical area; opinion within Congress is divided within and between political parties; and weak party discipline and the complexity of legislative procedures make it difficult for congressional reformers (and the interest groups that back them) to maintain momentum for reform.[19] But the strongest lever held by the president is his ability to argue that national security, and the requirements of foreign policy, demand a flexibility that is inconsistent with statutory mandates. Reinforcing the president's insistence that Congress defer to his constitutional preeminence in foreign policy has been the implied threat that he would take the matter directly to the American public if Congress did not give way. The few occasions when Congress has overruled the president arguably prove the dominance of the executive. The Nixon and Ford administrations, even with the support of business, failed to prevent passage of the Jackson-Vanik Amendment in 1974 (which stymied the administrations' efforts to promote détente by liberalizing trade with the Soviet Union) because of an extraordinary coincidence of events that weakened presidential authority generally, including Watergate and the Vietnam debacle.[20] The congressional imposition of sanctions against South Africa in 1986, over President Ronald Reagan's veto, is noteworthy precisely because it was so exceptional for a presidential veto in foreign policy to be overruled.[21] President George Bush's ability to get his veto sustained over congressional efforts in 1992 to link trade with China to improvements in China's human rights record, at a time when Bush had one of the lowest presidential approval ratings in recent history, testifies to the power of a president's appeal to his prerogatives in foreign policy.[22]

Simply because the president has succeeded in fending off congressional and interest group assaults on his authority to decide export policy, however, does not necessarily mean that he can command this foreign policy apparatus to do his bidding.[23] A bureaucratic apparatus that has enjoyed four decades of immunity from external encroachment on its practices, after all, might conceivably develop a political culture of resistance to presidents as well.

BUREAUCRATIC DEFIANCE

What the president needs from his bureaucracy to aid in his stewardship of foreign policy includes sufficient information about the problem at hand, adequately analyzed to give a proper diagnosis of its crux; appraisal of the major values and interests affected by the problem; search for a reasonable number of policy options and evaluation of the expected benefits and consequences of each; consideration of the obstacles that may arise in implementing each option; faithful implementation of the policy decided on; and sustained alertness to signs that adopted policies are not working as intended.[24]

What the president must anticipate is that his ability to get such assistance may be seriously impeded by the complex dynamics of organization behavior and bureaucratic politics, even when his subordinates are not engaging in deliberate chicanery. As the Tower Commission put the matter in its investigation of the Iran-Contra affair: "The policy innovation and creativity of the President encounters a natural resistance from the executing departments. While this resistance is a source of frustration to every President, it is inherent in the design of the government."[25]

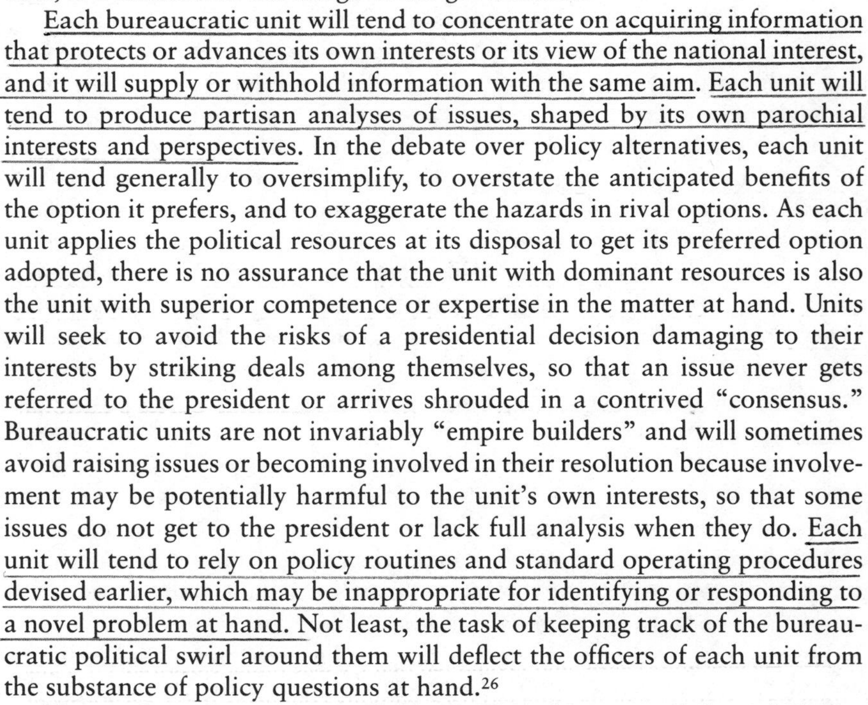

Each bureaucratic unit will tend to concentrate on acquiring information that protects or advances its own interests or its view of the national interest, and it will supply or withhold information with the same aim. Each unit will tend to produce partisan analyses of issues, shaped by its own parochial interests and perspectives. In the debate over policy alternatives, each unit will tend generally to oversimplify, to overstate the anticipated benefits of the option it prefers, and to exaggerate the hazards in rival options. As each unit applies the political resources at its disposal to get its preferred option adopted, there is no assurance that the unit with dominant resources is also the unit with superior competence or expertise in the matter at hand. Units will seek to avoid the risks of a presidential decision damaging to their interests by striking deals among themselves, so that an issue never gets referred to the president or arrives shrouded in a contrived "consensus." Bureaucratic units are not invariably "empire builders" and will sometimes avoid raising issues or becoming involved in their resolution because involvement may be potentially harmful to the unit's own interests, so that some issues do not get to the president or lack full analysis when they do. Each unit will tend to rely on policy routines and standard operating procedures devised earlier, which may be inappropriate for identifying or responding to a novel problem at hand. Not least, the task of keeping track of the bureaucratic political swirl around them will deflect the officers of each unit from the substance of policy questions at hand.[26]

The number of occasions in which presidents have actually been defied by the bureaucracies involved in export control is difficult to determine. Spicy tales of bureaucratic recalcitrance have become a common ingredient in presidential memoirs. Yet even if presidents were willing to confess all the instances in which they proved impotent in controlling their own subordinates, many successful acts of bureaucratic rebellion may have escaped presidential notice. Moreover, by law the Commerce Department is not obliged to tell an applicant why an export license has been approved or denied, and licensing decisions are exempt from judicial scrutiny, which might reveal the distance between presidential intent and bureaucratic action.

But a few cases are suggestive. In 1969, with the export control legislation adopted in 1949 due to expire, an alignment of prominent congressional Democrats and more conservative business groups sought to liberalize trade with the Soviet Union in the new Export Administration Act. Against the recommendations of his Commerce and State Departments, President Richard Nixon decreed at an NSC meeting devoted to the issue that his administration would oppose all liberalization, and he issued a formal directive on the matter to his cabinet a week later. Nixon prevailed in Congress, which voted to leave export liberalization at the discretion of the president, but he did not fare so well with his own bureaucracy. Henry Kissinger, who was then Nixon's national security adviser, noted the response: "No sooner were these instructions issued than the departments began to nibble away at them. Departments accept decisions which go against them only if vigilantly supervised. Otherwise the lower-level exegesis can be breathtaking in its effrontery." The Commerce and State Departments decided that Nixon had ruled only against new liberalizing legislation and that they were therefore free to loosen controls within existing law, under their own authority to establish export control lists and issue licenses. Repeatedly in the ensuing months, Kissinger had to block the Commerce and State Departments from decontrolling trade to the Soviet bloc on their own initiative, by bureaucratic fiat.[27]

President Jimmy Carter had his own experiences with bureaucratic defiance, provoking from him rather snappish behavior toward Secretary of State Cyrus Vance at a White House foreign policy breakfast where Carter was trying to hold his troops in line on a trade embargo imposed on the Soviet Union in 1978. According to Zbigniew Brzezinski, Carter's national security adviser:

> The President turned all of a sudden to Cy [Vance] and said that he doesn't want new trade initiatives started by Treasury, Commerce, or State, with the effect of going around his recent decision. . . . He doesn't want Marshall Shulman [of the State Department], Juanita Kreps [the Commerce Secretary], or others indicating to the Soviets in some fashion that this was just a little slap on the wrist. We want the Soviets to take this seriously, that we can do this to them, that it was meant to hurt. He was quite sharp, and I could tell that Cy was rather surprised. . . . Later that day, [Carter] sent me a memo back with my comments on [Treasury Secretary Michael] Blumenthal's objections, writing in the margin, "Tell Mike to support my policies."[28]

Blumenthal perhaps did stay in line, but within three weeks, the Commerce Department on its own granted the export license that Carter explicitly intended to block. As Brzezinski tells the tale, "I asked the President directly whether he had approved such a step, and he said he had not, but he told me not to push the issue too hard lest it precipitate the resignation of Juanita Kreps (who might feel thereby repudiated). On further investigation it turned out that State and Commerce had acted on their own."[29]

President Ronald Reagan's experience with bureaucratic defiance came from a Defense Department that apparently felt he was not sufficiently hawkish on export controls. In the Pentagon's view, if U.S. firms were subject to tighter export controls than other western countries imposed and thus lost export markets to foreign firms, that was just the price that Americans had to pay because of their global security responsibilities. The Commerce Department had a different view, and in 1984 it sought to remove limits on the U.S. export of wafering saws to western purchasers. (Wafering saws are used to slice silicon wafers for the manufacture of computer chips.) When the Defense Department opposed, the Commerce Department took the issue to the NSC and got a decision in its favor. The Defense Department, however, refused to give up on its opposition and exploited its role in the multilateral COCOM decontrol procedures to obstruct the decision there. Two years later, the Defense Department was still defying the NSC decision.[30]

In another instance, President Reagan approved the decontrol of semiconductor wire bonders (used in making computer chips), again because a foreign supplier was already exporting them. Despite Reagan's decision, the Defense Department the following year applied pressure on the foreign source of wire bonders, hoping to curtail their sales and thus justify continued U.S. export controls. When that effort failed, Defense fell back on interagency maneuvers to tie up the matter for another two years and prevent exports by U.S. producers.[31] Cases such as this are particularly delicate for the president, because by law he must promptly report to Congress if he modifies or overrules a Defense Department recommendation on export decontrol. No such report has ever been submitted by a president, which is not surprising since doing so would invite Congress to sit in judgment, treating the president as the mere equal of his own subordinates. No doubt the Pentagon understands this.

On another occasion, the Defense Department carried its demands for more authority over export controls directly to President Reagan and secured an executive order expanding the Pentagon's right to review exports to fifteen noncommunist countries that were suspected of transshipping controlled items to the Soviet Union. Then, in an act of bureaucratic entrepreneurship, Defense Secretary Caspar Weinberger decided he wanted more than just a presidential authorization, so he went directly to Senator Jake Garn (R-Utah) and proposed that the senator submit legislation giving the Pentagon control over exports to all COCOM countries. The president pulled his own Defense secretary back into line by sending Garn a letter rebuking Weinberger publicly for his initiative.[32]

The potential for bureaucratic defiance seems pervasive here. In any substantive debate on controls with the White House staff or the president, the advantage of technical expertise resides entirely with the agencies. The right to review control decisions, and thus the opportunity to engage in delay and harassment, is scattered among executive agencies by statute, making the president less than full master in his own house. The statute requiring the president to report to Congress when he overrules any Defense Department recommendation on decontrols also raises questions about who is master. Routine interaction and policy coordination with COCOM members potentially draws the loyalty of bureaucrats away from the White House and toward transnational groups of officials who develop their own institutional interests. Furthermore, in virtually every interagency forum in which decisions are to be made, a voting rule of unanimity prevails, fostering delay and lowest common denominator resolution of issues both among agencies and between the United States and COCOM members.

Indeed, an argument has been made that the consistency of U.S. export control policy is due precisely to the insulation of the bureaucracy, not only from societal pressures represented by Congress and interest groups but implicitly, as well, from changes in the foreign policy orientation of the president.[33] The complex export control apparatus embodies the presumptions of the cold war in its bureaucratic structure and statutory authority. Those Containment presumptions have continued to shape the bureaucracy's behavior, despite the ebb and flow of cold warriorism represented by presidents as diverse as Nixon, Carter, and Reagan. Without question, there are disputes among the bureaucracies. As one 1975 government study put it:

> The positions taken by each of the agencies appear almost a caricature: Defense officials vetoing any item they can get a handle on, if only to delay for a couple of years Communist acquisition of the technology; State . . . prepared to make an exception for almost any item, as long as it appears to contribute to detente; Commerce, making American firms' case that since technology is going to be sold in any case, the U.S. should at least reap the benefit of making the sale. In addition, CIA, the sole source of official judgment on "foreign availability" . . . continues to interpret "availability" and "equivalence" in the narrowest terms, preferring to delay trade whenever possible.[34]

Yet so long as they can manage their differences, these agencies also share an interest in perpetuating and insulating a bureaucratic "veto" system in which each enjoys significant voice and control.[35] It would seem that their insularity would make these agencies more dependent on the president. To the extent that the president keeps political competitors from intruding in foreign policy, he prevents the agencies from independently establishing partnerships with constituencies in the public or Congress who can offer to protect the agencies' interests, at budget time for instance, in exchange for favorable administrative decisions. In the absence of such partnerships, the agencies presumably must turn to the president for legitimacy and protection. And yet, even when they drift away from his purposes in pursuit of their own, the president does not dare withdraw his protection from these agencies because, if he did, he would be undermining his own grip over foreign policy.

EXPORT CONTROLS AS A MODEL FOR THE FUTURE?

Despite the potential for abuse here, from the president's standpoint the bureaucratic system for dealing with export controls has features that invite its use as a model for coping with foreign policy tasks of increasing issue complexity. To gauge from the recurrent complaints of Congress and business groups about the failure of the system to be more responsive to outside pressures for reform, this structure has done very well at resisting incursions and thus at least in principle has preserved the president's dominant role. The system brings to bear an uncommon range of technical expertise on a great volume of collaborative tasks in a manner that apparently does not excessively disrupt the agencies' performance of their other duties. The pervasive rules of unanimity mean that when the system confronts a problem judged by any agency as unusual, it shunts the matter onto a track that leads ultimately to the NSC. By reposing authority for actual decisions in a lead agency and granting other agencies only rights of consultation, the system potentially allows the president to prod or reverse decisions efficiently, by intervening at that focal point. And by incorporating broad rights of consultation, the system scatters information about pending decisions widely and thus raises the chance that controversial matters will reach the president's ear.

Moreover, this sort of consultative structure can be arranged to foster a particular hierarchy of values or goals in a foreign policy that must blend a complex array of competing values. For instance, when it comes to controls on U.S. exports, the general dominance enjoyed by the Defense Department in a bureaucratic system run on unanimity in essence establishes a "fail-safe" system in which conflicts between national security and the economic interests of U.S. firms are resolved in favor of national security, unless the president intervenes. On the other hand, when it comes to COCOM controls on firms in allied countries, the system is arranged so that the Defense Department draws up proposed COCOM limits but the State Department conducts the negotiations with allies, establishing a hierarchy in which harmonious relationships with allies are treated as more important in U.S. foreign policy than failure to control this or that export product. In essence, this complex scheme of lead agencies and consultative rights harnesses the parochial tendencies of each agency to foster wider presidential purposes.

But there are foreseeable problems. The president has defended the autonomy of the export control system against encroachment by groups outside the executive branch largely by his appeals to presidential prerogative and flexibility in matters of national security and foreign policy, appeals that depend for much of their political force on public and congressional fears of imminent foreign danger. Moreover, export controls have in the main been imposed for asserted reasons of national security, a realm in which other political actors have traditionally been less inclined to challenge the president than in broader foreign policy realms such as human rights or even import trade policy.[36]

Many of the new issue areas that must now be integrated into U.S. foreign policy, however, are not realms in which the president has established dominance over other political actors. Indeed, in most instances these are not "new" issues at all, simply ones that have been eclipsed in public attention by cold war security concerns. What is new is their rising prominence and the debate about which issues should take precedence over others in a new American foreign policy adapted to the post–cold war era. Only in rare cases will these be "new" issues in the sense that they have never been drawn into the public realm and therefore have never been the focus of politicking by groups in the public, Congress, or the bureaucracy. The Pentagon's foot dragging and objections in 1992 that it could not obey environmental regulations on the use of ozone-depleting chemicals without compromising the nation's military security, at a time when the Bush administration was under fire from other nations for its uncooperative attitude on global environmental issues, is an example of how "new" foreign policy problems may only be old policy realms thrust together in unaccustomed competition.[37] Compelled to hammer such disparate policy areas as military security and environmental protection into a coordinated foreign policy, the president will be forced to seize command back from the dominant political actors and coalitions that have entrenched themselves in these policy realms, coalitions built on both habit and statutory authority and marked by distinctive modes of political competition.[38] Moreover, the president will be attempting this at a time when the sense of cold war danger that once protected his autonomy in foreign policy no longer holds the nation in thrall.

"MANAGEMENT" IS NOT ENOUGH

It is difficult to see how a president can bring coherence to American foreign policy except by expanding his dominance into new policy realms. And it is difficult to see how he can push away those political competitors who at present have a stronger voice within his own bureaucracy in these realms except through raw personal force—through charismatic leadership that impresses upon the bureaucracy that *he* is their leader, that *his* personal vision must be their obligation, and that they must ignore the clamor of other voices and listen only to his. The episodes of defiance in export control policy related above, in instances involving the president's cabinet officers, suggest that a president cannot assure that his vision prevails if the only soldiers at his command are his top political lieutenants. Moreover, among the seven hundred or so political appointees in the executive branch whom the president can select, the average tenure in office is just two years. No president has seven hundred close and trusted friends, so few appointees will arrive with a sense of personal devotion to the president. All such appointments are intensely political matters, and even top officers may be selected for reasons that have little to do with their commitment to the president.[39] Since most appointees will abandon the president midway through his term, both their loyalty and their effectiveness in advancing his vision are open to doubt.[40] A president who seeks more pervasive and enduring influence will have to appeal for loyalty directly to those deeper in the bureaucratic ranks.

Presidents appear to spend surprisingly little time at such political mobilization aimed directly at the bureaucratic rank-and-file. I recall no instance during my two years in the Carter administration, for example, when the president came to the State Department to appear before his bureaucratic troops. Records of Carter's schedule in an average week confirm my own experience: National Security Adviser Brzezinski got six hours with the president during the week, or 9 percent of Carter's time; the secretaries of state and defense, along with all other cabinet officers, got 15 percent; members of Congress and foreign visitors got another 15 percent; and if there were any appearances before the bureaucratic troops in an average week, they were buried somewhere in the category of "people, ceremonial, and other," commanding 3.5 percent of Carter's hours—roughly the same amount of time he spent having lunch with his wife.[41]

Bureaucrats are, of course, also members of the public and thus audience to a president's general public appearances, press conferences, and televised speeches. But these make the president utterly dependent on the media, and principally television, where the natural dynamic is to dissipate and derogate the force of the president's message.[42] Evening sound bites selected by network editors are a pale substitute for those gestures of attention, grand and small, by which a president can appeal directly to his bureaucracy.

The suggestion that the president should infuse his leadership of the bureaucracy with charismatic elements, both rousing career bureaucrats from horseback and wandering among them on foot to rally them to his banner, runs counter to prevailing opinion among scholars and practitioners. A president who seeks cooperation by enlisting his bureaucrats as true believers, rather than by just appealing to their dispassionate professional ethics as career civil servants, will be challenging the distinction between politics and administration, between his role as partisan and his role as chief executive obliged under the Constitution to see that all laws are faithfully executed. In foreign policy especially, the president has become the dominant source of political energy and initiative in the American government, a fact that underscores the importance of ensuring that he is well advised and aided by public-spirited experts, not sycophants. Hence the objection, that if the president proselytizes among the bureaucracy, the inevitable result will be worse foreign policy. And some would add the argument that if the president drives all other political competitors and voices from the bureaucracy, the result would be an unrepresentative and irresponsible foreign policy as well.

Admittedly, there is a delicate balance to be struck here between charisma and prudence. A president who goes too far in imposing his vision on his subordinates risks undermining his own purposes in several ways. It is possible, for instance, that the narrower visions of the bureaucracy, however parochial, nevertheless contain ideas more promising than the president's own, and a president may lose this creative potential if pursuit of his own vision monopolizes the attention and resources of the bureaucracy. The president's proselytizing is not likely to be equally successful throughout each bureaucratic unit and all at once, and the resulting unevenness in mobilization may create divisions among professional staff and stresses among agencies that actually diminish coordination and efficient performance. The more effective the president is in mobilizing support, the more he may provoke countermobilization by those who oppose his vision, including efforts by those within the bureaucracy to secure protectors outside the executive branch. When the president forges a charismatic bond with those lower in the bureaucracy by mutual pledges of support, he ties himself to their decisions and actions in a way that may reduce his future room for maneuver. Where thousands of bureaucratic exegetes persuade themselves, on the basis of fragmentary clues from the Oval Office, that they understand exactly what the president wants, the sum of the bureaucratic parts will not necessarily be a cohesive foreign policy. Not least, if bureaucrats are encouraged to be true believers rather than dispassionate professionals, what happens to bureaucratic performance when one president gives way to another on Inauguration Day?[43]

These are all powerful arguments for caution. And yet they merely summarize problems that are inherent in bureaucracy. The theoretical distinction between politics and administration has already been blurred because the public and Congress have traditionally granted the president wide discretion in shaping foreign policy, at least in key issue areas, which he in turn has delegated to the bureaucracy. Seemingly technical bureaucratic decisions—should faster retargetting software be installed in the guidance computers of intercontinental ballistic missles? should a rural development aid project in the third world include birth control programs? should a free trade accord exclude pharmaceutical products?—often embody issues that can be resolved only by choices among core values, a situation that makes them as much politics as administration.[44] Unless the president rallies the executive agencies to his political vision, they will rally to their own when they make such decisions, with all the confusion and dissipation of energies which that implies. The president need not, and should not, rouse his subordinates to such partisanship that they abandon their professional objectivity; the general rouses the fervor of his troops so they will look to him and forget themselves, not so they will forget how to fight. The issue, therefore, is not whether the president should "disrupt" the bureaucracy by intruding in its daily business, but how best to do it so that coordinated policy results.

Mainstream advice from scholars and practitioners urges the president to apply adroit management and manipulation of the bureaucratic structure, not charismatic leadership, to achieve coordination. It seems to be taken for granted that the president can establish relationships of loyalty no further than his National Security Council staff, because alone among the foreign policy bureaucracies, the NSC staff is small, located within the White House compound, and uncorrupted by institutional interests or responsibilities other than to serve the president. Apparently "leadership" can reach as deep as the NSC staff; below that, "management" must prevail.

Wise management is certainly indispensable, but it is not sufficient. In their pursuit of effective management in foreign policy, presidents have oscillated between reliance on their cabinet secretaries and dependence on the National Security Council staff. By hard experience, most have concluded that the NSC staff is the better management tool. As a consequence, the NSC staff has grown apace, and although the professional staff still numbers fewer than a hundred, relationships of individual members with the president long ago became impersonal and bureaucratic. The NSC staff is already judged too small to coordinate policy properly, yet those who argue for an enlarged NSC staff concede that expansion may transmute the NSC into just another foreign policy bureaucracy with its own institutional dynamic.[45] Unless the president can raise the prospects of cooperation from the foreign policy bureaucracy in some manner that does not also raise the burdens of supervision, neither he nor his NSC staff will be able to cope. In the realm of foreign policy, often what the president needs will be as ephemeral as the intelligence analyst's alertness for the event that may never occur, the captain's steeled nerve for the battle no one wants, or the diplomat's stratagem for the crisis that has not yet happened. To gain such anticipatory cooperation, the president must arouse the kind of loyalty that yearns to serve and that supervises itself—the loyalty that springs from a shared vision.

And if the president woos and wins the bureaucracy, will winning lead to an imperial executive that is beyond the reach of the public and Congress and to policies that are unrepresentative and irresponsible? The escapades of President Reagan's National Security Adviser John Poindexter, NSC staff member Oliver North, CIA Director William Casey, and Assistant Secretary of State Elliott Abrams during the Iran-Contra affair in 1985–86 seem to give substance to the fear. Opinion polls showed consistent public opposition to Reagan's enthusiasm for the Contra rebels in their guerrilla war against the Sandinista government of Nicaragua, and Congress had passed legislation banning any executive agency from directly or indirectly supporting military operations in Nicaragua. Yet in their zeal to carry out their own interpretation of the president's wishes, Abrams, Casey, North, and Poindexter defied the public, Congress, and perhaps even the president by conducting and concealing a program of covert military supply for the Contras.

To be sure, the Iran-Contra episode illustrates the hazards of rogue behavior among zealous bureaucrats. But viewed in a wider context, it also underscores the substantial power wielded by Congress and the public over the foreign policy bureaucracy, even when bureaucrats try to be defiant. Article II of the Constitution vests the executive power in the president but remains utterly silent about what he is to be executive of (except to imply, by making the President commander in chief, that at least there would be an army and a navy). Hence, the entire bureaucratic apparatus of the modern executive is formally the creation of Congress, generated through legislation that Congress can extend or rescind (subject to the president's veto power). So long as Congress retains the constitutional authority to create or abolish executive agencies and positions, to set their policy mandates, to conduct oversight investigations, to regulate executive budgets, and ultimately to outlaw specific behavior by agencies and bureaucrats, and so long as the electorate subjects both Congress and the president to review at election time, the president can never completely drive political competitors out of "his" bureaucracy.

Nevertheless, as the president strives for policy coherence by asserting greater dominance within the executive branch, both Congress and the public will find fewer allies within the bureaucracy to help them thwart the president and thus will be pushed further away from the operational details of foreign policy. Arguably this is what the Framers intended in the Constitution. A feeble executive implies feeble government, Alexander Hamilton argued, which in practice means bad government. A sound republic requires a single energetic executive, a role the president can play within the bureaucracy only if his power is not dissipated by subjecting him "in whole or in part to the control and cooperation of others, in the capacity of counselors to him."[46]

CONCLUSION

Presidents have been warned for decades that an era was approaching in which the nation's foreign policy would of necessity become more complex, as new and more complex global issues clamored for attention. Yet for decades, presidents have held the new era in abeyance, keeping the nation's attention riveted on the Soviet threat and subordinating other foreign policy issues to the requirements of cold war military security. (Even President Carter's early efforts to end the U.S. "obsession" with the Soviet threat foundered on the Euro-missiles confrontation and the Soviet invasion of Afghanistan in 1979.) Now, the new era can no longer be postponed.

Fortunately, presidents will not be cast utterly adrift in this new era. The nation's experience with export controls over the past forty years shows that foreign policy issues of great complexity can indeed be brought together to yield a consistent and coherent policy, insulated from the transient political pressures of interest groups and Congress that can produce turbulence in policy, while accommodating the important consultative role that the public and Congress must play in setting the nation's foreign policy course.

Yet the experience with export controls also shows that to ensure continuity and strategic flexibility in foreign policy, the president must assert dominance within the executive branch, which means besting his political competitors in gaining the attention and cooperation of the bureaucracy. These competitors include Congress, which enjoys constitutional authority to establish, fund, and oversee executive agencies, as well as interest groups, which will also have prevailed on Congress to shape executive agencies and mandates in ways that favor their own particular views. Under the circumstances, the president is not likely to gain dominance over foreign policy by simply "managing" a bureaucratic structure built by his political competitors. Nor are his political competitors likely to yield the redoubts they have constructed for themselves within the bureaucracy simply because he asks them to retreat. To win this contest, the president must appeal directly to the bureaucratic ranks, persuading them to desert the camp of his competitors and regroup beneath his banner. And to keep his troops from scattering again, the president must understand that campaigning for the hearts and minds of his own subordinates is not a distraction from executive leadership but an indispensable element of it.

Campaigning among the bureaucratic ranks demands no more from a president than what it took to reach the Oval Office in the first place. No candidate for the presidency ever succeeded in winning the loyalty of voters in Iowa or New Hampshire by treating them in the way presidents are urged to handle the bureaucracy, by issuing directives from the Oval Office and leaving it to a few appointees to shake all the hands and kiss all the babies. So the problem is not an absence of the necessary political skills. The problem is a stunted conception of presidential leadership. If presidents hope to clear and then dominate the political space they need to forge a coherent foreign policy responsive to the challenges of a new era, they must first return to a more elemental understanding of the president as charismatic leader. It is a conception of leadership that the times demand.

Notes

I am grateful for generous and helpful comments on an earlier draft of this paper from Robert J. Art, Francis E. Rourke, John T. Tierney, and David A. Deese.

1. See Betty Glad, *Jimmy Carter: In Search of the Great White House* (New York: W. W. Norton, 1980), p. 476; I. M. Destler, "National Security II: The Rise of the Assistant (1961–1981)," in Hugh Heclo and Lester Salamon, eds., *The Illusion of Presidential Government* (Boulder, Colo.: Westview Press, 1981), pp. 277–78.

2. See I. M. Destler, Leslie Gelb, and Anthony Lake, *Our Own Worst Enemy: The Unmaking of American Foreign Policy* (New York: Simon and Schuster, 1984), pp. 11–30 and 261–288; Destler, "Rise of the Assistant," p. 279.

3. Although even then, I believed that the image of bureaucrats as saboteurs was vastly overdone in the bureaucratic politics literature, to the point where it encouraged presidential foolishness. See Donald L. Hafner, "Bureaucratic Politics and 'Those Friggin' Missiles': JFK, Cuba, and U.S. Missiles in Turkey," *Orbis* vol. 21, no. 2 (Summer 1977), pp. 307–33.

4. Paul Stares, *The Militarization of Space: U.S. Policy 1945–1984* (Ithaca, N.Y.: Cornell University Press, 1985), p. 199.

5. Max Weber, *On Charisma and Institution Building,* ed. S. N. Eisenstadt (Chicago: University of Chicago Press, 1968), esp. pp. 48–65. Richard Neustadt prefers the metaphor of the preacher, invoking Teddy Roosevelt's image of the presidency as a bully pulpit. But preachers generally call upon their listeners to look inward; a president needs the bureaucracy to look outward. See Richard Neustadt, "The Clerk against the Preacher," in James S. Young, ed., *Problems and Prospects of Presidential Leadership in the 1980s* (Lanham, Md.: University Press of America, 1982), pp. 1–36.

6. See, for instance, Destler, Gelb, and Lake, *Our Own Worst Enemy,* esp. pp. 11–30, 271–88.

7. Richard Neustadt, *Presidential Power: The Politics of Leadership* (New York: John Wiley and Sons, 1960), p. 34.

8. John Helmer, "The Presidential Office: Velvet Fist in an Iron Glove," in Heclo and Salamon, eds., *Illusion of Presidential Government,* p. 79.

9. See Aaron Friedberg, "Is the United States Capable of Acting Strategically?" *Washington Quarterly* 14 (Winter 1991): 5–23.

10. Terry Moe, "The Politics of Bureaucratic Structure," in John E. Chubb and Paul E. Peterson, eds., *Can the Government Govern?* (Washington, D.C.: Brookings Institution, 1989), pp. 267–329; Randall B. Ripley and Grace A. Franklin, *Congress, the Bureaucracy, and Public Policy,* 5th ed. (Pacific Grove, Calif.: Brooks/Cole, 1990).

11. See Christopher C. Shoemaker, *The NSC Staff: Counseling the Council* (Boulder, Colo.: Westview Press, 1991), pp. 6–19. The title of the special assistant has varied under successive presidents. Most recently, the position has come to be known as the national security adviser, which is rather misleading since the president has many advisers on national security, including all the formal members of the National Security Council.

12. Unless otherwise noted, the discussion of the bureaucratic structure here draws from John R. McIntyre, "The Distribution of Power and the Inter-Agency Politics of Licensing East-West High-Technology Trade," in Gary K. Bertsch, ed., *Controlling East-West Trade and Technology Transfer* (Durham, N.C.: Duke University Press, 1988), pp. 97–133; National Academy of Sciences, *Finding Common Ground: U.S. Export Controls in a Changed Global Environment* (Washington, D.C.: National Academy Press, 1991); John Heinz, *U.S. Strategic Trade: An Export Control System for the 1990s* (Boulder, Colo.: Westview Press, 1991).

13. The executive branch has never published the COCOM list, even though Congress mandated its publication in the Omnibus Trade Bill of 1988. (The British and Canadian governments have published the list, however, which makes the U.S. secrecy rather pointless.) In 1968, the Commerce Department's CCL apparently contained more than 1,200 categories of controlled items that were not on the COCOM list. By 1972, that gap had narrowed to 460 categories; by 1978, to 84 categories; and by 1979, to perhaps 38 categories. See William Long, *U.S. Export Control Policy: Executive Autonomy vs. Congressional Reform* (New York: Columbia University Press, 1989), pp. 31, 50.

14. In 1987, the Commodity Control List specifying the characteristics of each commodity subject to control contained 240 entries. The categories themselves may be quite narrow (e.g., "pulse modulators capable of providing electric impulses of peak power exceeding 20 MW or of a duration of less than 0.1 microsecond, or with a duty cycle in excess of 0.005") or exceedingly broad (e.g., "other electronic and precision instruments, including photographic equipment and film, not elsewhere specified, and parts and accessories, not elsewhere specified"). See National Academy of Sciences, *Balancing the National Interest* (Washington, D.C.: National Academy Press, 1987), pp. 81–82.

15. National Academy of Sciences, *Finding Common Ground,* pp. 80, 101; *Balancing the National Interest,* pp. 107, 116–21. The total trade activity covered by export licensing was vastly larger than the volume of direct trade with Soviet bloc countries because roughly 90

percent of all U.S. export licenses governed the export or reexport of goods from one western country to another to avert Soviet acquisition of controlled U.S. technologies through third parties. It is worth noting that the data gathered through the licensing procedure are a potential windfall for U.S. intelligence agencies trying to keep tabs generally on economic, military, and foreign policy developments in other countries, whether friend or foe. For this reason, extensive export controls are quite likely to survive the demise of the Soviet bloc. The formal justification will undoubtedly change—for instance, to prevent the spread of nuclear and ballistic missile weapons capabilities. I am grateful to Thomas W. Graham for this insight.

16. For a more detailed discussion of the politics of export controls, see Long, *U.S. Export Control Policy.*

17. Ibid., pp. 60–61, 76, 101–102.

18. On this point, see Chapters 1, 5. See also Robert Pastor, "The Cry-and-Sigh Syndrome: Congress and Trade Policy," in Allen Schick, ed., *Making Economic Policy in Congress* (Washington, D.C.: American Enterprise Institute, 1983), p. 160.

19. Long, *U.S. Export Control Policy,* pp. 40–41.

20. Steven Elliott, "Distribution of Power and the U.S. Politics of East-West Energy Trade Controls," in Bertsch, ed., *Controlling East-West Trade and Technology Transfer,* p. 94.

21. Long, *U.S. Export Control Policy,* pp. 42–43, 98–99.

22. President Bush's veto was sustained in the Senate, by a vote of 60–38, in March 1992. Bush was aided by farm-state senators from both parties, who feared that Beijing might retaliate by restricting sales of U.S. food exports to China. "China Will Keep Trade Privileges," *New York Times,* March 19, 1992.

23. It is conceivable that presidents have been able to protect export controls from outside meddling because Congress was never sincere in its efforts to interfere, preferring—as it did back in 1934 when it surrendered its role in tariff policy—to give the president wide authority and thereby make him take the heat from offended interest groups. However, the vigor with which Congress has tried to intrude in export policy, and with which presidents have fought to resist, does not seem to support this interpretation. In any case, the point remains that the withdrawal of Congress from the bureaucratic battle has not guaranteed the president dominance over his own subordinates. I am grateful to Robert J. Art for insights on this matter.

24. Alexander George, *Presidential Decisionmaking in Foreign Policy* (Boulder, Colo.: Westview Press, 1980), p. 10.

25. *Report of the President's Special Review Board,* February 26, 1987, p. V2.

26. George, *Presidential Decisionmaking in Foreign Policy,* pp. 112–13.

27. Henry Kissinger, *White House Years* (Boston: Little, Brown, 1979), pp. 152–55.

28. Zbigniew Brzezinski, *Power and Principle: Memoirs of the National Security Advisor* (New York: Farrar, Straus, Giroux, 1983), p. 323.

29. Ibid., p. 324. Brzezinski himself was not above engaging in a bit of independent foreign policy. Reportedly he established his own secret channel of communication with Iran during the hostage crisis, and when confronted on the matter by Secretary of State Cyrus Vance, "Brzezinski just flat out lied and denied it to the President." Walter Isaacson and Evan Thomas, *The Wise Men* (New York: Simon and Schuster, 1986), p. 727.

30. McIntyre, "Distribution of Power," p. 117. See also Long, *U.S. Export Control Policy,* p. 96.

31. Both Congress and presidents have been frustrated in their efforts to end U.S. export controls that prevent U.S. firms from selling products overseas that are already being exported by foreign firms. In 1979, Congress mandated that all export control categories should be reviewed to determine whether the products were being exported by foreign suppliers. But Congress failed to specify how promptly these reviews should be done. Consequently, in the first four years after Congress mandated them, only twenty reviews had been completed and only three items had been decontrolled. Apparently fearing that such interagency resistance to decontrols would provoke the allies into abandoning the entire COCOM system at the COCOM meeting of 1990, President George Bush was able to present a coherent decontrol plan to COCOM only by overriding bureaucratic process and applying pressure on the participating agencies to compel significant loosening of restrictions. See National Academy of Sciences, *Finding Common Ground,* pp. 96, 98; *Balancing the National Interest,* p. 175.

32. Bruce Jentleson, *Pipeline Politics* (Ithaca, N.Y.: Cornell University Press, 1986), p. 210.

33. See Long, *U.S. Export Control Policy,* p. 105: "As the organizational paradigm would

predict, fractionalized power in the export control administration encourages organizational parochialism. This attitude and approach, when coupled with standard operating procedures and negotiated solutions to policy questions, results in persistent, inflexible policies not easily disturbed by the intervention of government leaders and likely to change only incrementally, if at all."

34. *Commission on the Organization of the Government for the Conduct of Foreign Policy Report* (Washington, D.C.: Government Printing Office, June 1975), 4:447, quoted in McIntyre, "Distribution of Power," p. 121.

35. See Long, *U.S. Export Control Policy,* pp. 8–9, 38, 105, 106.

36. This distinction between the politics of national security policy and the politics of wider foreign policy is evident in a paradoxical way. Congress has allowed the president to embargo a wider variety of goods under general foreign policy arguments (President Carter's embargoes against the Soviets following their invasion of Afghanistan in 1979, for instance) but with greater restrictions on presidential action. The 1979 and 1985 laws on export controls, for instance, require annual review of all controls imposed for foreign policy reasons, and the 1985 legislation requires that the president consult with Congress and affected businesses before imposing such controls. Neither restraint applies if the controls are imposed for national security reasons. See *Finding Common Ground,* p. 64.

37. See "Military Is Seen Stalling on Ozone," *New York Times,* March 21, 1992. The transnational ties among environmentalist groups may be another example of the phenomenon David A. Deese discusses in Chapter 1: the increasing permeability of U.S. domestic politics to international influence and pressures.

38. For instance, environmental restraints fall in the broad realm of protective regulatory policy, where executive agencies, Congress, and private sector interest groups all exert roughly equal influence and the dominant political style is to resolve conflicts through bargaining and compromise. If environmental regulations compelled the Pentagon to curtail specific military operations, this would intrude on a policy realm in which the president and executive agencies bargain and compromise among themselves and Congress and outside groups exert little influence. If environmental regulations compelled the Pentagon to close military bases, however, this would intrude on a policy realm in which the Pentagon and Congress have traditionally shared influence and the dominant political style has been logrolling, an alliance of the parties so that all can get what they want. Scrambling these players and political styles together, one might imagine, could be like putting a football team and a baseball team together on a hockey rink and telling them to play soccer: unless someone serves as umpire and imposes new rules, chaos would reign. See Ripley and Franklin, *Congress, the Bureaucracy, and Public Policy.*

39. President William Clinton's efforts to appoint a cabinet that "looks like America" and to draw in "more than the usual white boys who run Washington" as departmental undersecretaries were laudable in many respects. Nevertheless, they apparently resulted in some appointments that would not otherwise have been the first choice of Clinton or his chief advisers. See "Diversity Pledge Slows Top Choices," *New York Times,* January 12, 1993. Once made, such appointments can be very difficult to undo, again for political reasons. Dean Rusk was not President John Kennedy's preferred choice as his secretary of state, but Kennedy allowed himself to be talked out of his preferences on political grounds. And reportedly, when Rusk's performance subsequently fell short of the president's expectations, Kennedy kept him on because (as McGeorge Bundy, Kennedy's security adviser, told friends) "you can't fire the Secretary of State, particularly if you hired him after only one meeting." See Arthur M. Schlesinger, Jr., *Robert Kennedy and His Times* (New York: Ballantine Books, 1978), pp. 238–40; David Halberstam, *The Best and the Brightest* (New York: Random House, 1972), p. 89.

40. See Richard Neustadt, *Presidential Power* (New York: Free Press, 1990), p. 235; Richard Rose, *The Postmodern President* (Chatham, N.J.: Chatham House Publishers, 1988), pp. 168–74. During the 1992 presidential campaign, candidate H. Ross Perot struck a responsive chord among the electorate with his criticisms of the Washington "revolving door," those instances in which a high official leaves government service to become a lobbyist or executive for a group, foreign government, or firm seeking to influence the agency with which the official once served. The more stringent ethics rules proposed by President Clinton to stop the revolving door suggest that presidents, too, share a suspicion that some high-level political appointees have their own future career interests more in mind than the president's programs.

41. Rose, *Postmodern President,* p. 152.

42. Neustadt, "Clerk against the Preacher," pp. 12–16.

43. Hugh Heclo, "An Executive's Success Can Have Costs," in Lester Salamon and Michael Lund, eds., *The Reagan Presidency and the Government of America* (Washington, D.C.: Urban Institute, 1984), pp. 371–74.

44. See Richard Nathan, "Political Administration Is Legitimate," in ibid., pp. 375–79.

45. See Carnes Lord, *The Presidency and the Management of National Security* (New York: Free Press, 1988); Shoemaker, *NSC Staff*, p. 117. I recall my surprise when the NSC staff member principally responsible for interagency coordination in arms control in the Carter administration told me that he had seen (but not met) the president only once, when his children had been invited to the annual White House Easter egg hunt. Presidential relationships even with senior NSC staff had been in decline since the Kennedy administration. Brzezinski tried to reverse the trend but succeeded in getting President Carter to attend only two NSC staff meetings in four years. Brzezinski fell back on weekly staff briefings in which he gave accounts of his own meetings with Carter, "so that vicariously, if not directly, they have a sense of engagement with a man for whom they are working so hard." See Dom Bonafede, "Brzezinski: Stepping Out of His Backstage Role," *National Journal*, October 14, 1977, pp. 1596–1601.

46. Alexander Hamilton, *The Federalist Papers*, no. 70. (New York: New American Library, 1961), pp. 423–31.

3 / Presidents, Opinion, and Institutional Leadership

BERT A. ROCKMAN

Because of constitutional interpretations of presidential prerogatives in foreign policy and the president's unique capacity to act, leadership in foreign policy is normally thought to be the particular responsibility of the president. Neither constitutionally nor realistically, however, is foreign policy an exclusive property right of the president. It is a shared responsibility wherein the president confronts considerable constraints. This chapter argues that in the 1990s the curbs placed on presidential power in making foreign policy by elite opinion, the media, and Congress are increasing as a result of fundamental changes in the international system and the character of American governance.

To be sure, Congress has never given absolute discretion to presidents. It always has been a participant, as is its constitutional right.[1] Similarly, the president is no foreign policy lamb. He has powerful constitutional prerogatives to initiate action and by constitutional interpretation is given the mantle of being "the sole organ of the nation in its external relations."[2]

Three major factors help shape the realities of the presidential role in foreign policy: the nature of the issues, the political context, and executive leadership. I lead off this chapter with a description and analysis of these factors. After identifying these shaping factors, I turn to a discussion of present trends in the context of U.S. foreign policy—trends that are moving the system of foreign policy making away from its presidency-centric base. Following this analysis are brief discussions of relative advantages that Congress and the president each bring to the foreign policy process and vis-à-vis each other. The conclusion identifies elements of constancy and fluidity in public opinion with respect to foreign policy and speculates about the ability of the country's political leadership, especially the president, to lead opinion in the new and more ambiguous international circumstances of the 1990s.

FACTORS SHAPING LEADERSHIP IN FOREIGN POLICY

As noted, three main factors tend to shape the relative influence and leadership roles of the president and Congress: (1) the nature of the issues being called into play; (2) the political context and policy antecedents in which these issues arise; and (3) the nature and style of executive leadership in foreign policy.

The Nature of the Issue

It has been a staple of the public policy literature in political science for some time that different types of policies evoke different sets of political and institutional actors. These, in turn, determine the political constituencies and elites that will continue to be attentive to the policies and presumably influential in shaping them.[3] There is some evidence that these fixed structural characteristics of policies and constituencies may be exaggerated,[4] but issues still seem a sensible place to begin to sketch a likely set of influences even if the sketch ultimately requires alteration.

A variety of issues fall under the broad rubric of foreign policy. The categories of foreign policy issues outlined below, if not wholly inclusive, reflect the broad variety. They are presented in a sequence that roughly measures the extent of presidential prerogative, or room to maneuver, beginning with the more secretive and crisis-oriented issues, such as intelligence operations, where presidential control generally can be expected to be greater. They then proceed through the more public, standing issues such as trade, foreign assistance, ethnic interventions, and human rights, in which policy is developed over an extended period and the roles of public opinion, the media, and Congress are likely to be more central. This section thus briefly analyzes intelligence operations and assessments; strategic policy; trade issues; foreign assistance; ethnic interventions; and human rights issues.

Intelligence Operations and Assessments By their nature, intelligence operations and assessments involve small circles of executive actors, operatives, and intelligence experts in and out of government. The scope for participation has expanded since the 1970s when the Senate and House created intelligence oversight committees. Congressional participation, however, is largely ex post facto, and it emerges typically after some event has brought it to selective public attention. The world of intelligence still remains a narrow one even if modestly larger than it had been previous to the formation of the intelligence committees.

Two related trends have helped widen attentiveness to intelligence activities and estimates. One is the breakdown in deference to executive authority, which Presidents Lyndon Johnson and Richard Nixon inadvertently helped promote and which divided government tends to exacerbate. The second is the development of a less deferential elite press attentive to and, above all, more willing to report on misguided intelligence operations or conflicts in intelligence estimates.

Except, however, for a monumental misdeed (such as Iran-Contra), intelligence operations and assessments remain a domain largely populated by interested experts. The range of experts so involved, however, seems to have expanded, which is perhaps a natural consequence of the more public nature of intelligence activity in the present period.

Strategic Policy There are two ways to think about strategic policy. One is conceptual, the other operational. The conceptual issues have been debated by defense intellectuals in think tanks, government agencies, universities, and specialists among congressional staffers. They have been the preserve of a highly informed elite—an elite that does not necessarily include the actual decision makers. Given the recently changed nature of the world and the role of strategic policy in it, this elite circle, like the defense industry itself, is becoming less influential.

Operationally, however, strategic policy is usually expressed in the context of defense procurement, including the size and structure of the defense budget and the targeting of weapons systems. Since in this case weapons systems programs define policy, the operational has more profound impact than the conceptual, whether or not the two are fully compatible with one another.

If the doctrinal elements of strategic policy are principally the preserve of defense intellectuals in government and out, strategic weapons procurement broadens the net of interested actors. Here, the president is likely to lead (and may be influenced by the bureaucracy and the services). However, Congress can, and often does, exert powerful influences on strategic procurement. Because weapons systems provide ample "pork," they understandably evoke strong interest lobbying and constituency tending on the part of key members of the authorizing committees and, especially, appropriations subcommittees. Strategic weapons decisions, while allocating goodly amounts of pork, are also decisions about national defense postures as well. Although, logically, the postures should determine the weapons systems, more frequently the weapons systems decisions induce debate about the postures or conceptual strategies. Usually, an even more focused group is involved at this level of debate, composed of a few key defense buffs in Congress and the actors associated with doctrinal definitions of strategic policy. All in all, strategic policy is largely an arcane area for public opinion, and the public has limited information. While the public is capable of reacting to broad themes such as more, less, or the same amount of spending for defense, it is rarely if ever able to differentiate debates about esoteric weapons systems or strategic doctrines. Indeed, to some extent that is even so in regard to rank-and-file members of Congress and an occasional president or two.

Because controversy over strategic weapons systems inevitably does bring into play controversies over strategic doctrine, congressional involvement in strategic weapons systems is more likely to follow doctrinal lines than purely constituency interest lines.[5]

Trade Issues Trade issues, perhaps more than any other set of foreign policy–related issues, attract intense interest group involvement. Producer groups and labor unions are especially involved, and regional interests also are called into play. Because of some of the interest groups that perceive themselves to be affected—labor unions for example—trade issues, on the surface, at least, carry partisan overtones. To some significant degree, this has been so throughout American history. Throughout the latter part of the nineteenth century, when the Democrats were the agrarian party and the Republicans were the party of industry (and also labor), the confluence of interests, region, and party came together. Then, Republicans were the party of protection, and Democrats were the party of free trade. In their public posturing, at least, these images are today reversed even though sentiment is, in fact, much further toward the free trade spectrum in both parties than it had been before the onset of World War II.

All of this means that trade issues are on the political agenda, and, as with many domestic policies, some groups gain visible benefits while others suffer visible losses. In fact, largely because interest and regional pressures can become so intense at the visible level, Congress has wound up seeking to shift direct responsibility for decision making to the executive. The logic of trade issues is much like that of tax bills. The temptation to provide ornaments for particular interests is compelling unless all parties find it in their interest to cede authority. That is why tax bills often have closed rules. Similarly, Congress also gives the executive a fair amount of slack to negotiate agreements (fast-tracking) or to seek redress for aggrieved industries (Section 301 of the 1974 Trade Act).[6]

However, there clearly seems to be some tendency to rein in the slack given to the executive as trade deficits mount, especially when labor-intensive industries seem to be the victims of the trade imbalances. Given the profound effects of trade policy on people's livelihoods and the rather abstract explanations to be made for the virtues of open markets, there is no doubt that trade issues will continue to exert pressure along domestic political fault lines. Particularly, perceptions of unreciprocated openings to markets will beget increased interest group activity, even if such activity is mainly to take advantage of these perceptions to protect existing markets. Such perceptions, in turn, will beget increased congressional intervention. But there are limits to the degree that Congress wants directly to manage trade policy and thus be responsible for it.

Foreign Assistance Is there anything as unpopular with the general public as foreign aid? Not likely. The general public resistance to foreign aid inevitably translates to some degree into elite resistance, especially, but not exclusively, in Congress. The extent to which Congress resists, however, tends to vary with the willingness of the president to expose himself to risks on the issue. In the absence of presidential leadership, members of Congress are unlikely to expose themselves to political risk. Although it is rare, occasionally Congress does actually provide leadership for foreign assistance. The case of aid to Poland and Hungary in 1990 fits such a circumstance.

If one problem is that assistance to anyone outside of the United States is unpopular, that is less true with regard to assistance to Israel (or any other country with a major internal constituency). The general public appears not to treat assistance to Israel as a special case, but backers of Israel in the United States clearly do. Israel is probably the only country for which support can translate into votes. Understandably, therefore, two countries, Israel and Egypt, soak up the bulk of U.S. bilateral assistance—and Egypt does so mainly because its aid is coupled to Israel's.

Ethnic Interventions The prospects of interventions in ethnic conflicts abroad are usually as unpopular as foreign aid to the general public, but, of course, they hold special salience for publics attached to those conflicts. Here, as in the Middle East or, more recently, in the Balkans, there is specific group mobilization that reflects intense interests. This mobilization is played out before a largely indifferent audience, however. Often such conflicts pit Congress, responding to the petitions of an aggrieved ethnic group, against an executive that is more likely to dismiss those grievances on the grounds of national interest or the dangers of intervention. The conflict between Greece and Turkey, for example, over Cyprus has a simple domestic political equation that fails to match the perceived international security needs of the United States. There are more Greek Americans than Turkish Americans, and the former tend to identify strongly with their heritage. Greek-American cohesiveness makes for effective lobbying and, therefore, for lots of congressional involvement, at least of a symbolic sort. On the other hand, Turkey, not Greece, was the key to the North Atlantic Treaty Organization's (NATO) southern flank strategy, and no American president was likely to risk alienating the Turkish government. Similar arguments could be made about the Israeli-Arab conflict and the particularly strong capacity of Jewish Americans to mobilize around what they perceive as threats to Israel. That capacity works better at the congressional level than the presidential one. Because Congress is especially susceptible to ethnic group lobbies that represent mobilized publics, there is normally a source of tension here between Congress and the president. Much of this tension, however, is played out at the symbolic level, where members of Congress have the opportunity to win plaudits while blaming the executive for frustrating their efforts.

Human Rights Issues These issues are largely the domain of attentive publics in the form of watchdog groups and of interested members of Congress and their staffs. In the executive, human rights as an issue inevitably faces difficulty with elements of the bureaucracy that are more attuned to considerations of Realpolitik. The extent to which human rights considerations are strengthened within the executive has a lot to do with the degree to which there is presidential commitment behind them. Jimmy Carter had much commitment to human rights, whereas the Nixon-Ford and Reagan-Bush administrations had rather little. Although, ironically, President Gerald Ford was pressured from the right wing of his party for seeming to ignore victims of repression in the Soviet Union—a campaign theme Ronald

Reagan used in his efforts to wrest the 1976 Republican presidential nomination from Ford—the human rights issue subsequently has cleaved, to some degree, along party lines. This cleavage is the result of two things. First, divided government, by usually pitting a Democratic Congress against a Republican White House, gives the Democrats in Congress an opportunity to score points against Republican administrations that appear insensitive to human rights. Second, the rise of détente in the Gorbachev years and the collapse of the cold war altogether after the disappearance of the Soviet Union removed the basis for a common position between the right and the left on human rights. Thus, aside from China, human rights became increasingly a partisan issue pitting a Democratic Congress against a Republican executive.

The Political Context and Policy Antecedents

What dictates the nature of congressional involvement in setting policy and in oversight? Alternatively, under what circumstances is presidential authority more likely to be given deference? In what ways do these involve public opinion? Three contexts seem to be especially important: (1) the depth and nature of partisan cleavages; (2) the level of presidential popularity; and (3) recent experiences in executive-legislative relations.

Partisan Fault Lines At the mass level, party comes into play in influencing support for the president's policies when the president is used as a source cue for approval or disapproval. This relationship seems to be especially true during periods of prolonged military engagement.[7] Under such conditions of protracted engagement, support or opposition tends to cleave along party lines. And when foreign policy divisions are salient, elites are especially likely to divide along party lines.

The key here is that party divisions on foreign policy are rarely stable for long periods of time. New issues come to be defined and older ones redefined along partisan fault lines. During the 1930s, and, to a degree, during the post-war period, some congressional Republicans sought to limit presidential discretion in foreign policy making. By the 1970s and into the 1980s, the shoe was on the other foot as other Democrats in Congress sought to inhibit the discretionary powers of the president. Between 1948 and 1964, a period that Aaron Wildavsky identifies as the "two presidencies" era when presidents presumably were granted greater deference from Congress than they have been subsequently, foreign policy was highly salient but not politically divisive.[8] The challenge of the cold war made foreign policy more salient. Yet there was also a good bit of bipartisanship. In the late 1940s, a Democratic president (Harry Truman) gained the support of key Republicans for a nonisolationist policy, for assistance to the Greek government under pressure from communist forces, and for the reconstruction of a

European continent that had been devastated by World War II. In the 1950s, a Republican president (Dwight Eisenhower) got the support of the Democratic leadership in both chambers for his foreign aid and collective security interests.

More frequently, however, foreign policy is divisive. As David W. Rohde points out in Chapter 4, the fault lines in foreign policy were increasingly revitalized along the partisan divide. Moreover, these partisan divisions also led to increased assertiveness on the part of Democratic majorities on the floor. The Reagan period especially brought sharp foreign policy divisions to the fore and generated considerable party-based conflict in foreign policy.[9]

Presidential Popularity The relationship between presidential popularity and presidential discretion in foreign policy is abundantly controversial. The issues have to do with the extent to and ways in which presidential popularity is mediated through Congress—if it is mediated at all. Charles W. Ostrom, Jr., and Brian L. Job, for example, argue that presidential popularity is related to the use of force in international disputes: the more popular a president is, the more likely he is to use force.[10]

How much discretion a president has, of course, also depends on the nature of the issue. The issue may determine whether congressional action is needed or whether presidential action can be initiated subject only to legislative review. Popularity may lead presidents to take risks where they can initiate action. But where presidents are dependent on Congress, they are more likely to be affected by the favorable or unfavorable distribution of congressional seats and the controversy that attaches to a proposed policy alteration.

Executive-Legislative Relations Consensus breeds trust; conflict breeds distrust. The cold war perceptions of the 1950s largely created an environment of deference to the president and cooperation between the branches despite the existence of divided government for most of that period. Conflicting perceptions of the country's foreign policy needs in the midst of divided government, however, are certainly likely to lead to a less harmonious relationship.

Over the past two decades, divisions on foreign policy have been sharp, and divided government has prevailed most of the time. Not very surprisingly, then, the pattern has been similar to U.S. domestic policy making, namely, efforts by the legislature to establish restraints on the executive and efforts on the part of the president to act as though there was no legislature. The presence of a unified government should help diminish these divisions, but it by no means ensures smooth sailing for any president. The fact is that Congress constitutionally is a vital element in the foreign policy process. Crisis and bipartisanship plus judicial interpretations of the presidential role in foreign policy over the years weakened congressional assertiveness. Distrust of the executive and growing partisanship reinvigorated congressional assertiveness, and it is probable that there will be some decline in its expression once the party divide between Congress and the executive is closed. Even so, Congress does not seem ready to give up its claims to policy making in foreign affairs regardless of the political character of the executive.

Executive Leadership

What are the essential ingredients of presidential leadership in foreign policy? They might be characterized as goals, clarity, commitment, and consensus building. Most presidencies tend to fall short on consensus building. The close vote on a resolution to empower the president to commit U.S. forces to combat in the Persian Gulf at the outset of 1991 illustrates that the consensus on the Gulf War was created post hoc by its success rather than in its political preparations. The reason presidents often eschew consensus building is that they tend to identify foreign policy as strictly within their domain. They justify that conception through the belief that members of Congress are mere constituency tenders and advocates of particularistic interests, whereas the presidency alone embodies the national interest. Among post–World War II presidents, Truman and Eisenhower benefited from the widespread perception of communist threat, which made their consensus-building tasks easier.

Leaders cannot be committed or attentive to everything. That generalization surely applies to U.S. presidents. As a result, particular issues become especially salient to particular presidents. To look briefly at the last three presidents, for example, Jimmy Carter was committed to ratification of the Panama Canal Treaty and to the Camp David Accords; Ronald Reagan was committed to increased defense spending and relentlessly pressuring the Soviet Union, but not much else; George Bush was committed to a "no new settlements" policy in the West Bank and to maintaining favorable relations with the Chinese government. These seem to be almost wholly idiosyncratic sets of commitments, characterized mainly by presidential preferences and temperaments. Few of these particular commitments attracted great attention from the public before they became presidential commitments. Few of them could be said to be wildly popular. Indeed, some, such as the Panama Canal Treaty and the continuation of cordial relations with the Chinese government in the aftermath of the Tiananmen Square massacre were positively unpopular. While President Reagan's vast increases in defense spending had support in surveys taken before his election in 1980 and for almost a year afterward, that support ceased shortly after he began to flood the Pentagon with gold. Additionally, President Bush's commitment to holding up loan guarantees to the Israeli government until it stopped building new settlements on the West Bank was popular with the general public mainly because of the unpopularity of foreign aid implied in the loan guarantee package. But for the group with intense interests and significant political resources, namely American Jews, Bush's actions in withholding the loan guarantees were regarded as an unfriendly sign.

In sum, there seems to be no particular pattern as to what presidents choose to commit their presidencies to in the way of foreign policy initiatives. Opportunities help define these initiatives. A president's preferences and calculations may allow him to seize certain opportunities. But there seems little reason to believe that presidential commitments are driven by public opinion.

For the most part, presidents tend to be inheritors as well as progenitors of ideas. The mix is different across different presidencies. The consuming conception of the cold war began with the Truman administration and came to an end during the latter stages of the Reagan administration and the early stages of the Bush administration, despite vestigial remains of cold war ideas in the defense policies of the Bush administration. Nonetheless, the cold war was the overarching concept behind U.S. foreign policy for more than four decades. It provided predictability and policy inheritances. Its absence makes for a less predictable world and also a less predictable U.S. foreign policy.

Some presidents sharpen existing premises. Reagan was one who did so. His foreign policy goals were relatively clear, and he was remarkably clear in articulating them. His focus was to pressure the Soviet Union everywhere in the world and in the arms race, forcing a retreat in the face of relentless pressure and superior resources.

Some presidents seek subtle redirections of policy within the existing premises. President John Kennedy was one, moving from a strategic focus on the Soviet Union to a tactical focus, seeking to engage perceived Soviet surrogates across the world through the use of counterinsurgency tactics. A shift in military doctrines was implied by the shift in political goals. On a somewhat different plane, that of diplomacy, President Nixon moved U.S. policy toward normalization with the Soviet Union and China but continued to battle the Soviets in peripheral theaters. The policy, however delicately crafted, appealed neither to doves nor to hawks among the elite and thus eventually lost support from cold war advocates after Ford took over from Nixon. Doves liked détente but disliked peripheral engagements. Hawks liked peripheral engagement but disliked détente.

A few presidents have sought to alter the existing premises and policy inheritances or, perhaps, to create new ones co-existing alongside the old ones. President Carter's emphasis on north-south issues was readily overtaken by the intensification of cold war pressures arising from both cold war advocates domestically and seemingly hostile Soviet responses externally. Signs of Soviet hostility such as the abrupt dismissal of early zero-zero arms control proposals, the invasion of Afghanistan, and so forth, confirmed the cold war perspective of the Soviet Union as an untrustworthy enemy. Carter's goals often lacked clarity in their articulation, were in conflict within his administration, and frequently seemed at odds with events.

In foreign policy, then, presidential leadership is a variable. How important the variable of presidential leadership becomes rests to a significant degree on the effort that presidents exert and the clarity with which they articulate their goals. But even high levels of presidential commitment do not guarantee that presidents will get their way (at least, legally) in the face of division within the country. Obviously, Ronald Reagan tried time and again to rally the country on behalf of the Contras in Nicaragua. But the policy lacked decisive support because of a general reluctance among the public to get involved and because opposition among liberal Democrats in the Congress was especially fierce.

Presidential leadership, of course, brings to the fore the issue of who wields what levers in influencing foreign policy.

LEVERS OF INFLUENCE

In the relationship between the president and Congress over foreign policy, each branch holds certain levers. Here, we explore the comparative advantages that each of the branches holds in trying to influence the behavior of the other.

The Congressional Advantage

The main arrows that lie in the congressional quiver to influence foreign policy are those of legislation, appropriations, and oversight, whether formal or informal. Congressional influence is maximized when presidents need something that they cannot obtain by acting alone. Typically, this situation occurs when the president requires legislation or, at least, prefers it.

Divided government has led to frequent unilateral assertions of presidential prerogative on contested turf, which, in turn, have led to congressional efforts to restrict presidential and executive behavior. The now-famous Boland Amendment that was the centerpiece of the Iran-Contra investigation was passed to deter intelligence agencies from using public funds to support the Contras in Nicaragua—a clear expression of congressional distrust of the Reagan administration. The use of a National Security Council staffer, Oliver North, and the sale of arms to Iran from which huge profits were made allowed the Reagan administration to violate the spirit, but not the exact letter, of the law.

The Persian Gulf War mobilization of forces also brought forth a potential constitutional conflict. We cannot know what President Bush would have done had at least one chamber voted against the resolution permitting the use of U.S. forces to engage in combat. Popular lore has it that the president would have proceeded regardless.[11] Notably, the keynote speaker at the 1992 Republican National Convention, Senator Phil Gramm of Texas, proclaimed that the president would have gone ahead even without congressional authorization on the basis of the commander in chief role. Whether such a unilateral commitment would have brought forth a full-blown constitutional crisis probably rested on the outcome of the war. Had things gone badly, the president would have been in serious trouble, but in all likelihood, even with an authorization, the president would have had deep difficulties anyway.

Although most roads to getting things done require some form of congressional approval when Congress exercises an ex ante check on the executive, the ability of the president to take the initiative frequently leaves Congress in the position of being an ex post critic of executive actions and processes. Given the temptation for unilateral behavior on the part of presidents operating under divided government, Congress is placed increasingly in precisely the role of ex post critic. Much, therefore, depends on whether the outcomes are perceived to be successes or failures.

A great deal of my emphasis has been on what presidents can get Congress to accept and on presidential circumvention of congressional authority when they fail. But that is only part of the story. For Congress itself can launch major, as well as minor, initiatives in fulfilling its role as a foreign policy making partner. As much attention needs to be paid to the congressional role as a carrier of initiatives as is given to the presidential role.

An important difference between Congress and the presidency, though, is that unless both chambers have veto-proof majorities, Congress needs presidential consent more than presidents, realistically, need congressional consent.

The Presidential Advantage

When words translate into foreign policy—as they frequently do—the president has unique advantages because he is perceived abroad as the most authoritative spokesperson for U.S. policy. Among other things, this perception enables presidents or their agents to issue threats of a military, economic, or diplomatic nature. The "take care" clause of the Constitution and the commander-in-chief role specified in the Constitution give a president unique advantages in committing U.S. military forces. Such advantages, as yet, have not been much inhibited by legislation such as the War Powers Resolution, and certainly not by the constitutional authority given to Congress to declare war.

Initiative rests with the president; the burden of reacting to that lies with Congress. How much Congress is willing to accede to the president rests upon a number of factors including the extent to which a president has party majorities in Congress, the level of the president's popularity, and the degree to which the undertaking is perceived to have been necessary but, above all, successful.

Public opinion is largely a backdrop to conflicts over policy by official policy makers. It provides parameters around which actual policy needs to show some minimal responsiveness. Policy makers typically understand these broad bounds (especially those in Congress) but are hardly slavishly responsive to them.[12] The more constant are public views, however, the more these bounds are apt to have some influence either on policy making or on the ways in which policies actually are justified.

CONSTANTS AMONG THE PUBLIC

Three major constants in public views are noted here. They tend to provide loosely defined boundaries for foreign policy makers.

Noninterventionism

The taste for military intervention has historically been weak among the mass public, typically for isolationist reasons. Indeed, the taste for military intervention commonly is strongest among attentive publics and elites.[13] Even in the events leading to the Persian Gulf War, an analysis by John Mueller of popular opinion indicates that "popular opinion was not driving the country to war; the public was being led to it."[14] (See Chapter 8.)

Protracted interventions have notable difficulties sustaining popular support. As with domestic policy, presidents have a hard time selling sacrifices for indefinite and vague goals. Korea was one lesson, Vietnam another, and, unfortunately, the introduction of U.S. marines into Lebanon yet a third. Notably, after such lessons resistance to sending troops or even air power into the Balkans has been powerful, regardless of the compelling moral case to be made.

Ambivalence about Protectionism

A maxim applicable to positions individuals take in complex organizations—namely, where you stand depends upon where you sit—is equally appropriate to public attitudes regarding trade and protectionism. A great deal depends upon perceived winners and losers. Most industrial labor unions veer toward protectionism mainly because, while capital is mobile, labor is not. Producer groups, however, are not uniformly protrade. Agricultural interests, historically and currently, are export oriented; heavy industrial interests, also historically and currently, are oriented to blocking imports. So-called rust belt interests, being more industrially focused, are also more protectionist, whereas so-called sun belt interests tend to be more export dependent.

Trade issues are functional issues, meaning that for the most part the only publics likely to get involved are those who see their interests directly affected. Since threats to unprotected markets are likely to be perceived as being greater than are opportunities to exploit markets, there is always a latent tendency toward a protectionist response.

The policy-making pattern involves a combination of congressional credit claiming and blame avoidance. Members of Congress are in a position to support the grievances of key constituent industries, while, at the same time, Congress grants quite a bit of deference to the executive on trade.

Nonsupport for Foreign Aid

While international assistance programs always have difficulty generating support, the close of the cold war makes support even harder to get. A hard sell is now an even harder sell. As with trade, however, there is no powerful reason to believe that actual policy, as distinct from considerable posturing en route to policy, will be strongly affected by isolationist sentiments.

On the whole, despite public sentiment that resonates with isolationist postures, there is not a great deal of evidence to suggest that these sentiments directly dictate foreign policy making. In a system as loosely jointed as is the American political system, advocates for protectionism and for blocking international assistance programs can readily find spokespeople on Capitol Hill. But that does not mean that they can win their case. Whatever their political persuasion, presidents tend to act more alike on these issues than differently. Uniformly, presidents have sustained their support for both trade agreements and assistance programs.

FLUIDITY: PRESENT TRENDS MOVING AWAY FROM A PRESIDENCY-CENTRIC SYSTEM

Contemporary trends will provide stark challenges to presidents' conceptions of themselves as the nation's preeminent foreign policy maker. Some trends reflect changes in the international system; others reflect changes in the nature of U.S. internal politics.

Divided Government

Despite the present respite from divided government, there is no compelling reason to believe that the era of divided government is over. The electoral results of 1992 are anything but straightforward. Although virtually every trend was running in favor of the Democratic challenger, Bill Clinton won by less than 5 percent of the popular vote.

In the longer run, the U.S. electoral process is as likely to continue coughing up Republican presidents and Democratic Congresses as not. When it does, the policy conflicts of the 1970s and 1980s likely will be revisited. Further, if past is prelude, when conflicts between the branches mount and presidents and Congresses resort to other means, such as public hearings and embarrassment, distrust between them will rise.[15] For presidents to assert what they believe to be their proprietary rights in foreign policy, unilateral actions and administrative edicts (abetted by courts favorably disposed to executive power) will likely be the favored instruments. In short, there should be no reason to believe that foreign policy will look any different from domestic policy behavior.

With the onset of unified government under President Bill Clinton, however, these conflicts should diminish. It would be foolish to expect them to disappear altogether, but they should generate less intense conflict between Congress and the president and less ill will and distrust. How long-lasting the condition of unified government will be, however, remains to be seen. Betting on it would be unwise.

The End of the Cold War

The termination of the cold war has reduced an obvious presidential rationale for peripheral interventions as well as activities, legitimate or illegitimate, that could be cloaked under the veil of the cold war. Of course, the end of the cold war also reduces the rationale on which much foreign assistance, especially but not exclusively arms transfers, had been based. It also diminishes the advantage of Republican presidential candidates who, more than their Democratic foes, focus on the presidential foreign policy role and on the necessity of strong stances.

In actual fact, the world may now be a more dangerous place. It is certainly a more confusing place. But the strategic need for American involvement has been, like the Berlin Wall, reduced to rubble—or is at least so perceived. Under these conditions, whatever role the United States should be playing, presidents are less likely to be given deference, and there is likely to be more, if less intense, conflict over what U.S. foreign policy should be. This conflict is likely to mean a widened debate among elites amid a relatively high degree of mass disinterest, a condition apt to be sustained until a crisis such as that which occurred in the Persian Gulf looms.

Growing Prominence of Trade, Environmental, and International Regulatory Issues

Issues revolving around both collective and distributive goods in the world system and in regional communities will become increasingly prominent. Global issues, such as the environment, are certainly likely to gain greater attention. So, too, will trade issues that may become more acrimonious in the absence of the cold war. Such issues will bring domestic interests into play, especially interests that are faced with bearing greater costs. These issues often involve commitments to regulations that are themselves a source of domestic contention. The 1992 environmental summit in Rio de Janeiro reflects the intrusion of domestic political conflicts onto the global scene. One may expect this intrusion in other areas, such as public health, as well. Trade agreements such as the North American Free Trade Agreement

(NAFTA) will bring about redistributions among domestic economic interests that also will produce a fair amount of domestic conflict. Although policy requires a presidential role, that role will surely be contentious, and while the president is likely to be a prominent player (if he so chooses), he will lack the same advantages that accrued to his former role as unofficial "leader of the free world."

Conclusion

Two trends overall seem to be driving change: (1) exogenous changes in the world system, and (2) changes in the character of governance in the American political system. The first factor reduces the rationale for centralized policy making and presidential discretion. The second factor leads to the contentiousness of choice. The impact of both factors together is likely to reduce the role that presidents have assumed is theirs as the exclusive shapers of foreign policy, because the always-fragile separation between domestic and foreign policy will become increasingly permeable.

Contained within these changes are conflicts about institutional prerogative. These conflicts will diminish but not disappear with the reappearance of unified government. They will sharpen under conditions of divided government. Foreign policy, in other words, will become "politics as usual."

"Politics as usual" tends to constrain presidential prerogative, and the disappearance of a singular overriding perception of threat does so as well. It may be that our first post–cold war president and politician par excellence, Bill Clinton, finds that prospect less threatening than his predecessor had. Clearly, nonmilitary issues will be more at the forefront than during the cold war days. Yet, ironically, U.S. military engagements, or at least shows of force, may be more frequent in a world in which armed disputes within and between states in the third world and the formerly held communist territories will generate diplomatic pressure for outside intervention. Such disputes no doubt will test the extent to which the president and Congress will be able to cooperate and, indeed, to work together to determine when and how and under whose command U.S. forces should be deployed. The end of the cold war has eroded a political fault line running through U.S. foreign policy making. Whether the erosion of that fault line will prompt more cooperation across institutions in figuring out a doctrine for military intervention in a less predictable and possibly even more dangerous world remains to be seen.

Notes

1. For an extensive discussion of the general question of congressional deference to presidents in foreign policy making, see Steven A. Shull, ed., *The Two Presidencies: A Quarter Century Assessment* (Chicago: Nelson-Hall, 1991).

2. Note the frequently cited Supreme Court decision of 1936, *U.S. v. Curtiss-Wright,* in which Associate Justice George Sutherland, writing on behalf of the Court, approvingly quoted this antecedent view of Chief Justice John Marshall. See Christopher H. Pyle and Richard M.

Pious, *The President, Congress, and the Constitution: Power and Legitimacy in American Politics* (New York: Free Press, 1984), p. 238.

3. Classic statements of this proposition are, among others, Theodore J. Lowi, "American Business, Public Policy, Case Studies, and Political Theory," *World Politics* 16 (1964): 677–715; Theodore J. Lowi, "Four Systems of Policy, Politics, and Choice," *Public Administration Review* 32 (1972): 298–310, Randall B. Ripley and Grace A. Franklin, *Congress, the Bureaucracy, and Public Policy,* 5th ed. (Pacific Grove, Calif.: Brooks/Cole, 1990).

4. See Frank R. Baumgartner, *Conflict and Rhetoric in French Policymaking* (Pittsburgh: University of Pittsburgh Press, 1989).

5. For example, see James M. Lindsay, "Parochialism, Policy, and Constituency Constraints: Congressional Voting on Strategic Weapons Systems," *American Journal of Political Science* 34 (1990): 936–60. However, unlike strategic doctrines, the relationship between tactical arms and strategies often is less clear. The struggle for "pork" is even more competitive. Tactical arms procurement typically involves a thick tangle of bureaucratic and service interests, contractors and constituency interests, and, hence, a heavy influence of the defense subcommittees on the appropriations committees.

6. For a general overview of trade policies and the role of congressional involvement, interests, and executive discretion, see David B. Yoffie, "American Trade Policy: An Obsolete Bargain," in John E. Chubb and Paul E. Peterson, eds., *Can the Government Govern?* (Washington, D.C.: Brookings Institution, 1989), pp. 100–138.

7. John E. Mueller, *War, Presidents and Public Opinion* (New York: John Wiley, 1973).

8. Aaron Wildavsky, "The Two Presidencies," *Society* 4 (December 1966): 7–14.

9. See George C. Edwards III, "The Two Presidencies: A Reevaluation," *American Politics Quarterly* 14 (1986): 247–63.

10. Charles W. Ostrom, Jr., and Brian L. Job, "The President and the Political Use of Force," *American Political Science Review* 80 (1986): 541–66.

11. Bob Woodward, *The Commanders* (New York: Simon and Schuster, 1991).

12. On the role of congressional responsiveness, especially, see Barry M. Blechman, "The Congressional Role in U.S. Military Policy," *Political Science Quarterly* 106 (1991): 17–32.

13. See Bruce Russett, "Doves, Hawks, and U.S. Public Opinion," *Political Science Quarterly* 105 (1990–91): 515–38.

14. John Mueller, "American Public Opinion and the Gulf War: Trends and Historical Comparisons," paper presented at the conference "The Political Consequences of War," sponsored by the National Election Studies, Center for Political Studies, and Brookings Institution, Washington, D.C., February 28, 1992.

15. See, for example, Benjamin Ginsberg and Martin Shefter, "After the Reagan Revolution: A Postelectoral Politics?" in Larry Berman, ed., *Looking Back on the Reagan Presidency* (Baltimore: Johns Hopkins University Press, 1990), pp. 241–67.

Part III / The Congress

4 / Partisanship Leadership and Congressional Assertiveness in Foreign and Defense Policy

DAVID W. ROHDE

Until the mid-1960s, bipartisanship on matters related to international affairs[1] was both a goal and a fact of presidential-congressional politics. According to Dean Acheson, the purpose of bipartisanship in foreign policy was reaching "agreement . . . [that] will have the support of the American public and thus lift foreign policy issues from the area of partisan consideration."[2] Even with Acheson as secretary of state from 1949 to 1953, when bipartisanship involved consultation between the White House and both parties in Congress, most issues of international affairs exhibited presidential leadership and congressional deference. As Cecil V. Crabb, Jr., and Pat M. Holt argued:

> For the first two decades after World War II Congress was largely content to leave the management of foreign relations to the executive branch—thereby providing impetus for the emergence of the imperial presidency; when it did play a significant role in foreign relations, Congress normally supported the diplomatic policies and programs advocated by the White House.[3]

This pattern of consensus and deference was broken by conflict over the Vietnam War and other international affairs issues during the Johnson and Nixon administrations in the late 1960s and early 1970s. The conflict over the war was, however, largely between the president and Democrats in Congress; the support of most Republicans remained firm. When Richard Nixon became president, there was frequent disagreement between the White House and northern Democrats on foreign and defense policy, but Nixon was usually able to rally a bipartisan conservative coalition of Republicans and southern Democrats to support his position.

This chapter demonstrates that conflict over international affairs issues has increased and become more partisan during the last two decades and that in the process Congress has become more assertive in these matters—that is, more willing to substitute its collective judgment for the president's. It also offers an explanation for those changes—that electoral developments have made preferences on policy more similar within the two parties and more different between them. Meanwhile, reforms adopted within the House of Representatives in the 1970s reduced the legislative clout of the conservative coalition that had usually supported the president and strengthened the powers of the Democratic majority's leadership. The argument and evidence focus on the House, but key contrasts are drawn with the Senate. After outlining these developments, the chapter discusses how they affected legislative politics in the arena of international affairs. The final section considers specific policy conflicts regarding some foreign and defense policy issues during the Reagan-Bush years.

POLITICAL CHANGE AND CONGRESSIONAL POLITICS

Electoral Politics and Congressional Parties

Since the New Deal, the Democratic party in Congress had exhibited a great deal of heterogeneity on policy, with the principal cleavage being geographic, along North-South lines.[4] Intraparty conflicts, however, usually involved domestic issues. Disagreement on foreign and defense policy was comparatively rare until the Vietnam War shattered the cold war consensus on these matters. Conflict over Vietnam spread to other aspects of American involvement abroad and to questions involving levels of defense spending and the choice of weapons for the military. In addition, disagreement on a range of domestic issues grew larger.

These policy differences among Democrats in Congress were rooted in their differing electoral circumstances. Their constituencies differed demographically; in the early 1960s, 79 percent of southern Democrats came from overwhelmingly rural districts, while nearly half of northerners came from heavily urban ones.[5] Southern districts were, moreover, overwhelmingly Democratic and predominantly conservative in sentiment. Southern Democratic voters tended to support the Vietnam War and heavy defense spending, while opposing much of the domestic agenda favored by northern Democrats, particularly regarding civil rights. Thus northern and southern Democrats in the House were pulled by electoral pressures in opposite directions on most major issues. Disagreement was so frequent that by the 92d Congress (1971–72), on roll call votes that found a majority of Democrats opposed by a majority of Republicans, southern Democrats' support for their party's positions averaged only 46 percent. That is, *most* southerners voted more often with Republicans than with their Democratic colleagues.

Yet, just as the 1960s saw the deepening of intraparty differences for Democrats, the seeds of future change were also sown. First, the Supreme Court's one-man, one-vote rulings on congressional districting undermined the rural bias of southern districts. Over the next two decades, moreover, migration to and within the South expanded the urban-suburban population relative to rural areas.

Second, the Voting Rights Act of 1965 led to a significant alteration of the composition of the electorate in the South. Before this law, southern congressional constituencies—measured in terms of persons who actually participated in elections—were overwhelmingly Democratic, predominantly white, and largely conservative. Under the Voting Rights Act, black voter registration and electoral participation increased enormously. The effects of this expansion of the electorate were primarily felt within the Democratic party, with which most of the new voters identified. Because most black voters held relatively liberal policy views, the result was a gradual shift in the sentiments of Democratic primary electorates and of the electoral coalitions of Democratic candidates in a liberal direction over time.

Third, at the same time that black enfranchisement was occurring, another change was taking place within the southern electorate: the decline of white conservative strength among Democratic identifiers and voters, and a corresponding increase within the Republican party. Due to opposition to the national Democratic party's positions on civil rights, social welfare, and national defense, many white southern Democrats began to depart from their traditional political roots. This shift began with Republican votes in presidential elections and abstention from Democratic primaries, but for many it further developed into support for Republican candidates below the presidential level as well. These developments also had the effect of liberalizing the southern Democratic electorate, reinforcing the impact of black enfranchisement. In addition, they created a southern Republican electorate that was extremely conservative in makeup.

The combined impact of these trends was substantial. The dual liberalization of the Democratic electorate altered the incentives for potential congressional candidates based on their ideological orientations. Comparatively liberal candidates encountered real possibilities for successful nomination fights where their chances had previously been bleak. Conservative candidates, on the other hand, found Democratic primaries to be less hospitable, and indeed substantial numbers concluded that it was in their best interests to switch their party allegiance and compete for a Republican nomination. Whether they were former Democrats or lifelong Republicans, the winners of GOP primaries almost always presented very conservative platforms to the electorate.

These developments among voters and in the nomination process replaced the old one-party factional politics described by V. O. Key[6] with real two-party competition in many districts and altered one-party politics in others by the 1980s. Some southern congressional constituencies were predominantly urban with substantial minority populations, making them and their Democratic representatives virtually indistinguishable from their counterparts in northern cities. The majority of congressional constituencies in the South were heterogeneous in character, where it was plausible to imagine either party winning. However, it was frequently the case that an extremely conservative GOP nominee would be pitted against a less extremely liberal Democrat, producing a Democratic victory. Finally, a significant number of districts were created with a pronounced Republican advantage.

These changes produced a mixed pattern of electoral outcomes. First, Republicans gradually won more southern House seats, in many instances replacing the most conservative southern Democrats. The GOP share of southern seats increased from 8 percent (9 of 120) in 1960 to 38 percent (52 of 137) in 1992. Second, where Democrats were succeeded by other Democrats, the changed nature of party coalitions within the districts usually produced representatives whose electoral incentives and personal inclinations led them to be significantly more supportive of party positions than their predecessors. Finally, those Democrats who remained in office were induced, partly by electoral change and partly by the reforms discussed below, to side more often with their northern colleagues. These outcomes substantially reduced the policy split between northern and southern Democrats. By the 100th Congress in 1987–89, southern Democratic support for the Democratic party's positions averaged 76 percent, up 30 points from the 92d Congress (1971–73).

The greater homogeneity among Democrats might have been a hollow gain if the declining share of southern seats had cost them majority control of the House. The southern losses, however, were offset by northern gains, and in many instances (particularly in the Northeast) the losses resulted from moderate Republicans being replaced by liberal Democrats. These losses, coupled with the addition of the very conservative southerners, shifted the ideological balance of the GOP's House delegation to the right.

Thus changes in the Democratic and Republican parties in the House produced memberships that were more internally homogeneous and more different from one another. The same electoral developments (with the exception, of course, of redrawn district lines) produced parallel, albeit less pronounced, effects within the Senate membership, where average party loyalty for southern Democrats increased from 53 percent in the 92d Congress to 79 percent in the 100th. There, however, the loss of southern Democratic seats did play a role in producing a Republican majority from 1981 through 1987.[7]

House Reform and Its Impact

As the policy divisions among Democrats deepened during the late 1960s, liberals in the House became increasingly concerned that institutional arrangements were biased against their interests. Within most committees, chairs had substantial powers to influence policy outcomes, and because of the automatic application of the seniority rule, they did not need to take account of the preferences of other Democrats when making decisions. Party leaders had few powers to offer incentives or to impose sanctions in order to influence committee behavior. Moreover, procedural rules for floor consideration made it difficult to change committee proposals, further enhancing committee independence. This situation was disadvantageous for House liberals because the patterns of seniority gave a disproportionate share of committee leadership positions to conservative southern Democrats. These conservative Democrats often allied with Republicans in committee and on the floor and used their institutional powers to produce policy outcomes that were opposed by a majority of Democrats.

Democratic reformers altered this perceived institutional bias in a variety of ways. First, by 1975 they struck directly at the powers of committee chairs through the "Subcommittee Bill of Rights" and related rules changes. Subcommittees were given specific jurisdictions in committee rules and their own budgets and staff. Assignments of subcommittee chairs and subcommittee members were made based on seniority. Control of all of these aspects of committee operation was removed from the independent control of committee chairs. The reformers, however, sought to guarantee that the decentralization of power to subcommittees would not make them autonomous, as full committees had previously been. Rather, they vested final control over most committee operations in the Democratic Caucus on each committee. Even more important, those caucuses could remove subcommittee chairs if their behavior was unacceptable to committee Democrats. These rules changes created a more democratic decision-making environment. The expectation was that this environment would result in committee policies that better reflected overall committee preferences.

Second, the full Democratic Caucus mandated automatic, secret ballot votes in the caucus on committee chairs at the beginning of every Congress. This arrangement, coupled with the creation of committee caucuses, made committee leaders responsible to the majority of the majority party to a degree unheard of since early in this century. Committee chairs could be removed by the Democratic Caucus if their stewardship was retrospectively judged to be unsatisfactory, as could subcommittee chairs by committee caucuses. These arrangements were intended to make sure that committee leaders would not pursue policy options that strayed too far from the outcomes favored by a majority of Democrats. The new procedures were, moreover, not a hollow threat. They were used to remove a number of committee and subcommittee chairs during the 1970s and 1980s. Without the *automatic* protection of seniority, committee leaders had to consider the wishes of rank-and-file Democrats.

Third, while reducing the power of committee leaders, the reformers chose to strengthen Democratic party leaders. The Speaker was empowered to refer bills to more than one committee and to set deadlines for consideration of those bills. Control over Democratic committee assignments was transferred to the party's Steering and Policy Committee, the membership of which was dominated by party leaders and appointees of the Speaker. By the 1980s it was made clear to members that the degree of support for party positions would play a significant role in determining assignments to the most desired committees.

Perhaps the most important change was the decision to place the Rules Committee—which controlled the procedures under which bills are considered on the floor—firmly under the control of the Speaker, now empowered to appoint and remove its chair and the Democratic members.[8] These reforms enhanced the ability of the leadership to influence individual members and to affect the fate of legislation reported by committees. The aim here was to reinforce the responsiveness of committees to a consensus among Democratic members when one existed and to increase the leadership's ability to protect satisfactory committee policies on the floor.

Fourth, the relationship between committees and the floor was altered by new procedures that permitted recorded votes on amendments on the House floor and that set up an electronic voting system, reducing the amount of time it took to hold a floor vote. Previously, voting on amendments had been anonymous, reducing the impact of constituency forces on members' behavior and increasing the leverage committee leaders had. The new rules reversed that balance, and electronic voting made it feasible to consider a large number of amendments on the floor. Analysis by Steven S. Smith shows that amendment activity expanded greatly in the wake of these new procedures.[9] The changes were double-edged. The reformers anticipated that the threat of this capability would induce committees to be responsive to rank-and-file sentiments when drafting legislation. If, however, a committee was not responsive, then members could override its decision and enact their preferences through the amendment process.

In the Senate, formal rules changes like those above were not necessary. Committees had always been less autonomous, and floor procedures had always permitted amendment activity. The changes here instead involved the dissolution of informal norms that had constrained junior liberals from full, active participation in Senate activity, thus leaving disproportionate influence over policy to senior, more conservative members. During the 1970s, junior members threw off these limits, and the institutional influence of committees was further reduced. Members could freely rewrite legislation on the floor. Actually, institutional arrangements in the Senate limited the effects of electoral change on partisanship. Individual senators had substantial power under the rules, and the minority frequently had to be accommodated in a way that was unnecessary in the House.

THE CONSEQUENCES OF CONGRESSIONAL CHANGE

The preceding section discussed the ways in which shifting electoral forces produced changes in the makeup of the Congress's membership and outlined the pattern of institutional reform. Now I will examine the effects of these changes on the relationship among members, committees, and party leaders—first in general and then specifically in the realm of foreign and defense policy.

THE GENERAL EFFECTS OF A CHANGING CONGRESS ON POLICY CONFLICTS

The Democratic reformers of the 1970s enacted changes designed to weaken the power of conservatives within the committee system and to enhance the ability of the party leadership to protect and advance legislation widely favored by House Democrats. Because most of the reforms involved changes in the Democratic Caucus's rules—requiring only majority support within the party—they had limited effects while Democratic divisions remained substantial. As, however, intraparty homogeneity and interparty differences increased in the late 1970s and after, the impact of these changes on the patterns of conflict over policy became more pronounced.

The Speaker's power of multiple referral of bills "has allowed the leadership to capitalize on existing sources of leverage, and in so doing extend its role and influence."[10] Multiple referral enhances the Speaker's power over scheduling and is used for most legislation important to the leadership. In addition, the Speaker sometimes becomes the arbiter of intercommittee differences, negotiating compromises among contending positions.

The influence of the Speaker with committees has been enhanced particularly by control over the Rules Committee. This Committee determines, for most bills, the terms of floor debate (specified in what are called "special rules"): how much time will be allotted, what amendments will be permitted and how they will be considered, and whether the House's usual procedural requirements will be waived or enforced. Before reform, the Rules Committee was an independent, generally conservative committee that leaders of other committees had to negotiate with individually. Under the new rules, the Democratic leadership could control the Rules Committee's decisions when it chose to. Thus committee leaders knew that special rules could be crafted to enhance the chances of policies the leadership favored or to undermine proposals that were deemed unsatisfactory. As special rules were more frequently designed to protect Democrat-favored committee policies, conflict over those rules became more frequent and more deeply divided along partisan lines.

The effects of other reforms that were more directly targeted on committees and their leaders can be seen through the changing patterns of conflict over floor amendments. With the curbs in the powers of chairs and the possibility of removal of recalcitrant committee leaders, the reformers expected that committees would be more likely to adopt policies that were satisfactory to rank-and-file Democrats. It appears that their expectations were borne out. During the late 1970s and the 1980s, the frequency of floor amendments that were predominantly favored by Democrats decreased, while those favored by Republicans increased, indicating a shift in partisan satisfaction with committee proposals. Moreover, Democrat-favored amendments were adopted more frequently and Republican-favored ones were passed less often. Thus when policy shifts away from committee positions did occur on the floor, they went more often in the Democrats' direction. Finally, the degree of interparty conflict on amendment votes increased substantially.

The combination of greater committee responsiveness to Democrats' preferences with greater ability to alter committee proposals on the floor yielded bills that—in final form, as passed by the House—were increasingly satisfactory to the Democratic majority and increasingly unsatisfactory to the GOP. In the 91st Congress (1969–71), barely 60 percent of the bills were more satisfactory to the Democrats than to the Republicans on initial passage by the House. By the 100th Congress (1987–89), the corresponding proportion was 95 percent. On these votes, too, as with amendments, the degree of interparty disagreement increased substantially over time. Thus the evidence indicates that by the 1980s, the combination of changes in members' preferences and in the House's institutional arrangements altered the relationship among committees, party leaders, and rank-and-file members, resulting in policy outcomes that were increasingly favorable to the Democratic majority.

The Impact on International Affairs Issues

The description above outlined the general effects of congressional change. Certain policies and certain committees, however, are different from the general pattern, and some are more different than others. This is particularly true with respect to foreign and defense policy.

As noted above, the period before the mid-1960s was generally marked by congressional deference to and bipartisan support for the presidential actions in international affairs. It was not that this arena was entirely free of presidential-congressional conflict or that conflict was never partisan; such situations were, however, relatively rare. Indeed the recognition of presidential strengths in foreign and defense policy compared to domestic issues led Aaron Wildavsky to propound a systematic advantage held by the president in his "two presidencies" hypothesis. In his assessment of the evidence, Wildavsky concluded that in "the realm of foreign policy there has not been a single major issue on which Presidents, where they were serious and determined, have failed."[11]

The president's dominant position with respect to international affairs rested in major part on a widespread agreement in Congress and in the electorate on certain principles underlying the nation's policies. As Crabb and Holt stated, from World War II until the late 1960s, "a national consensus existed on two ideas: that communist expansionism threatened the security of the United States and other independent nations, and that it had to be resisted by the strategy of containment."[12] With shared agreement on goals and broad strategy, the public and Congress usually were willing to let the president take the lead on the means to be employed.

The president's leading position was reflected in the makeup and operation of the two principal House committees that dealt with foreign and defense policy—Foreign Affairs and Armed Services—although the two were markedly different from one another in certain respects. The Armed Services Committee was a bastion of support for the military and for presidential requests to increase defense spending. Members of Congress seek committee assignments to advance the achievement of their goals, and the Armed Services Committee attracted representatives from districts with substantial interests in defense, either because of defense industries or military bases. For its members, the committee primarily fostered their interest in reelection, and they were relatively unconcerned about policy questions. As one member said during the mid-1950s, "We mostly reflect what the military people recommend; military policy is made by the Department of Defense. Our committee is a real estate committee."[13]

The preexisting commitment to high defense spending within their districts made most committee members automatic supporters of high spending and vigorous use of the military internationally. This orientation continued, moreover, into the 1960s and 1970s, as Chair Mendel Rivers's (D-S.C.) comment to Nixon's new secretary of defense in 1969 indicates: "Mr. Secretary . . . I am sure you know, better than anybody, that this committee is the only official spokesman for the DOD [Department of Defense] on the floor."[14] The committee's Democratic members were unrepresentative of (i.e., more prodefense than) House Democrats. In the 92d Congress (1971–73), for example, 44 percent of committee members were from the South, while only 35 percent of House Democrats were from the South.

The Foreign Affairs Committee was also characterized by strong support for presidential leadership. For example, Richard F. Fenno's analysis of the committee in the early 1970s described its policy coalitions as "executive-led" and, indeed, "monolithic."[15] Presidential leadership on policy was primary, and both the committee and the House were inclined to accede to it on most matters: "As late as 1970 the Chairman of the Foreign Affairs Committee regarded it as a junior partner in foreign policy decision making; in his opinion the committee should normally support the executive branch's foreign policy positions."[16]

On the one issue with some partisan-ideological conflict—foreign aid—the committee contingents of both parties were dominated by aid-supporters. The committee's Democratic contingent was also sectionally unrepresentative, but in the opposite direction from that of the Armed Services Committee. The Foreign Affairs Committee was more liberal than the caucus, with only 24 percent southerners compared to 35 percent southerners among all House Democrats.

By the 1970s, the cold war consensus within the electorate and within the House that had served as the basis for presidential dominance of international affairs issues first fractured, then dissolved. The inclination of the committees to respond to the president continued, however. As House

Democrats from the North became less acceptant of presidential leadership in foreign and defense policy and more opposed to the substance of presidential positions, the unrepresentativeness of these committees (especially Armed Services) became an increasing problem. Indeed, its policy implications provided a major impetus for the reform effort.

In the 1980s, as noted earlier, the sectional split among Democrats was mitigated, and interparty conflict increased. These developments happened, however, at different rates in different policy areas, generally occurring first in domestic matters, later in foreign policy, and latest (and least) on defense. Moreover, Democratic party contingents on the committees became, if anything, even less representative of the caucus. By the beginning of the 101st Congress (1989), 33 percent of House Democrats were from the South, while on the Foreign Affairs Commitee 15 percent, and on the Armed Services Committee 52 percent, of the Democrats were from the South. In each case the unrepresentativeness of the committee created potential problems. Because the Armed Services Democrats were much more supportive of the military than their caucus was, it was likely that committee proposals would be unsatisfactory to Democrats. The Foreign Affairs Committee, on the other hand, would be more likely to produce policies that were satisfactory to the dominant liberal faction of the Democratic party, but its proposals might be vulnerable to opposition from a coalition of Republicans and more conservative Democrats on the floor.

The differences between the committees, and their corresponding different relationships to the Democratic Caucus and to the House floor, led to varying effects from the increases in intraparty homogeneity and interparty conflict and from the reforms of the 1970s. First, growing Democratic homogeneity produced strong pressures for party leaders to act vigorously on behalf of party positions for which there was widespread support. I have termed this situation "conditional party government"—that is, party leaders were not important or active on every issue but were expected to be active when there was significant party consensus for a position.

One implication of this situation was that Democratic leaders were less likely to provide automatic support for presidential positions in international affairs and were more constrained to stand for the views of their party members. An analysis by Steven S. Smith shows that from the 1950s through the 1980s, Democratic leaders were progressively more likely to oppose the foreign policy positions of Republican presidents. He concluded from his research that "in the foreign policy arena, leaders show increasingly partisan patterns in their support for presidential positions, increasing activity as party spokesmen, increasingly partisan patterns of consultation with presidents, and increasing activity as policy leaders."[17]

The changed circumstances also altered the way that the Speaker employed reform-granted powers over the Rules Committee relative to foreign and defense policy. This change meant shaping the floor environment to protect committee proposals that Democrats supported or (as we will see below in the case of Armed Services) to undermine proposals that they opposed. In either case this shift resulted in progressively more partisan responses on floor votes on special rules for international affairs bills, as the data in Figure 4-1 show. It portrays the average party difference (i.e., the average absolute difference between the percentage of Republicans and Democrats voting aye) on votes on special rules from 1969 to 1990, grouped by presidents.[18] Partisan disagreement was quite low in the Nixon-Ford years but increased to very high levels under Reagan and Bush.

Next we can examine evidence on amendment activity. Amendments are efforts to change the content of the policy a committee has recommended. Thus by looking at the patterns of who supports amendments and which amendments are successful, we can draw inferences about the locus of dissatisfaction with committee policies and about the relationship between committees and the floor. Figure 4-2 shows the proportion of amendments that were Democrat-favored (i.e., received a higher percentage of support from Democrats than from Republicans) in each policy area.[19] First, we can see that the proportion of Democrat-favored amendments is consistently higher on defense than on foreign policy matters, reflecting dissatisfaction resulting from the unrepresentativeness of the Armed Services Committee.

Second, there are some variations over time. Through the early 1980s, the Foreign Affairs Committee remained fairly responsive to the president. This responsiveness meant greater satisfaction for Republicans under Richard Nixon and Gerald Ford and in Ronald Reagan's first term than under Jimmy Carter. Thus there was a lower proportion of Democrat-favored amendments in the Carter years. Later in the decade the responsiveness declined as the Democrats became more cohesive, and so Democrats had less reason to seek changes in committee policies and Republicans had more. As a result, the proportion of Democrat-favored amendments declined and the Republican-favored proportion increased. On defense issues, Republican incentives to seek change on the floor increased only under Carter (when the committee responded positively to some of the president's strategic policies) and in Reagan's second term.

Table 4-1 presents data on the proportion of amendments that were successful. They clearly show the changing relationship between committees and the floor. In both foreign and defense policy, the percentage of Democrat-favored amendments that were successful increased over time, while the success rate for Republican amendments generally declined. The only exception to the trend was for GOP amendments in foreign policy under Bush.

The interaction between the committee's initial judgments and floor action results in a final bill containing the collective decision of the House. Until the mid-1980s, the inclination of Congress to defer to the president and of a majority on the floor to defer to committees usually resulted in foreign policy bills that were more supported by Democrats and defense bills

Figure 4-1 / Average Party Difference on Votes on Special Rules in Foreign and Defense Policy, 1969–1990, by Presidential Administration (consensual votes excluded)

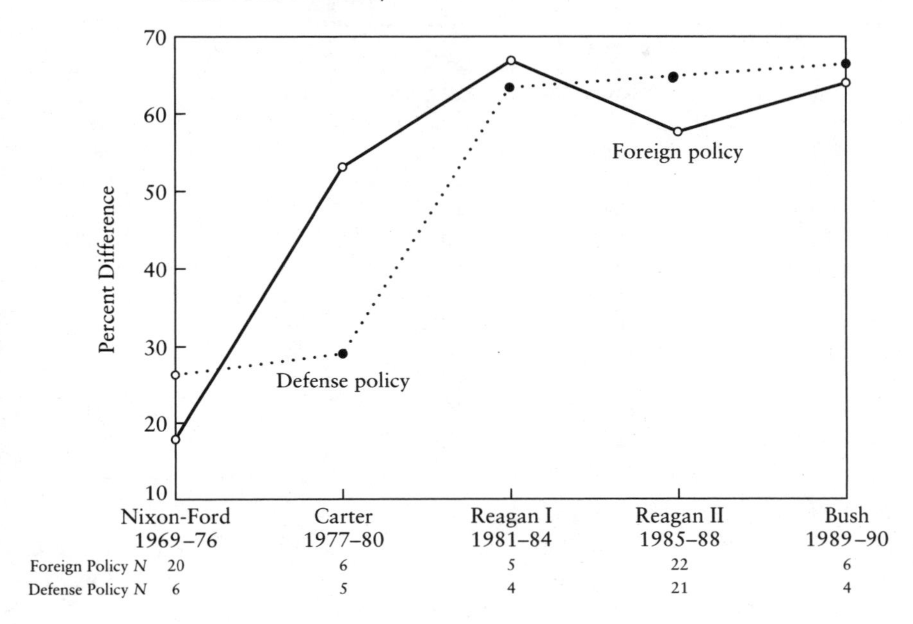

Figure 4-2 / Proportion of Amendments That Were Democrat-Favored in Foreign and Defense Policy, 1971–1990, by Presidential Administration (consensual votes excluded)

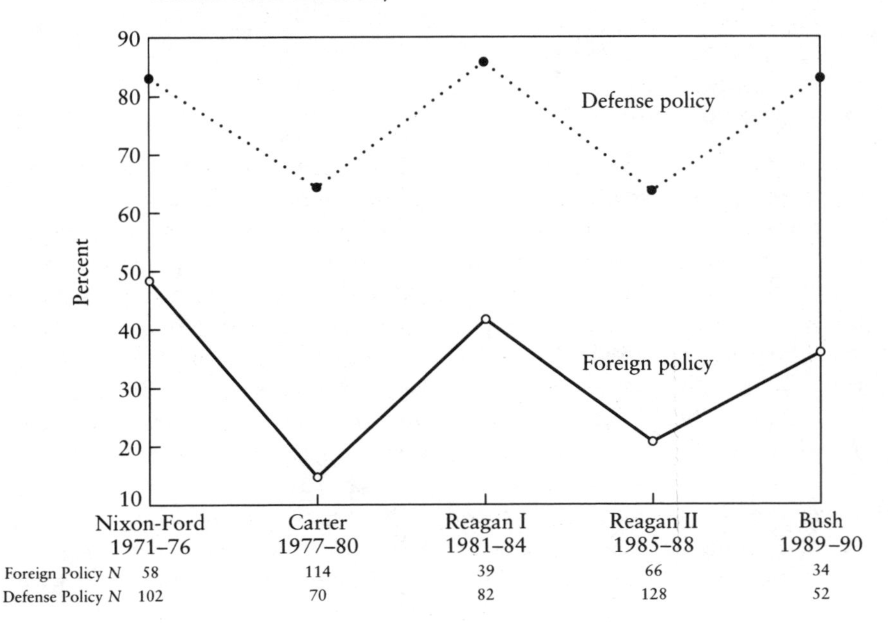

Table 4-1 / Proportion of Amendments Passed in Foreign and Defense Policy, 1971–1990, by Presidential Administration, Controlling for Which Party Favored the Amendment (consensual votes excluded)

	Foreign Policy		*Defense Policy*	
ADMINISTRATION (YEARS)	DEMOCRATIC-FAVORED (N)	REPUBLICAN-FAVORED (N)	DEMOCRATIC-FAVORED (N)	REPUBLICAN-FAVORED (N)
Nixon-Ford (1971–76)	35.7% (28)	43.3% (30)	15.6% (83)	36.8% (19)
Carter (1977–80)	44.4% (18)	40.6% (96)	11.1% (45)	56.0% (25)
Reagan I (1981–84)	56.3% (16)	30.4% (23)	25.7% (70)	50.0% (12)
Reagan II (1985–88)	50.0% (14)	30.8% (52)	40.7% (81)	27.8% (47)
Bush (1989–90)	58.3% (12)	45.4% (22)	51.2% (43)	11.1% (9)

that were more favored by Republicans (and their conservative Democratic allies). In both areas, partisan disagreement on the bills was fairly low, as Figure 4-3 shows. In Reagan's second term and under Bush, however, both foreign policy bills and defense bills were more likely to be Democrat favored. Moreover, as Figure 4-3 shows, partisan disagreement on these bills was markedly greater, especially on defense matters.

Partisan Resurgence and Policy Conflict in International Affairs

The aggregate evidence presented above is consistent with the theoretical arguments about the changing relationship among members of the House, the president, and committees in the wake of electoral change and reform. However, aggregate voting data can tell us only so much. Roll call results are the product of the interaction of members' preferences and the nature of the agenda. Reliable inferences about one of these aspects require firm *prior and independent* knowledge about the other. Roll call analysis usually involves making inferences about members' preferences by making the heroic (and almost always implicit) assumption that the nature of the roll call agenda is the same over time. Certainly such an assumption is unwarranted with respect to this analysis, as it is in most cases. While the discussion above does involve some independent knowledge about changes in the agenda, aggregate results must be treated gingerly and regarded as tentative. Therefore, to reinforce the reliability of our inferences, in this section I will

Figure 4-3 / **Average Party Difference on Initial Passage of Bills in Foreign and Defense Policy, 1955–1990 (consensual votes excluded)**

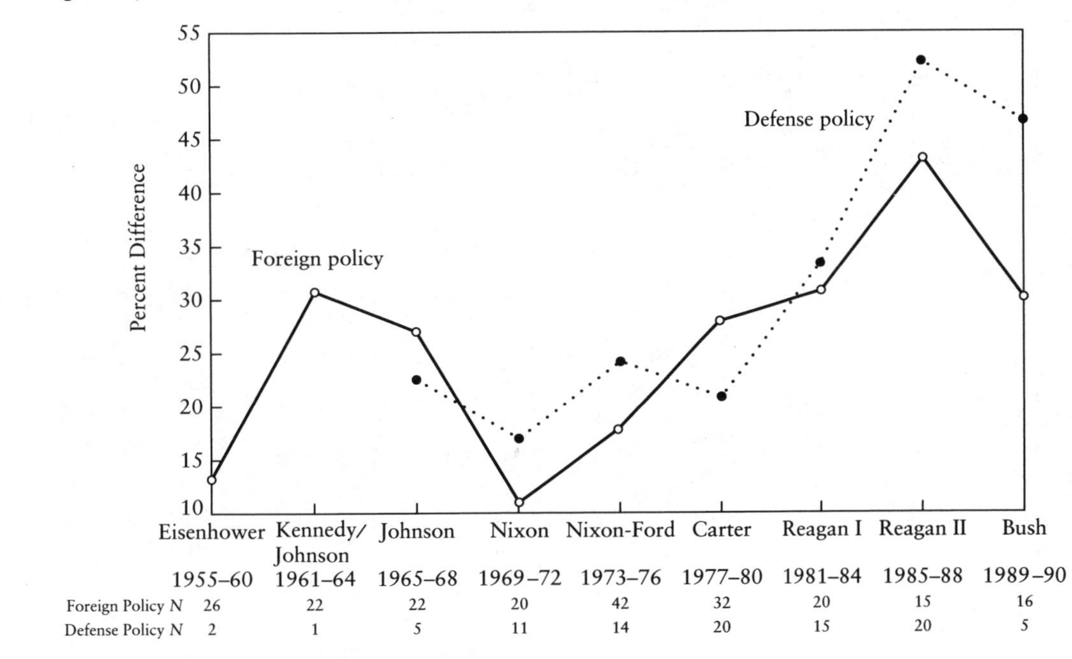

present specific discussions of a number of the major foreign and defense policy issues of the Reagan-Bush years.

Defense Authorization

Of all the issues in the international affairs area, none better illustrates the transition from bipartisan deference to the president to partisan congressional assertiveness than the annual authorization of defense procurement. Before the Vietnam War, defense authorization bills were low-conflict items. From 1961 through 1968, only once were there more than 15 votes against final passage. That was in 1963, when 33 members (all but one Republicans) voted nay because the bill exceeded President John Kennedy's request by almost $500 million.

During the Nixon administration, the procurement bill became a vehicle for conflict over Vietnam. In addition, disagreements began to develop over weapons systems and the overall level of defense spending. In Nixon's first term, some northern Democrats opposed initial passage of the procurement bill, but the proportion never reached even 40 percent of that group. As the war waned, the other issues began to predominate, but again almost all the opposition to the Armed Services Committee's recommendations came from northern Democrats. From 1971 through 1976, 23 amendments dealing with weapons systems, nuclear policy, or overall spending received roll call votes. All were Democrat-favored; 18 were conservative coalition votes.[20] On every one, the party difference was smaller than the North-South Democratic difference.[21] Moreover, only one of these amendments passed.

These patterns of conflict continued through the Carter administration and Reagan's first term. Liberal Democrats were generally opposed to Reagan's massive defense buildup and disagreed with the president on much of his nuclear weapons policy. Between 1981 and 1984, 33 Democrat-favored amendments were proposed on these issues. Of these, 24 produced conservative coalition votes, and on 29 the party difference exceeded the North-South difference. Only 3 amendments passed, but Democratic support for them was generally higher than in the past.

The real evidence of change came during the second Reagan term. It began with a shift in the leadership on the Armed Services Committee, when at the opening of the 99th Congress in 1985 the Democratic Caucus removed Melvin Price (D-Ill.) as chair and replaced him with Les Aspin (D-Wis.). Much of the vote against Price stemmed from the fact that he was eighty and infirm, and many members believed that he was not able adequately to lead the committee. However, in choosing Aspin the caucus passed over a number of senior Democrats who were strong supporters of Reagan's policies to choose the seventh-ranking member of the committee and a Pentagon critic.

Conflict was moderate over the authorization bill in 1985. Many amendments were adopted, but most were of minor significance and few related to the kind of military policy issues we have been discussing. One notable exception was an amendment that barred testing of antisatellite (ASAT) weapons. The House passed the amendment, but Aspin accepted a compromise in conference that favored conservative preferences on that issue as well as a number of others.

In 1986, however, conflict over defense authorization was much more widespread and revealed the effects of the resurgence of partisanship. The House floor ratified a full-scale assault on Reagan's defense policies and the recommendations of the Armed Services Committee. Five significant amendments that changed various specific policies were adopted: (1) bar nuclear tests over one kiloton; (2) ban full-scale tests of ASAT weapons; (3) prohibit production of new (binary) chemical weapons; (4) require U.S. observance of the limits of the Strategic Arms Limitation Treaty (SALT) II; and (5) cut funding for the Strategic Defense Initiative (SDI) by 40 percent. In addition, the House passed an amendment that reduced spending 11 percent below Reagan's proposed budget. These six amendments were supported by a coalition of liberal and moderate Democrats, including Aspin and Majority Leader Jim Wright (D-Tex.). On all the amendments except the one on chemical weapons, a majority of southern Democrats joined with a majority of their northern colleagues, while in every instance at least 78 percent of Republicans voted against the amendments. The unrepresentative nature of the Armed Services Committee is illustrated by the fact that the vote on the budget cut amendment among southern Democrats who were not on the committee was 49–11 in favor, while among committee southerners it was 7–8 against.

The support of the party leadership was instrumental in passing the package of amendments, because the leaders structured the special rule for the bill to facilitate victory. First, the amendments were scheduled at three different points during the nearly two weeks of debate. This scheduling permitted Democrats who favored the package but who had substantial conservative elements in their districts to mix these votes with pro-Reagan votes on other amendments, thus avoiding what John W. Kingdon called a "string of votes" that could undermine their reelection.[22] Second, the special rule barred amendments to the liberal-moderate coalition's amendments that could have made them less potent. Finally, the special rule employed "king-of-the-mountain" procedures under which amendments dealing with the same subject were grouped together and voted on in a prespecified order. Only the last one in a group that was passed was included in the bill, regardless of the number that received majority support or which one received the most votes. The coalition-supported amendments were, of course, scheduled last.

Another example of the combined impact of procedural change and changes in the Democratic membership occurred at the opening of the 100th Congress in 1987. The compromises that Les Aspin had engineered on the authorization bill in 1985, particularly those regarding ASAT weapons and the M-X missile, angered many Democrats, who believed that he had failed adequately to defend House positions. In 1986 further hostility to Aspin developed after he supported additional aid to the Nicaraguan Contras. As a result of these actions, the Democratic Caucus voted to remove Aspin as committee chair. In the subsequent election, Aspin ran again, opposed by three other committee members. On the third ballot, Aspin won reinstatement by a 133–116 vote. This contest made Aspin realize that he had to be responsive to the sentiments of the Democratic Caucus. As one of the leading liberals, Barney Frank (D-Mass.), said about Aspin's campaign: "I know one of the things [Aspin] said was, 'Look, I understand what you are saying when you said I did not sufficiently represent the caucus's position. I will be much more firm in sticking to those positions.' "[23]

In 1987 the pressure from the defense views of the caucus was reflected in a more responsive set of policy recommendations in the Armed Services Committee's authorization bill. As a consequence, not only were there liberal Democrats who wanted to offer amendments to restrict the Reagan administration further, but there were also conservative Republicans who wished to move the committee bill more toward the president's positions. Again there was a Democratic coalition behind a set of amendments, and again the Democratic leadership crafted a series of special rules that were designed to advantage the coalition and to disadvantage the Republicans. The new Speaker, Jim Wright, played a significant role in planning the strategy on the bill. Five of the Democrat-favored amendments were adopted, and each one was supported not only by a majority of northern Democrats but also by a majority of southerners. Among the Republican-favored amendments, on the other hand, not one passed and none received a majority of southern Democrats' votes. Les Aspin's summary judgment reflected the results of the conflict: "It's becoming a much more Democratic defense bill."[24] Similar events and similar outcomes occurred in the debate on the 1988 bill.

The transition to the "kinder, gentler," and supposedly more moderate Bush administration did not halt partisan conflict between the House and the president on defense policy. In 1989, the Armed Services Committee bill reflected to a degree Democratic preferences (it cut, for example, $800 million from Bush's request for the B-2 Stealth bomber). Democrats went further, however, during the floor debate. Amendments were adopted that reduced SDI funding, further cut spending for the B-2, cut in half the authorization for rail-garrison basing of the M-X missile and limited total M-X deployment to fifty missiles, and prohibited the production of binary chemical munitions. On every one of these amendments, a majority of northern Democrats *and* a majority of southerners were in agreement, and opposed to a majority of GOP members.

In 1990, the transition to a post–cold war world led both the president and Congress to support reductions in overall defense spending, but there were still disagreements regarding how the funds should be allocated. The

Armed Services Committee, following Aspin's lead and the Democratic Caucus's preferences, cut billions from Bush's request for the B-2, M-X, SDI, and other strategic programs and allocated the money instead to certain conventional weapons programs. Then, on the floor, GOP efforts to increase SDI funding were resisted, and an amendment to cut the program further (supported by northern and southern Democrats) was passed.

The extremity of the change toward partisan conflict and congressional assertiveness on defense policy is shown by House action on the 1991 bill. In the wake of the military victory over Iraq, and with Bush's approval ratings at historic levels, one might have expected the president's influence on this issue to be maximized. Instead, the Armed Services Committee again slashed funds from SDI and the B-2, and shifted the money to conventional weapons. When Minority Leader Bob Michel (R-Ill.) offered an amendment to substitute Bush's defense proposals for the committee's, it received only 5 Democratic votes, and on the bill's final passage 92 percent of Democrats voted in favor, while the president and 84 percent of Republicans were opposed.

I close this discussion by examining the trends in approval rates among Republicans and northern Democrats on final passage of the House's initial version of the defense bill. These data are displayed in Figure 4-4 for 1969 through 1991.[25] During the Nixon and Carter years, one-fourth to one-third of northern Democrats registered their dissatisfaction first with the Vietnam War and then with defense policy more generally by opposing the bill. However, opposition was heightened by the Reagan defense buildup, peaking in 1983 when a majority voted against passage. Then, as Democratic preferences began to have more influence on the floor and in committee, support began to rise again, never falling below 75 percent after 1985. Republicans, on the other hand, had support levels above 90 percent every year until 1983, when policy outcomes began to fall short of presidential requests. From 1986 on, when Democratic homogeneity permitted the party leadership to take an active role in changing committee proposals on the floor, a heavy majority of Republicans opposed final passage. Defense policy had become an arena of partisan conflict, and a solid majority of Democrats was quite prepared to substitute its judgment on many important matters for the recommendations of the president and the Armed Services Committee.

Reagan's Contra Aid Policy

Unfortunately, space will not permit as detailed a consideration of other issues, so the remaining discussion is confined to highlights, in order to illustrate the variety of matters that have exhibited evidence of change. The first of these involves President Reagan's efforts to secure aid for the Nicaraguan Contras. This hotly contested issue, on which the sentiments of a closely divided Congress seesawed back and forth, exhibits a number of features that are important for our purposes.[26]

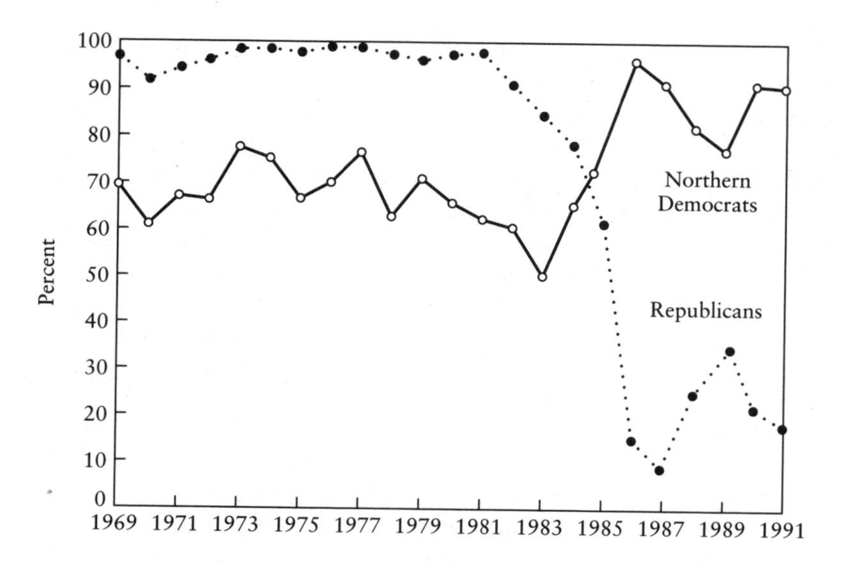

Figure 4-4 / Percentage of Republicans and Northern Democrats Supporting Initial Passage of the Defense Authorization Bill, 1969–1991

First, the issue of aid to the Contras offers a clear, specific example of the general theme of increased congressional partisanship and assertiveness on international affairs issues. While the president did win a few rounds during the battle that went on during most of his two terms, the eventual outcome was a defeat for Reagan and his congressional allies, blocking what he wanted to do. The president received consistently strong support from GOP members; the opposition to his policies was orchestrated by the Democrats. They reached a contrary independent judgment on the issue and were willing to substitute that judgment for the president's, despite his public appeals for support and the emotional attacks against them by GOP representatives. For example, in the 1987 debate on a resolution to delay $40 million in aid while previous assistance was accounted for, Minority Whip Trent Lott (R-Miss.) claimed that its effect would be "handing Central America over to the Soviets."[27] It has long been argued that one reason Congress traditionally deferred to the president in foreign policy was that members wanted to avoid being blamed if things went wrong. Here was one instance where sufficient numbers were willing to take the risk.

Second, this issue illustrates the independent role that party leaders have come to play in foreign policy. Speaker Jim Wright was largely responsible for crafting the peace proposal that was announced by the White House in August 1987. Liberal critics of Contra aid believed that this joint effort was really a Republican trap laid for the Speaker and the Democrats, and Wright risked his prestige and perhaps even leadership position by supporting it. Wright was also interacting independently with the heads of the Central

American governments, and he was able to announce the peace plan they had agreed on only two days after the White House proposal.

Third, we see here again the effects of the majority party leadership's ability to structure the legislative agenda to advantage party policies. The special rule for consideration of the aid moratorium in March 1987 barred any amendments and prohibited any instructions to be attached to a motion to recommit from the Republicans. During the Rules Committee's consideration of the rule, Trent Lott said, "I view this whole process as a sham. You're slam-dunking us, and you've got the votes to do it."[28] Then in March 1988, debate on the proposal that sealed the fate of Contra aid authorized consideration of a GOP plan and a Democratic substitute for it. Amendments to these proposals were banned, as were instructions in a motion to recommit. Because the substitute would be voted on first, its adoption (which is what occurred) would bar the Republicans from a separate up-or-down vote on their proposal. Minority Leader Bob Michel characterized the rule as "abuse of the legislative process."[29]

Finally, this issue illustrates the importance of the change in the political leanings of southern Democrats. To be sure, some of them stuck with the president throughout the battles over aid to the Contras, but others did not. Despite virtually unanimous opposition to Contra aid among northern Democrats on every test vote from 1983 on, their view never prevailed unless it got the support of at least one-half of the southerners. On the final major test (the vote on the Democratic substitute for Michel's resolution in March 1988), southerners voted with their party and against President Reagan's preferences 60–18.

Economic Sanctions against South Africa

The magnitude of the change in policy preferences among southern Democrats is demonstrated by consideration of another major foreign policy dispute from the Reagan years: whether to impose economic sanctions on South Africa because of its apartheid policy. American policy toward southern Africa had always been influenced by attitudes on domestic racial issues. Efforts during the 1970s to get the United States to participate in United Nations–sponsored sanctions against the white-ruled Rhodesian government were resisted by the conservative coalition. For example, in 1972 the Foreign Affairs Committee sought to authorize joining in the sanctions, but the effort was blocked by a floor amendment that southern Democrats supported 65–7. In 1977, U.S. participation was finally authorized with the vigorous support of President Carter, but southern Democrats opposed the bill 37–44.

By 1985, things had changed substantially. That year the Foreign Affairs Committee reported a bill imposing modest economic sanctions on South Africa. Six weakening amendments were offered on the floor by Republicans, but none passed and none received the support of more than 8 southern Democrats. They supported passage of the bill by 73–6 over Reagan's opposition. To stave off Senate action on the conference report, the president announced his own, more-limited set of sanctions. Then in 1986 congressional supporters of sanctions argued that Reagan's sanctions were far too weak, and the House and Senate agreed on a stronger set. The president vetoed the bill, but the House overrode the veto with virtually unanimous Democratic support (232–4), and the Senate followed suit.

By 1988, liberal Democrats claimed that the previous sanctions were having little effect because there were too many loopholes. In May, the Foreign Affairs Committee reported a bill barring almost all trade with South Africa, and then the far-reaching bill had to be referred to seven other committees that had jurisdiction. Using his powers relating to multiple referrals, Speaker Jim Wright imposed deadlines for consideration of the bill. The need to act on other priority legislation kept the deadlines from being met, but the leadership kept up the pressure and committee action was completed by August.

Under the Speaker's direction, the Rules Committee stripped from the bill a Republican-sponsored amendment that had been adopted in the Banking, Finance and Urban Affairs Committee, and then it drafted a restrictive special rule permitting only three amendments to be offered (all sponsored by Republicans). On the floor, all three amendments received overwhelming support from the GOP, but none got more than 23 southern Democratic votes. The southerners then endorsed passage of the bill 66–7. Senate consideration of the issue stalled because it was clear that sufficient support was not available to override a threatened veto by President Reagan. Action on new sanctions in the Bush administration was blocked by movements to reform the South African system, although there was disagreement over whether sufficient change had occurred to justify removing the previous sanctions.

This issue clearly demonstrates the way changed electoral circumstances have altered the domestic influences on conflict over foreign policy. When conservative whites dominated the constituencies of most southern Democrats, the majority of them could not join with their northern colleagues in efforts to use U.S. economic influence to penalize discriminatory foreign governments. In the wake of the major electoral changes outlined above, support for vigorous use of sanctions against South Africa became almost as strong among southern Democrats as among Northerners, even in the face of determined opposition from President Reagan.

Other Issues from the Bush Administration

In the discussion of defense authorization bills, we saw that conflict between the president and House Democrats persisted after the transition from Ronald Reagan to George Bush. In this section, I will consider a few other issues from the Bush years in order to assess further the effects of that transition.

Certainly the most salient international affairs issue of the Bush administration was the debate over whether Congress should authorize the president to use force against Iraq. This issue was partly a conflict over institutional prerogatives that has caused tension between Congress and the president throughout the nation's history, but particularly since the Vietnam War and the passage of the War Powers Resolution in 1973. Did the president have inherent authority to use force, as Bush spokespersons claimed, or was congressional endorsement in advance required? The latter was the view of most Democrats in Congress and of a number of prominent Republicans.

In the event, Congress did vote on authorizing the use of force and did endorse it, but the decision still split along partisan lines. The Democratic leadership in both houses opposed the president, although the vote lacked the usual leadership pressure on individual members. As House Majority Whip Richard Gephardt (D-Mo.) stated, "We expect and want all of the Members to vote their conscience."[30] In the House, Bush got virtually unanimous support from the GOP (only 3 defectors), but he was opposed by more than two-thirds of the Democrats. Not surprisingly, the president's support was greater among southern Democrats than northerners, but even there 32 of the 85 members opposed him.[31]

What is so remarkable about this vote is not that the president won. Rather it is that significant opposition developed before the fact, that it occurred so sharply along partisan lines, and that the opposition attracted such substantial support. One has only to think back to the Gulf of Tonkin resolution of 1964—unanimous in the House and only 2 opponents in the Senate—to see how much things have changed. Again, in order to oppose the president many Democrats were willing to accept the electoral risks that they would be blamed if their policy went wrong. Republicans have used the war vote to rally opposition and recruit opponents against Democrats who voted "wrong." This strategy appeared to be ineffective in the 1992 congressional elections.

Another sharp dispute between Congress and Bush involved responses to China's suppression of dissent in June 1989. Shortly thereafter the House unanimously passed sanctions against China, but the bill to which they were attached was not enacted. Later that year Congress passed a bill designed to prevent deportation of Chinese students whose visas had expired. Bush vetoed the bill but promised that the students would not be deported. The House voted 390–25 to override the veto, but enough Republicans rallied behind the president in the Senate to sustain him.

Later, the dispute revolved around whether China's most-favored-nation (MFN) trade status should be renewed. The Ways and Means Committee reported a bill in 1991 that would impose stiff conditions on China regarding moves toward democracy and individual freedom before renewing MFN status. In July, House Republicans sought to recommit the bill in order to grant the president wide discretion on the matter, but the move failed and the bill was passed. On both votes Bush was opposed by heavy majorities of northern and southern Democrats. A less harsh set of conditions was included in the final conference report, which the House endorsed in November by 409–21 against Bush's wishes. The president vetoed the bill, but the House voted to override 347–61, with all but 10 Democrats and two-thirds of the GOP supporting.

CONCLUSIONS

This chapter argues that over the last two decades Congress has grown increasingly assertive in foreign and defense policy and that conflict over these issues has grown increasingly partisan. The main domestic cause of these developments was electoral change: alterations in the patterns of political participation and voter loyalty, particularly in the South, resulted in increased homogeneity of preferences on policy among members of Congress in both parties and in greater differences between the parties. Intraparty agreement and interparty conflict was then reinforced in the House by changes in institutional procedures that undermined the independence of committees, made committee leaders responsible to majority party caucuses, and strengthened the powers of party leaders.

The evidence presented above indicates that committees did become more responsive to majority party preferences on international affairs issues, that the Democratic leadership used its power over the agenda to protect policies preferred by the party, and that greater opportunities for amendments on the floor further moved policy in the direction of majority party preferences. More specifically, we saw that sectional differences among Democrats have been significantly reduced on defense policy, on the involvement of the United States in the internal conflicts of other countries, and on foreign policy issues that are closely related to domestic racial issues. Moreover, we saw that partisan conflict on international affairs between congressional Democrats and the president has persisted beyond the transition from Reagan to Bush.

The analysis of specific issues has been useful in reinforcing and clarifying some aspects of the analysis. For example, the Democratic leadership's use of its agenda powers on the defense authorization bills during Reagan's second term shows clearly that this capability is not only used to *protect* committee products from floor attacks. In this instance the leadership joined with Democratic members to undermine and alter committee judgments because they did not reflect the dominant preferences of the majority party. The temporary removal of Les Aspin as chair of the Armed Services Committee also illustrates that the application of reform procedures is often firmly rooted in policy conflicts. The altered institutional arrangements of the House can and do have a powerful impact on policy outcomes when the precondition of a consensus within the majority party is met.

The persistence of partisan conflicts into the Bush administration serves to demonstrate that the sharp disagreements of the previous eight years were not simply the consequence of typically extreme ideological views on the part of the president, behind which GOP members of Congress felt compelled to rally. To be sure, conflicts were moderated during the new president's first two years, but that was due more to an altered world context. As Barbara Sinclair has said, during that period "geopolitical events were the most important determinant of relatively tranquil relationships. . . . Certainly Congress shows no sign of retreating from its assertive foreign and defense policy role."[32] Rather, the persistence of disagreement and its reinvigoration in the 102d Congress reinforce the conclusion that conflict over defense and foreign policy is rooted in basic electoral forces. Changed electoral alignments have shaped the recruitment of new members of Congress and altered the influences on the preferences of those that have survived from the time before the changes took effect.

In closing, it is important to be clear about what is *not* being claimed here. First, not all international affairs issues are now conflictual, either within Congress or between Congress and the president. Some issues demonstrate widespread consensus. Consider, for example, the bill to authorize sanctions against countries and companies determined to be contributing to the spread of chemical and biological weapons—co-sponsored by ideological opposites Senator Claiborne Pell (D-R.I.) and Senator Jesse Helms (R-N.C.) and endorsed by President Bush in 1991.

Second, not all issues are now partisan, nor were all issues nonpartisan or bipartisan before. The China sanctions matter shows that bipartisan consensus within Congress can develop against the president's views. On the other hand, the issue involving the interaction of new loan guarantees to Israel and Israeli policy regarding settlements on the West Bank has created uncertainty and division within both parties. Indeed, the multiple cross-cutting disagreements in the foreign aid area, generally, have meant that only one independent authorization bill for these programs has become law since 1981.

Third, there are still many aspects of international affairs policy in which congressional efforts to assert positions different from those of the president are of limited or no effect. The president retains broad ability to commit American troops to actual or potential combat situations, despite the War Powers Resolution. In addition, the Senate's advice-and-consent authority regarding treaties is often bypassed by the president's authority to conclude executive agreements. Furthermore, the courts have tended to uphold executive interpretations of congressional delegations of power unless they clearly violated statutory language.

Finally, the sectional disagreements within the Democratic party have not entirely disappeared. As we saw, a majority of southerners in the House opposed their party to endorse the authorization of military action against Iraq. More generally, southerners are still less likely to support defense cuts.

Thus, with regard to congressional assertiveness and partisanship in international affairs, we are not dealing with an either-or situation. The changes that have occurred are matters of degree. But sometimes differences of degree are so extensive that they become differences in kind. That is precisely the case here. Electoral and institutional changes have produced a significantly different context that has led to very different outcomes with regard to foreign and defense policy in the 1980s and 1990s. We can expect this new situation to persist as long as the conditions that produced it persist.

Notes

The major support for the project of which this paper is a part came from the National Science Foundation through grant SES 89-09884. Support for interviews with members of the House of Representatives was received from the Dirksen Congressional Center. Roll call data were supplied by the Interuniversity Consortium for Political and Social Research, which bears no responsibility for the analyses or interpretations presented here. I am grateful for the research assistance provided by Renee Smith and James Meernik, and to Paul Peterson for his comments.

1. Some analyses of these policies use the term "foreign policy" in juxtaposition to defense policy, while others mean that term to include defense matters. For the sake of clarity and simplicity, I will use the terms separately. "Defense policy" includes weapons procurement, war powers, defense organization, and military security issues like foreign military aid (but not matters like military construction and veterans' benefits, which are classified as domestic issues). "Foreign policy" includes issues like nonmilitary foreign aid and diplomatic and trade relationships with other countries. Thus the distinction between the two hinges on military versus nonmilitary matters. I will use the term "international affairs" to denote defense and foreign policy issues generically.

2. Dean Acheson, quoted in John Rourke, *Congress and the Presidency in U.S. Foreign Policymaking* (Boulder, Colo.: Westview Press, 1983), p. 80.

3. Cecil V. Crabb, Jr., and Pat M. Holt, *Invitation to Struggle: Congress, the President and Foreign Policy,* 2d ed. (Washington, D.C.: CQ Press, 1984), p. 58.

4. The detailed analysis underlying this section and the next has been presented in a number of previous works. On electoral change, see David W. Rohde, "The Electoral Roots of the Resurgence of Partisanship among Southern Democrats in the House of Representatives," paper presented at the Annual Meeting of the American Political Science Association, 1991; David W. Rohde, *Parties and Leaders in the Postreform House* (Chicago: University of Chicago Press, 1991). The latter work deals extensively with the House reforms and the activities of party leaders. On the House Democratic leadership, see Barbara Sinclair, particularly "House Majority Leadership in the Late 1980s," in Lawrence C. Dodd and Bruce I. Oppenheimer, eds., *Congress Reconsidered,* 4th ed. (Washington, D.C.: CQ Press, 1989), pp. 307–33; Barbara Sinclair, "House Majority Leadership in an Era of Legislative Constraint," in Roger H. Davidson, ed., *The Postreform Congress* (New York: St. Martin's Press, 1992), pp. 91–111. The changed relationship among committees, floor politics, and legislative outcomes is considered in David W. Rohde, "Agenda Change and Partisan Resurgence in the House of Representatives," in Alan D. Hertzke and Ronald M. Peters, Jr., eds., *The Atomistic Congress* (Armonk, N.Y.: M. E. Sharpe, 1991), pp. 231–58. On presidential-congressional conflict, see Rohde, *Parties and Leaders,* chap. 5; David W. Rohde, "Divided Government, Agenda Change, and Variations in Presidential Support in the House," in Paul Peterson, ed., *Congress and the Making of Foreign Policy,* forthcoming. Senate-House contrasts are discussed in David W. Rohde, "Electoral Forces, Political Agendas, and Partisanship in the House and Senate," in Davidson, ed., *Postreform Congress,* pp. 27–47.

5. Southern states include the eleven former Confederate states, plus Kentucky and Oklahoma.

6. V. O. Key, Jr., *Southern Politics* (New York: Vintage, 1949).

7. The floor was not the only arena in which Democratic party unity demonstrated an increase in the House and Senate. Daniel S. Ward showed that Democratic unity grew between the 96th and 100th Congresses on roll calls within every House and Senate committee he examined. See "The Life of the Party: The Individual Basis of Partisanship in U.S. House and Senate Committees," Rice University Institute for Policy Analysis working paper, 1992.

8. In addition to the works cited above, the changed role of the Rules Committee and its relationship with the leadership are dealt with in Stanley Bach and Steven S. Smith, *Managing Uncertainty in the House of Representatives* (Washington, D.C.: Brookings Institution, 1988).

9. Steven S. Smith, *Call to Order* (Washington, D.C.: Brookings Institution, 1989), chap. 2.

10. Melissa P. Collie and Joseph Cooper, "Multiple Referral and the 'New' Committee System in the House of Representatives," in Dodd and Oppenheimer, eds., *Congress Reconsidered,* p. 265.

11. Aaron Wildavsky, "The Two Presidencies," *Society* 4 (December 1966): 7–14, quotation on p. 8.

12. Crabb and Holt, *Invitation to Struggle,* p. 62.

13. Quoted in Lewis A. Dexter, "Congressmen and the Making of Military Policy," in Robert L. Peabody and Nelson W. Polsby, eds., *New Perspectives on the House of Representatives* (Chicago: Rand McNally, 1963), p. 311.

14. Mendel Rivers, quoted in Steven S. Smith and Christopher J. Deering, *Committees in Congress* (Washington, D.C.: CQ Press, 1984), p. 107.

15. Richard F. Fenno, Jr., *Congressmen in Committees* (Boston: Little, Brown, 1973), pp. 27, 30.

16. Crabb and Holt, *Invitation to Struggle,* p. 44, describing the stated views of Chair Thomas Morgan (D-Pa.).

17. Steven S. Smith, "Congressional Leaders and Foreign Policy", in Peterson, ed., *Congress and the Making of Foreign Policy* (forthcoming).

18. Consensual votes (i.e., votes receiving 90 percent or greater support) are excluded from the analysis. There are not enough votes in earlier years to extend the series back.

19. The time series begins with 1971 because that is when new rules on recorded votes on amendments went into effect. The analysis involves only first-degree amendments; second-degree amendments (i.e., amendments to amendments) do not necessarily indicate dissatisfaction with the committee's judgments.

20. A conservative coalition vote is one on which a majority of northern Democrats votes in opposition to a majority of southern Democrats and a majority of Republicans.

21. This difference is measured similarly to party difference: the absolute difference between the percentage of northern Democrats and the percentage of southern Democrats voting aye.

22. John W. Kingdon, *Congressmen's Voting Decisions,* 3d ed. (Ann Arbor: University of Michigan Press, 1989), pp. 41–43.

23. Barney Frank, quoted in *Insight,* February 16, 1987, p. 19.

24. Les Aspin, quoted in *Congressional Quarterly Weekly Report,* May 23, 1987, p. 1066.

25. Southern Democrats are not included because their support level never fell below 93 percent.

26. For an account of the battles over this issue, see Philip Brenner and William M. Leo Grande, "Congress and Nicaragua: The Limits of Alternative Policy Making," in James A. Thurber, ed., *Divided Democracy* (Washington, D.C.: CQ Press, 1991), pp. 219–53.

27. Trent Lott, quoted in *Congressional Record,* March 11, 1987, p. H1191.

28. Trent Lott, quoted in *Congressional Quarterly Weekly Report,* March 14, 1987, p. 460.

29. Bob Michel, quoted in *Congressional Record,* March 3, 1988, p. H644.

30. Richard Gephardt, quoted in *National Journal,* January 19, 1991, p. 194.

31. It was surprising, perhaps, that the usually more hawkish Senate provided less support for the president. The vote there was only 52–47, and the authorization of force was opposed by 82 percent of the Democrats, including 10 of the 17 southerners.

32. Barbara Sinclair, "Governing Unheroically (and Sometimes Unappetizingly): Bush and the 101st Congress," in Colin Campbell, S.J., and Bert A. Rockman, eds., *The Bush Presidency: First Appraisals* (Chatham, N.J.: Chatham House, 1991), p. 173.

5 / Congressional Activism in Foreign Policy: Its Varied Forms and Stimuli

JOHN T. TIERNEY

Virtually all observers of American foreign and defense policy agree that the long-standing structures of power and decision making in these policy realms have been undergoing dramatic changes, with Congress figuring more and more prominently over time. Until the Vietnam War era, foreign policy was controlled to a great extent by the president and an exclusive, executive-centered foreign policy elite. And although the president and the national security bureaucracy still remain in the driver's seat on many issues, Congress plays a far more important role now than it did a couple of decades ago. But how do we describe congressional involvement? How do we specify the forms and dimensions of Congress's growing assertiveness in the realm of foreign and defense policy?

The preceding chapter by David W. Rohde provides us with one answer, an incisive and compelling one: we understand Congress's growing assertiveness in foreign and defense policy as a function of increasing partisanship, bred of gradual electoral changes and a consequent sharpening both of intraparty agreement and interparty conflict. In an era dominated by persistently divided government—Republican control of the White House and (mostly) Democratic control of Congress—partisanship has become the driving force in congressional differences with the executive over foreign and defense policy. The political parties and their leaders are the important aggregating force in Congress, without which it is impossible for Congress to assert itself in the way Rohde specifies—that is, showing a willingness "to substitute its collective judgment for the president's."

Rohde's chapter is the clearest elucidation yet available of the ways in which party differences define congressional activism in foreign policy, but it highlights only part of the overall picture of changing congressional involvement in the formulation of foreign and defense policy: it focuses on Congress as a lawmaking institution and locates the roots of its policy contrariness in growing partisanship.

This second chapter on Congress and foreign policy is meant to illuminate other features of congressional involvement in foreign policy making. It looks beyond the narrow function of lawmaking (which is, after all, only a small part of what Congress does) and beyond partisanship and party leadership (Congress's centripetal, aggregating forces) to take stock of the more centrifugal forces in the congressional political environment: individual legislators' political incentives and opportunities, the fragmentation of congressional committees, the diversification of organized interests with a stake in foreign policy, and the multiplication of foreign policy "idea brokers." This chapter shows that while the congressional parties of today may be speaking with louder and clearer voices on foreign policy than they used to, these are not the only policy-relevant voices being heard in the cacophony on the Hill. Moreover, this chapter demonstrates that there are many different ways members of Congress can affect policy, so we need to take account of more than just lawmaking by majorities if we are to have a more complete picture of congressional involvement in foreign policy.

AN ALTERNATE VIEW OF CONGRESS AND ITS INFLUENCE

It is a commonplace observation among scholarly observers of Congress—indeed, it is a theme of one of the leading college textbooks on the national legislature—that "there really are two Congresses, not just one."[1] One Congress is the lawmaking body that was the focus of the preceding chapter—the institutional Congress that acts as a collegial body of shifting majorities, carrying out constitutional responsibilities, challenging the executive branch, passing legislation, making policy. In this Congress, the party and leadership forces that organize and orchestrate legislative behavior—planning for favorable rules and other legislative conditions, marshaling unruly partisans, assembling successive majorities for passing a law—are of paramount importance and, as Rohde shows, help us to understand and explain the changing nature of congressional behavior on foreign policy and other issues.

The Congress of Individual Members

The other Congress is the collection of individual legislators—the political arena in which separately elected and differently inclined representatives work out the assorted tensions among their views of the national interest, their individual constituencies' interests, and their own personal political interests. Although Congress frequently acts as one institutional entity—when its two houses agree to pass a law, for example—it is not a single, monolithic institution. Indeed, it is about as far from that as something that calls itself an organization can get: it is an assemblage of 535 individually elected representatives and senators (and thousands more of their unelected staffers), all stretched across the normal range of human personality and proclivities, across a spectrum of legislative experience and years of service in Congress, across 535 different constituencies, across the ideological spectrum, across two political parties, across the two chambers of Congress, across scores of congressional committees, and across a labyrinthine legislative process with many different points of opportunity for shaping political outcomes. In this Congress, what is important is the individual legislator, each marching to his or her own peculiar political drumbeat, each responsive to a different mix of organized and unorganized constituencies, each animated by his or her own unique combination of motives, ambitions, opportunities, pressures, and incentives.

One of the most important consequences of the reform upheaval in Congress in the 1970s (discussed in the previous chapter) is that in its aftermath individual members of the House counted for more than they did in the past and came to have more opportunities to engage in the sort of policy entrepreneurialism for which they came to be known.[2] The formal channels and norms structuring the distribution of influence in the chamber (strong chairs, rigid jurisdictional boundaries of committees, norms of apprenticeship, and so forth) eroded, leaving individual members freer to pursue their own policy interests and objectives. It gradually reached the point where one did not even have to have a seat on a relevant subcommittee to have influence (although that always helps): today it is often sufficient simply to have a strong interest in a region or an issue and to care more about it than anyone else.

This individualization of policy influence is possible because the post-reform Congress makes it relatively easy to push an issue, even for those who are not committee leaders or senior members. Again, some of this influence is simply a matter of demonstrated interest and expertise. A good example is the case of Representative James Oberstar (D-Minn.), who holds no seat on a committee related to foreign affairs but is the universally recognized congressional expert on Haiti (having lived there for several years in the early 1960s) and who has earned the privilege of being consulted whenever United States actions concerning Haiti are considered.

Another example involves the efforts of Representative Joe Moakley (D-Mass.), chair of the Rules Committee, who has had tremendous influence in cutting military aid to El Salvador because he got interested in human rights abuses there, became quite expert on the issue, and earned the respect and hearing of his colleagues. Moakley has also worked to secure refugee status to Salvadorans who have come to the United States and helped to rewrite the 1990 immigration law to allow more Salvadorans to immigrate legally to the United States.[3]

Techniques of Influence

Just as this individualized Congress is highly varied and richly textured in its members' different purposes, motives, and interests, so also are those members lavishly equipped with different techniques for influencing or affecting foreign policy. When it comes to trying to affect foreign and defense policy, members of Congress have many arrows in their quivers. As the preceding chapter by Rohde suggests, one very important form of congressional policy influence comes from the institution's aggregates, the parties, which assemble majorities to pass substantive policy legislation in order to make Congress's collective will prevail. Legislation surely is the most striking and conspicuous means by which Congress tries to shape foreign policy, and often it is also the most direct and certain in its impact. But if we focus only on the passage of substantive legislation (such as defense authorizations, Contra aid policy, economic sanctions against South Africa), we miss a wide variety of other ways in which Congress institutionally and its members individually influence policy.

Thus it is worthwhile to consider some of the *other* ways members of Congress can influence foreign policy. The following discussion is by no means exhaustive but merely is meant to provide some sense of the varied techniques and resources for influence that members have at their disposal.

Amendments and Other Statutory Techniques Some of these other techniques in fact still fall under the broad category of legislative or statutory efforts. For instance, members frequently offer amendments to legislation in an effort to advance their own policy preferences or those of interest groups. Sometimes these amendments have important policy consequences. For example, in 1975 Congress stated its commitment to human rights by adopting the so-called Harkin Amendment (Section 116a of the Foreign Assistance Act), forbidding U.S. economic assistance to regimes in gross violation of internationally recognized human rights unless the aid directly benefits needy people. Sponsored by Representative (later Senator) Tom Harkin (D-Iowa), the amendment had been drafted by staffers of the Quakers and the Washington Office on Latin America and then received additional crucial lobbying support from the Americans for Democratic Action.[4]

Of course, it is important to point out that the ability of House members to make or affect foreign policy directly by passing laws or amendments was greater in the years immediately following the major reforms in the House—the mid- and late 1970s—than it is in the early 1990s.[5] In the immediate postreform period, rank-and-file members had unprecedented access to the floor. So those members wishing to make law—or sometimes just mischief—in foreign policy would use the foreign aid authorization bill (or sometimes the foreign operations appropriations bill) as their legislative vehicle. But by the late 1980s the majority party leadership had come to control access to the floor much more tightly, especially by controlling the floor schedule: bills tend not to get to the floor unless the leadership agrees to schedule them. In addition, the leadership to a considerable extent controls the amending process as well. Most major legislation now is considered under restrictive rules, so there are few amending free-for-alls of the sort that occurred frequently in the mid-1970s. Members today are more likely than they were in the late 1970s to see their pet amendments disallowed if the party leadership considers them to be bad policy, bad politics, bad timing, or simply a waste of time.

But none of this is to say that amendments are now closed in the House as an avenue of policy making, only that it is a more difficult path than it once was.

Even if entrepreneurial lawmaking is thus more difficult now in the House, the Senate of course continues to provide an environment much friendlier to policy entrepreneurs because of the ways in which the legislative process in that chamber confers great power on senators as individuals. Particularly important in this regard are unrestricted debate, the absence of any real limits on floor amendments, and the weakening of the norms that used to limit senators' exploitation of these powers. When combined with the fact that senators generally have greater media access than House members and that most majority party senators chair several subcommittees, the Senate environment gives the individual senator a considerably greater shot at influencing policy on whatever issue he or she happens to find interesting.

And even in the House, members still see amending activity as a useful technique for influencing the direction of policy. Part of its attractiveness, of course, is that, as any observer of legislative politics knows, an amendment does not have to be adopted in order to have an effect. For one thing, an amendment, especially if it has considerable support, can force accommodations and compromises in the broader bill. Moreover, sometimes amendments are offered merely for their symbolic value, useful as a way of appeasing organized interest groups, sending a message to the executive branch, or even sending a signal abroad. As Robert A. Pastor has noted:

> A congressman introduces an amendment with 50 cosponsors that threatens Japanese products with tariffs if Japan does not remove its barriers to U.S. exports. A senator introduces an amendment that threatens to cut aid to Mexico if that government does not cooperate more in the pursuit of drug traffickers. The intent of both resolutions is to send signals in three directions, each with a different purpose: to the foreign government—Japan, Mexico—to open up its markets or cooperate on drugs; to constituents, to indicate that their congressman is acting on their concerns; and to the State Department, to push it to negotiate more forcefully.[6]

Whatever the purpose of members' amending activity—whether to attract attention to themselves, to send signals, or to influence coalitions or compromises—the amount of activity is increasing, with the number of amendments offered and the number of hours devoted to consideration of the annual defense authorization bill increasing steadily during the 1970s and 1980s.[7]

Committee Activities Another way members of Congress affect foreign policy is through their activities as members of committees. Part of their committee work involves considering new legislation. This is a prime opportunity for members, even quite junior ones, to have a say in the shaping of policy proposals through their participation in the bargaining, compromising, and accommodating that characterize legislative deliberation. Even when proposed legislation does not pass, it still can have policy impacts of the sort noted above for amendments.

But most committee work has less to do with shaping new programs than with reconsidering and overseeing existing ones. The routine processes of authorization and appropriation focus committee attention on one of Congress's most important functions—oversight (or monitoring) of executive agencies and the programs they administer. In defense policy, in particular, Congress's oversight efforts have undergone dramatic changes in the past several decades with the gradual adoption of annual authorizations for most portions of the defense budget. Members of the Armed Services Committee now authorize almost every element of the defense budget each year, down to almost the last rifle and uniform.

From an institutional or policy point of view, the occasional advantages of this sort of close financial and programmatic oversight may be outweighed by the disadvantages of constricted vision. As Senator Sam Nunn (D-Ga.), chair of the Senate Armed Services Committee, has said: "At its worst . . . this trend to micromanagement has the staff and members focusing on the beach while we should be looking over the broad ocean and beyond the horizon."[8] Some academic observers such as Robert J. Art agree, lamenting that "powerful political realities" push Congress into looking "mostly at the details of defense spending but rarely at the big picture."[9]

But to say that members of Congress are more inclined to pay attention to—and can have an effect on—the details of policy than on the broad outlines is not to trivialize this form of influence. On the contrary, all the best case studies of policy formulation and implementation demonstrate an axiom central to both politics and policy analysis: to know what a measure does, it is important to look beyond its broad purposes to the particulars. After all, the details are what specify such critical matters as how much is to be spent, on what, and under what constraints.

Moreover, from the standpoint of individual members of Congress, this microscopic attention to the details of defense spending has great political utility. For one thing, it is easier for a member to understand and talk about line-items in the defense budget (aircraft, tracked combat vehicles, naval vessels, and the like) than to engage in "policy oversight" by, for example, considering seapower strategy or the trade-offs between force and diplomacy necessary to achieve national security goals. Second, by attending to the details, members can look out for the interests of constituents in their districts or states, protecting weapons systems, military bases, supply contracts, and the like. Third, with some frequency members find in the budget opportunities for grandstanding by making some splashy revelation that attracts lots of media attention their way, as Senator Charles Grassley (R-Iowa) did in the mid-1980s with his famous press release about the navy paying $435 for a hammer and the air force paying $3,000 for a coffeepot.[10] In that case, of course, an individual member's political grandstanding had lasting policy consequences, as Congress responded (to this and other news about widespread improprieties in the relationships between military contractors and procurement officials at the Pentagon) by embracing more ardently its plans to cut defense spending and by enacting a variety of measures under the rubric of "procurement reform" that include numerous technical modifications in the way the government pays defense contractors.

In the Senate, committee members have another avenue of influence—their constitutionally conferred right to "advise and consent" on presidential nominations to high-level positions in the executive branch. This power came to play quite forcefully in the 1989 decision by members of the Senate Armed Services Committee to reject President George Bush's nomination of former Senator John Tower (R-Tex.) to become secretary of defense. Originally intended by the framers of the Constitution to enable the Senate to check on the character and qualifications of those whom the president selects for major policy positions, the Senate's confirmation power has come to be used much more broadly by members as a technique for congressional control over the policy and administrative processes. As G. Calvin Mackenzie has noted, after having studied the politics of presidential appointments: "Confirmation is a process, not simply a vote. And . . . the confirmation process provides ample opportunity for the Senate to influence the President, his nominees, and their subsequent policy decisions [even] without rejecting any substantial number of nominations."[11] Mackenzie shows that senators use the committee hearing stage of the confirmation process to instruct, to warn, and even to humiliate nominees whom the Senate ultimately plans to confirm. By these means senators can achieve a considerable degree of influence over nominees, who will be likely to give careful thought before taking any subsequent policy actions that might violate the views expressed by important groups of senators in their confirmation hearings and to actions that violate promises or strong expressions of opinion that they themselves have offered in those hearings.

Procedural Innovations James M. Lindsay and Randall B. Ripley have pointed to the various ways Congress uses "procedural innovations" to alter the structures and processes by which decisions are made in the executive branch—to affect either the mix of people and viewpoints involved in decision making or the processes by which decisions are made. Examples include the congressional decision in 1977 to create the Bureau of Human Rights and Humanitarian Affairs in the State Department to institutionalize attention to human rights in the executive's formulation of foreign policy, and the provision incorporated in the Trade Act of 1974 enabling members of Congress to serve as official advisers in international trade negotiations. As Lindsay and Ripley point out, "The underlying premise is that changing the process changes the policy," which is especially appealing to Congress since the seemingly neutral character of such procedural innovations "makes it easier to build a winning coalition around procedural changes than around substantive policy changes."[12]

Another important kind of procedural innovation Lindsay and Ripley identify is the reporting requirement, by which Congress instructs the executive to prepare reports for Congress to inform the legislature of executive actions or decisions, as in the case of Central Intelligence Agency reports to congressional committees about covert operations.[13] Congress uses these reports for a variety of purposes, alone or in combination: administrative oversight, legislative action, public education. Legislators may request reports to help them initiate legislation in the absence of any presidential effort or to put pressure on the executive. And sometimes members turn to reporting requirements as a substitute for substantive lawmaking when there is insufficient support or consensus to pass an enforceable law. One advantage of reporting requirements is that one does not even have to get them included in statutes; they can simply be included in committee reports that accompany legislation.[14] The number of reports has gone up astronomically in recent years: the Defense Department's required reports to Congress went from 79 in 1967 to 861 in 1990, of which 211 are recurring reports.[15]

From the standpoint of individual members, required reports are attractive in a variety of ways. First, they provide legislators with information they need to deliberate. Second, being forced to prepare reports can have the effect of inducing executive branch officials to anticipate congressional reactions and thus, perhaps, to craft policies and programs with a sensitivity to congressional concerns and preferences. And third, legislators can use the information from the reports for their own tactical political purposes. As Robert Gates, former deputy director of the Central Intelligence Agency, has put it, this information is "often used to criticize and challenge policy, to set one executive agency against another, and to expose disagreements within the administration."[16]

Direct Diplomacy or Action Another nonstatutory way members of Congress can influence foreign policy is by engaging in direct diplomacy. This action takes several forms. Perhaps the most conspicuous form is what James M. Lindsay has called "Lone Ranger diplomacy," a phrase he coined to refer to "the penchant some members of Congress have to conduct their own foreign policy."[17] The most famous recent episode of Lone Ranger diplomacy involved House Speaker Jim Wright's (D-Tex.) 1987 incursion into the Nicaraguan peace process, when he met over a three-day period with Nicaraguan President Daniel Ortega, Contra leaders, and Nicaraguan Cardinal Miguel Obando y Bravo in an effort to pursue a diplomatic resolution to the ongoing conflict between the Sandinistas and the Contras.[18]

Whereas Lone Ranger diplomacy is quite rare, many members of Congress have routine, direct contacts with foreign governments—when visiting foreign dignitaries or heads of state make trips to Capitol Hill to talk with legislators, when foreign embassy personnel stationed in Washington head to the Hill to lobby members of Congress on foreign policy, and when members of Congress make visits abroad. As Lindsay points out, one way in which these congressional contacts with foreign governments can be important is by reducing the president's autonomy and unilateral ability to get what he wants: "If Congress and the foreign government know what each other considers to be an acceptable outcome, they can push the president away from his ideal outcome and closer to their own."[19]

Going Public Just as presidents can change the dynamics (and often the direction) of public policy by "going public," members of Congress can also shape public opinion on foreign and defense policy matters by "framing" issues, "packaging them in a way that attracts the attention of the media and the executive branch, places the issue on the agenda, and puts the administration on the defensive."[20] The processes and actions by which members of Congress influence public opinion include holding committee hearings, making emotional floor speeches that get clipped for the network news programs, issuing press releases and staff reports, writing op-ed columns in national newspapers, appearing on Sunday morning news shows, helping to mobilize constituencies, and so forth. Any such action can have the effect of pushing the executive branch (or other policy actors or, for that matter, other countries) to reconsider some policy or proposal and to refine or revise it to accommodate new views or reconcile new demands.

Often, of course, one finds these various techniques being used in combination with one another and for dual purposes—affecting policy as well as advancing a legislator's own reputation or electoral fortunes. For example, a member may launch an informal caucus and use the initiative to demonstrate mastery of an issue's intricacies. Or a member may commission investigations from the General Accounting Office or the Congressional Research Service and attract attention by releasing findings through the press on a slow news day. This latter approach—working the media, attracting publicity, hustling to make a name—is such a common part of life on Capitol Hill these days as to be second nature to most legislators. But this is not to suggest that this behavior is deplorable or undesirable: broadening the scope of the conflict via the media may be one of the few effective weapons members of Congress can bring to bear when the White House seems immovable on an issue like Contra aid or South Africa. Even if they have few hopes of producing immediate, tangible accomplishments, members know that they can raise issues, reshuffle agendas, and frame alternatives through skillful use of the media.[21]

Examples of members who have shown skill at going public in order to influence foreign policy are Senator Chris Dodd (D-Conn.) and Representative Stephen Solarz (D-N.Y.). Dodd enjoys considerable influence on Central American issues by virtue of his visibility as a senator, his stint as a Peace Corps volunteer in the Dominican Republic, his fluency in Spanish, and his demonstrated expertise on the issues. This combination of attributes enables him to command publicity virtually at will.[22]

Solarz, one of Congress's most relentless headline hunters, is fond of playing to the folks back home as well as to the press and television galleries. He was among the members of Congress (Senators Richard Lugar and Paul Laxalt were others) credited with "bringing down" Philippine President Ferdinand Marcos in the mid-1980s by reframing public opinion toward Marcos and his government. By his staff's count, the ubiquitous Solarz appeared on thirty-four radio and television shows and was quoted in eighty articles in national newspapers in a five-month period in 1986.[23] The efforts of Solarz and others played an important role in reshaping public opinion and ultimately leading President Ronald Reagan to withdraw his long-standing support for the embattled dictator.

In sum, in the contemporary Congress political maneuvers intended to influence the consideration of foreign policy can spring up virtually anywhere: members of Congress seem not to regard such niceties as seniority or committee membership as necessary attributes qualifying one to hold forth on matters of foreign policy or to aspire to be a mover and shaker in this once-elevated policy domain. It is true that some members are more likely to play such a role because of their institutional or partisan responsibilities (the congressional leadership, for example) and some because of their expertise or sustained access to experts and large reservoirs of information (the chairs and members of the relevant congressional committees). But many others (individual legislators with no leadership role or foreign policy committee assignment) simply have opinions, interests, or preferences and manage to participate by knowing how to seize available opportunities for purposes of advancing their own agendas or career considerations.

When one looks at this broader, more varied congressional political environment, there is so much to keep an eye on—so many different views or scenes to take in—that what observers focus on is bound to vary widely. Because of the complexity and richness of congressional political activity, any particular view of it is likely to be incomplete. The best we can do is point to a variety of changes in the congressional political environment that affect both the opportunities and the incentives for enhanced involvement by individual representatives in the politics of American foreign policy. One important change in the opportunity structure for participation in foreign policy is that Congress no longer marches to the tune of a few committee chairs (or even of committees), so foreign policy issues have become opportunities for members of varying stature and visibility to stand out by virtue of their persistence, expertise, or cleverness.

THE CHANGING STRUCTURE OF COMMITTEE POWER

Nowhere are the changing structures and patterns of congressional involvement in foreign and defense policy making more apparent than in the parts played by the relevant standing committees. As the recognized workhorses of

Congress—the centers of policy making and of oversight of the executive—committees are the institutional mechanism through which Congress flexes much of what muscle it has in the foreign and defense policy arenas. Perhaps the most remarkable change in the structure of committee power is the extent to which it has become fragmented, with more and more committees getting into the act on foreign policy issues. Part of this fragmentation is the result of larger institutional changes in Congress in the 1970s, discussed in more detail in the previous chapter. Part of it also stems from the fact that the influence of the principal committees traditionally responsible for formulating the national legislature's positions on these matters—the Senate Foreign Relations Committee and the House Foreign Affairs Committee—has declined noticeably over the past quarter-century (the former more so than the latter). The Senate committee's downward slide seems far more conspicuous and more important—perhaps because its previous stature was so much greater and because it has a substantial zone of constitutionally conferred decision-making responsibility (reviewing treaties and presidential nominations) from which the House committee is excluded.

In a penetrating analysis of the Senate Foreign Relations Committee's decline, James McCormick sees several features of the committee's internal dynamics as contributing to its deterioration: (1) substantial turnover in its membership has left it composed of "members with more limited service and more limited political influence within the institution"; (2) a widening ideological gulf within the committee (with the Democratic and Republican members becoming more and more polarized ideologically) has reduced the chances of compromise and thus the prospects for effective decision making; and (3) a precipitous decline in the quality of leadership from the committee chair. This last factor may well be the central reason for the deterioration of the committee's effectiveness since 1974, when J. William Fulbright (D-Ark.) left the Senate after serving for a full quarter-century as chair. Not only has there been frequent turnover in the Senate committee chair in the years since then (there have been five chairs in less than two decades), but these individuals—John Sparkman, Frank Church, Charles Percy, Richard Lugar, and Claiborne Pell—"have not been recognized as forceful leaders with well conceived foreign policy agendas."[24] The decline of the Senate Foreign Relations Committee created a vacuum into which individual senators and other committees moved. Where once the Foreign Relations Committee was the premiere institutional platform for congressional influence in foreign policy, its decomposition has encouraged new Senate aspirants for foreign policy primacy.

On the House side, the decentralization of committee power took a different form. Whereas the chorus of voices in the Senate enlarged as the Foreign Relations Committee gradually withered from decay, the chorus of voices in the House enlarged as a consequence of deliberate steps to push power and responsibility down from the level of the full Committee on Foreign Affairs to its subcommittees. McCormick's analysis demonstrates how the increased powers of the subcommittees and the subcommittee chairs—and the growth of staff at both the committee and subcommittee levels—have spread out the power base in the House on foreign policy matters. One consequence of this enlarged power base has been a conspicuous increase in the number of oversight hearings held by House Foreign Affairs subcommittees since the 1970s. And according to McCormick, the House subcommittees' monitoring of the policy behavior of the executive branch has been routine, focused, and aggressive—another point of contrast between the House and the Senate committees.[25]

While the principal foreign policy committees in both houses have been undergoing dramatic changes, other developments have led to a further fragmentation of power on Capitol Hill with respect to foreign policy. The most conspicuous of these developments has been the emergence in the 1980s and early 1990s of the Armed Services Committees in each chamber as important forums for handling key foreign policy issues. McCormick notes that this change occurred in part because the nature of the policy issues dominating the agenda during this period (debate over the U.S. defense budget in general and over weapons systems and arms control issues) fell more clearly within the jurisdiction of the Armed Services Committees than that of the foreign policy committees. Another reason for the Armed Services Committees' new primacy was the "vacuum left by the inaction of the foreign policy committees"—a void eagerly filled by the armed services panels, as evidenced in the 1987 debate over the Reagan administration's attempted reinterpretation of the Antiballistic Missile (ABM) Treaty and again in late 1990 and early 1991 when the hearings held by the armed services panels involving the Persian Gulf crisis clearly overshadowed the foreign relations committees' deliberations.[26] The Armed Services Committees in both chambers also have benefited from assertive and skillful leadership in the persons of Representative Les Aspin (D-Wis.) and Senator Sam Nunn (D-Ga.).[27]

The House and Senate Appropriations Committees also have come to play a prominent role, especially in foreign aid policy. Since the House Foreign Affairs and the Senate Foreign Relations Committees have managed to secure the passage of only two foreign aid authorization bills in the past twelve years (in 1981 and 1985), the Appropriations Committees have taken the leading role, seeking waivers of the requisite authorization measures in order to appropriate funds. As McCormick notes, the practical effect of this trend has been to lessen the influence of the foreign policy committees and to increase the power of the Foreign Operations Subcommittees of the two Appropriations Committees.[28] Whereas in the past the appropriators did what the authorizers told them to do, that no longer holds true when the Senate Foreign Relations Committee is virtually dead on the vine and the House Foreign Affairs Committee is splintered into autonomous subcommittees.

A similar sort of fragmentation has occurred in the area of trade policy.

At one time not long ago, control over the contents of trade bills was in the hands of the chairs of the Senate Finance Committee and the House Ways and Means Committee. But just as the structure of committee power has been fragmented in the more traditional foreign policy and defense domains, in trade policy making as well participation has now spread to include the House Energy and Commerce Committee, the Armed Services Committees, the banking committees, and the Judiciary Committees.[29]

In sum, many institutional changes on Capitol Hill since the early 1970s have scattered power and responsibility and opportunities for influence. The decentralization and fragmentation of congressional power have multiplied the number of people on the Hill who may have a say on foreign policy and defense matters.

But these institutional developments, though very important, constitute only part of the changing field of forces drawing more and more members of Congress into the broad arena of foreign affairs. The changing character of the policy agenda and the presence of new external demands, interests, and policy ideas are also important stimuli to involvement by members of Congress.

THE CHANGING CHARACTER OF THE POLICY AGENDA

Christopher Matthews, the author and journalist who acquired much of his firsthand political experience working as chief spokesperson for House Speaker Thomas P. ("Tip") O'Neill, Jr. (D-Mass.), once wrote of him:

> His long rise to one of the country's most contested positions through a half century of successful elections was built on something hard and elemental. It is the nugget of wisdom prized by all great political figures: to understand and influence your fellow man, don't focus too much on the grand, intangible issues: keep a tight watch on what matters most to him or her personally.[30]

The Speaker himself is well known for his own brief formulation of this principle: "All politics is local," he was fond of saying. For years, members of Congress seemed to embrace and follow the implications of this maxim for how they should balance their attentions between global difficulties and local problems: the latter should clearly come first; electoral success comes from attention to highway and water projects, not to the terms of grand alliances or broad diplomatic initiatives. And political scientists have busied themselves for years, dutifully describing and remarking on the localistic perspectives of national legislators.

While there is of course still much to be said for Tip O'Neill's understanding of how to retain the electoral loyalty of the folks back home, this general posture no longer has quite the inviolability it once had, largely because policy issues themselves are quite different now from what they were in previous decades. More and more of the policy challenges that fill Congress's agenda have a distinct "intermestic" character—that is, they are issues on which old distinctions between domestic and international affairs are now clearly irrelevant: global warming, deforestation, pollution control, fishing rights, nuclear and chemical weapons proliferation in the third world, drug trafficking, terrorism, human rights, new trade interdependencies and imbalances, the global integration of capital markets, and the exchange value of the dollar. Noting the way these challenges "underscore the growing irrelevancy of distinctions between domestic and international affairs," Walter J. Oleszek argues that many of the policy problems and demands that come before Congress these days are beyond the control of any single nation. If solutions are to be found, it will only be through international cooperation and coordination.[31]

All of this is not to say that representatives and senators now have less incentive to heed traditionally domestic concerns such as inflation, unemployment, health care, environmental protection, transportation, and the like. Rather, they have incentive to regard all these issues with enhanced sensitivity to their international dimensions, as Oleszek notes: "Simply put, no member can politically afford to remain insulated from global developments because no district or state is isolated from their effects."[32]

Similarly, members of Congress have incentive to pay attention to the varied implications of issues such as foreign aid and trade, to see how legislation on such matters might be used as vehicles for achieving other political objectives such as environmental protection or the advancement of human rights (or used simply as opportunities for symbolic posturing or "position taking"). Moreover, paying attention to these issues can work to the electoral advantage of members who are skillful enough to frame for their constituents the domestic importance of these matters. Explaining why corporate and business groups working the "other side" of the South African antiapartheid issue in the mid-1980s made so little headway, even among congressional Republicans, Bruce W. Jentleson noted Senator Majority Leader Robert Dole's (R-Kans.) observation that the issue had become a "domestic civil rights issue." Jentleson went on to explain:

> Prime Minister Pieter Botha was Governor George Wallace; Bishop Desmond Tutu was the Reverend Dr. Martin Luther King, Jr.; and the Anti-Apartheid Act was the Civil Rights Act. Republicans, looking more than ever to recruit blacks into their party, thus voted for the Anti-Apartheid Acts of 1985 and 1986 by margins almost as huge as the Democrats.[33]

In short, it is apparent that the changing character of the policy agenda has altered the incentive structure for members of Congress on foreign policy issues. Once largely ignored by legislators who regarded them as irrelevant to their districts and detached from the kinds of interests to which they are accustomed to responding, foreign policy issues now have taken on more of the flavor of domestic policy issues. Foreign policy issues are increasingly being prosecuted politically by a broad constellation of organized interests that behave similarly to the ways they behave in domestic policy arenas—using their varied political resources to press legislators to carry their cause to victory, a path that increasingly takes legislators beyond the water's edge.

NEW EXTERNAL DEMANDS, INTERESTS, AND IDEAS

Members of Congress are increasingly drawn to foreign policy matters by the growing number and increased efforts of organized interests intent on influencing the direction of congressional decisions over foreign and defense policy matters. The reasons for the increased presence of organized interests in the foreign policy realm are numerous. Some have to do with unfolding events and changing conditions such as the globalization of the American economy, changes in the strategic security interests of the United States, and the growing federal budget deficit.

Other reasons have to do with the changing mix of groups in society having concerns that are touched by foreign and defense policy. For example, there are the traditional interests—such as corporations, trade associations, labor unions, and ethnic lobbies—that mobilize politically around some economic interest at stake, such as a trade agreement, a foreign aid appropriation, or a weapons contract. But recent years have also seen the proliferation of groups (such as nuclear freeze and human rights organizations) contending over nonmaterial interests and values.

Still other reasons for the increasing role of organized interests in foreign and defense policy have to do with the changing structure of power and decision making on these issues. When foreign policy was controlled primarily by the president and an exclusive foreign policy elite, the number of access points open to organized interests was small. But by the mid-1970s, the institutional and political environments for making foreign and defense policy had changed dramatically, in ways that offered organized interests new opportunities to participate in policy deliberations and encouraged their active involvement. The attentive public was becoming more ideologically polarized on foreign policy issues and the mass public was becoming less passive, more distrustful, and more volatile and unpredictable as it hankered simultaneously for peace and strength.[34] Moreover, the press was newly aggressive and skeptical in its coverage of foreign and defense policy and thus was a potential new ally for those on the outside eager to shape policy agendas.

Finally, Congress was carving out a formidable institutional presence for itself in the development of foreign and defense policy: the national legislature, as we have seen, was becoming more assertive, more resourceful, more decentralized—and thus more permeable to those eager to have some input in foreign and defense policy-making processes.

As the old executive-centered foreign policy elite is weakened by the inroads of a more dynamic and inclusive mix of policy actors, those who wish to understand the politics of foreign and defense policy need to have a better handle on the ways that Congress and interest groups are linked on these issues.[35] One place to begin such a consideration is by exploring the diversification of foreign policy interests and considering the ways in which these changing interests implicate congressional politics.

The Diversification of Foreign Policy Interests

The organized interests trying to influence congressional decisions on issues of foreign and defense policy are neither as numerous nor as wide ranging in scope as one finds in the broader arena of domestic politics, where both the number and diversity of interests represented in the political process are truly striking. Still, foreign and defense policy issues engage hundreds of organizations representing the broad mix of interests active on Capitol Hill, especially ethnic groups, advocacy and cause groups, foreign governments and foreign economic interests, and American economic interests.

Ethnic Groups Among the most noticeable of the private organizations active on issues of foreign and defense policy are ethnic interest groups. Surely there has been more written about ethnic lobbying organizations than about any other particular kind of organized interests in the foreign policy arena.[36] Ethnic lobbies get involved in foreign policy issues on behalf of many different racial, cultural, and religious subgroups in the American population. There are sizable and active organizations representing Irish, Italian, Polish, African, and Asian Americans, to name a few. The phalanx of organizations active in Washington on behalf of particular ethnic interests is large and growing rapidly to include new organizations such as the Cuban American Foundation and assorted new eastern European ethnic organizations such as the Ukrainian National Association, the Polish American Congress, and the Armenian Assembly of America.

Of all the ethnic groups, the so-called Jewish lobby is widely regarded as the most important and powerful throughout most of the past forty years, its political influence stemming chiefly from two factors: (1) the extraordinary issue attentiveness and high voting participation rates of American Jews, and (2) skillful assemblage in Washington of several Jewish-American organizations that lobby Congress on issues affecting Israel, the best known of which is the American Israel Public Affairs Committee (AIPAC). Since its founding in the early 1950s, AIPAC has prospered and built itself into one of the most formidable lobbying organizations in the United States.

Between 1980 and 1990, AIPAC's staff quadrupled to more than one hundred, its membership quintupled to fifty-five thousand households, and its member-financed budget increased almost tenfold, from $1.4 million in

1980 to $12 million in 1991.[37] AIPAC tries to work its will on Capitol Hill by orchestrating aggressive grass-roots pressure campaigns, cultivating key members and staffers in the House and the Senate who become strategic allies and sources of timely information and political intelligence, and showing its muscle often enough to reinforce its own image as a heavyweight that has to be heeded, as when AIPAC successfully targeted Senator Charles Percy (R-Ill.) for defeat in 1984.

Although the Jewish lobby's formidable resources failed in 1991 to translate into a policy victory on Israel's request for $10 billion in loan guarantees for refugee resettlement, its record of success over time has been impressive, especially in persuading Congress to appropriate huge amounts of foreign economic and military aid to Israel, which receives far more foreign aid from the United States than does any other country.[38]

And, of course, American Jews are not alone among ethnic groups in being able to point to successes in the foreign policy realm. Perhaps the other most notable instance of direct policy influence by an ethnic lobby involved the Greek-American community's response to the Turkish invasion of Cyprus in 1974. The American Hellenic Institute Public Affairs Committee (AHIPAC) mobilized to coordinate the activities of various Greek-American groups and succeeded, despite White House resistance, in getting Congress to impose an embargo on American arms to Turkey.[39]

Generally speaking, however, when one moves away from AIPAC and the politically active constituency of American Jews, the record of ethnic group lobbying success is far less imposing, to the point where most analysts seem to agree that the impact of such groups on American foreign policy is minimal.[40]

But it is clear that changing political circumstances can affect an ethnic group's political opportunities. The Cuban-American lobby, for example, has been enjoying some enhanced clout recently, benefiting from changing world events such as the breakdown of communist regimes in eastern Europe, the emergence of a new government in Nicaragua, and the end of the cold war. As Cuba becomes more and more isolated as one of the few remaining outposts of hard-line communism, members of Congress are attracted by the symbolic appeal of taking action against Castro's totalitarian regime.[41] This changed situation makes a difference to the political fortunes of the Cuban American National Foundation (CANF), created in 1981 to lobby and frame opinion on Cuba and Cuban Americans. Between 1981 and 1992, CANF's directors and the political action committee they control, the Free Cuba PAC, Inc., donated more than $1 million directly to candidates for Congress and the White House and helped to raise even more through their sponsorship of fund-raising dinners. This sort of effort paid off for CANF in October 1992 when President George Bush signed the Cuban Democracy Act, which extends the economic embargo that the United States began when Fidel Castro took power in 1960 and imposes penalties on foreign subsidiaries of American companies trading with Cuba.

CANF had pushed this legislation hard in Congress, arguing that it would hasten the downfall of Castro.[42]

Moreover, there is no doubt that, just as in domestic politics, groups that do not rank high in the overall pecking order of Washington power politics may nevertheless enjoy advantages in particular instances by virtue of having strategically placed allies in Congress. The foreign policy concerns of Polish Americans, for example, are advanced in Congress by such formidable advocates as Dan Rostenkowski (D-Ill.), chair of the House Ways and Means Committee, and John Dingell (D-Mich.), chair of the House Energy and Commerce Committee. Similarly, although Armenian Americans find their relatively small numbers to be a political handicap, they have as champion of their cause Senator Claiborne Pell (D-R.I.), chair of the Senate Foreign Relations Committee. Pell has a large Armenian constituency in Rhode Island, and he uses his institutional position to watch out for its interests, such as in the late 1980s when he was able to get the Senate Foreign Relations Committee and later the full Senate to approve a resolution condemning violence against ethnic Armenians in the Soviet Azerbaijani region of Nagorno-Karabakhj.[43] Although the policy impact of such a resolution is typically minimal (in this case it got the Soviet Foreign Ministry so upset that U.S. Ambassador Jack Matlock was called to the Kremlin and formally rebuked), it illustrates one of the ways in which even politically frail ethnic interests can affect the politics of foreign policy—by raising issues in the press and in Congress and thereby affecting the agenda.

Advocacy and Cause Groups There is nothing particularly new about the presence in foreign and defense politics of citizens' groups organized around common concerns or political causes of a noneconomic or nonoccupational nature. Although such groups have been around for decades, the number of these organizations politically active on foreign and defense issues is much greater than ever before, and their interests range far beyond war and peace issues to include such matters as global environmental degradation and human rights.

Some citizens' groups that get involved in foreign policy ordinarily have domestic policy as their primary focus but are induced by circumstance or opportunity to broaden their perspectives to an international or global perspective. This is the case, for example, with respect to many environmental groups that have gotten involved in various areas of foreign policy, such as trade, as global environmental issues have loomed larger and larger. National environmental groups, which once consigned trade issues to meagerly staffed "international" departments, if they dealt with them at all, devoted considerable resources in the early 1990s to injecting an ecological perspective into the debate over the proposed North American Free Trade Agreement and protecting environmental standards threatened in the most recent round of negotiations on the General Agreement on Tariffs and Trade (GATT).[44]

Other citizens' groups active on foreign policy issues are advocacy organizations that seek selective benefits on behalf of persons who are in some way politically incapacitated or otherwise unable to represent their own interests effectively. Among such groups are Amnesty International, the Friends Committee on National Legislation (the Quaker lobby), the United Church of Christ, and the International League for Human Rights—all organizations that lobby in Washington on human rights issues, pressing the United States to do something to help persons suffering from conditions that range from torture in Paraguay, to slavery in Mauritania, to starvation and malnutrition in East Africa and Southeast Asia.[45]

Citizens' groups and cause groups have also been playing a more and more important role over time in foreign aid policy, in part perhaps because this is a policy area tailor-made for interest group intervention. Members of Congress engineer the details of foreign aid by earmarking (setting aside) funds for specific countries or particular projects. This process of earmarking has at least two advantages from the standpoint of legislators and the proponents of particular interests. First, it is a clear and direct way of forcing the executive branch to pay attention to their wishes with respect to foreign policy, including such particular concerns as environmental protection and human rights. Second, it addresses the political needs of legislators on foreign assistance issues, which often are, understandably, the lowest priority to most members' constituents: after all, foreign aid does not provide visible benefits for very many home states and districts. But earmarking gives legislators a way to claim political credit with whatever narrow constituencies, such as Greek Americans or pro-Israel lobbies, are pushing for earmarks.

But congressional micromanagement of foreign aid expenditures goes beyond cordoning off dollars for specific countries. Congress also stipulates the details of development assistance, specifying how much will be spent in each of eight different functional categories of development aid: agriculture, education, health, child survival, AIDS, population, energy, and the environment. Needless to say, the efforts of organized interests are central to the politics of these earmarks, engaging a wide spectrum of citizens' groups from A to Z—from human rights organizations such as Amnesty International to population control groups like Zero Population Growth. For example, there has been a continuing struggle in recent years about earmarks for population programs aimed at lowering birthrates in underdeveloped countries. Population control organizations tended to win support for these earmarks on the Hill, but only after bitter struggle with members supported by prolife groups that do not like the earmark because some of the money goes to China, which has been condemned for its abortion practices.

Foreign Governments and Foreign Economic Interests Another prominent constellation of organizations active on Capitol Hill includes foreign governments and foreign businesses, which lobby on behalf of their own interests or hire Washington lobbyists to do it for them. There is nothing particularly surprising or unnatural about this. After all, the American national government adopts many different policies—on defense arrangements, arms sales, foreign aid, immigration, import restrictions and other trade practices, and so on—that affect the economic strength, political stability, and national security of other countries.

Most foreign governments get involved in congressional politics directly through their embassies. Embassy personnel monitor unfolding events on the Hill with an eye toward protecting national interests. Visits to the Hill by embassy staff are now so commonplace that members of Congress consider foreign emissaries a staple part of the retinue of lobbyists trooping through the corridors of Congress each week. But a more conspicuous way that foreign governments and corporations now seek to influence congressional decision making is by retaining highly regarded American lobbyists as their agents in Washington. Many former cabinet secretaries, House members and senators, White House staffers, and trade officials, once out of office, hire themselves out as foreign agents, representing the interests of other nations on Capitol Hill. The long roster of those employed in this way (now or recently) reads like a who's who of the modern American political elite, including Clark Clifford, Lloyd Cutler, William Rogers, William Ruckelshaus, Stuart Eizenstat, and Paul Laxalt. These foreign agents represent their clients' interests in the same ways they would for domestic clients. They forge links with legislators from districts where their clients have facilities. They help their clients create and manage political action committees.[46] In short, they use the legislative and political networks they have been nurturing for years and make the most of their access, political skills, and credibility.

But there is rising concern in Washington and among observers of national politics that foreign-owned corporations are throwing around too much weight as they penetrate the American political system in ways that may not be in the national interests of the United States. Although foreign economic interests with real clout in Washington hail from points around the globe (from Britain and Canada, Switzerland and Germany, the Netherlands and South Korea, and countless other countries), in the view of many observers the most disturbing specter of foreign political power has Japanese features. Critics of "the Japan lobby" assert that Japanese companies (and the Japanese government) spend tens of millions of dollars each year on Washington lobbyists, consultants, and public relations firms, infiltrating Washington's fragmented decision-making apparatus to such an extent that the Japanese hold considerable sway, especially over American trade policy.[47]

Defenders of Japan's involvement in Washington politics argue that the case against the Japan lobby is vastly overstated; that Japanese lobbying is not much different from other lobbying on Capitol Hill, where competing interests often cancel each other out and have only marginal impact on policy; and that, in fact, much of the Japanese lobbying has been counterproductive because of the negative publicity it has generated.[48]

While scholars disagree about the importance of foreign economic lobbying efforts, it is sufficient for our purposes to note that the Japanese and others have established in Washington formidable networks of highly regarded American advocates who have the kinds of skills, insider knowledge, strategically placed contacts, and far-reaching financial resources that typically spell access and influence in American politics.

American Economic Interests It is not surprising that in a capitalistic economy such as that of the United States the overwhelming majority of organizations actively engaged in representing their interests to the government are economic organizations of one sort or another—labor unions, corporations, trade associations, business alliances, and the like. This is as true in foreign and defense policy as in other arenas of American public policy. American businesses and other economic concerns clearly have strong interests in a broad range of foreign and defense issues, including weapons procurement, arms sales, foreign economic assistance, and trade policy. The last few years have brought some changes in the scope of their involvement, however—particularly noticeable in trade policy where business interests figure prominently.

In the realm of trade policy, the range of business interests looking to Congress for help or relief has expanded considerably from the period following World War II, when the prevailing pattern of interest group involvement on trade issues was of a small number of industries seeking protection, each pretty much watching out for itself. Smaller industries were encouraged by congressional leaders to use the quasi-judicial trade remedy procedures of the Tariff Commission (later the International Trade Commission). But as I. M. Destler has noted, larger industries, such as textiles, sometimes managed to negotiate separate deals for themselves—deals that usually involved promises not to stand in the way of broader trade-liberalizing initiatives.[49] There were no significant groups staking out a broad protectionist position, nor were those groups that stood to benefit from open trade, such as consumers, active on the issue.

But by the mid-1980s, the confluence of important new economic developments around the world gave the politics of trade some new twists and began anew to attract the attention of Congress to trade issues. Among the more salient economic developments were the dramatic worsening of the U.S. trade deficit; the looming emergence in 1992 of the European Community as an economic giant whose movements would send economic shock waves around the world; the emergence of Canada and Mexico as ever more important trading partners of the United States; and the economic awakening of the Pacific rim.

In addition to changes affecting the politics of interests were changes in the politics of symbols and ideas. In the slogan of "fair trade," protectionists finally found an effective symbolic counterweight to the "free trade" theme that so long had played so well for their opponents. Moreover, economic nationalism was once again becoming a potent political force in American politics (and especially on Capitol Hill), as more citizens and political leaders came to perceive that it was not the Soviet "evil empire" of Ronald Reagan's legendary rhetoric that the United States had fallen behind, but Japan and the capitalist powers of the West. Many members of Congress found themselves drawn to trade policy as they saw more evidence for the proposition that the United States had been fighting in the wrong trenches for the past thirty-five years. Noting the general changes in the political and intellectual climate on the trade issue, Destler concludes:

> Trade politics has become more partisan. The elite has grown less committed to liberal trade. Intellectual challenges to open-market policies have grown. Patterns of trade politics have become more complex. All these changes have weakened the old system for diverting and managing trade policy pressures, and most have increased the political weight of those backing trade restrictions.[50]

And the growing number and range of U.S. industries seeking trade protection (or at least a government-secured "level playing field") has now come to include not just mature, relatively low-tech industries whose manufacturing processes can be readily duplicated in other countries but also the high-tech makers of products like semiconductors and telecommunications equipment.

None of this means, of course, that narrow interests looking for protection always win. Sometimes organized interests (such as the steel industry in the mid-1980s) are unsuccessful in pushing their parochial trade demands, even when other countervailing societal pressures are weak, because government policy makers (or "state actors," as statist theorists would say) resist those demands out of regard for the proposal's widespread political and economic implications.[51] Sometimes organized interests may fail even in a vigorous bid to win a protectionist measure such as an import quota if legislators fear the effects of retrospective voting by a generally inattentive public that is nevertheless capable of reacting to large and immediate costs traceable to legislative actions.[52]

But frequently legislators do choose to serve attentive and organized interests on trade matters. In some instances, doing so poses no readily discernible costs to domestic interests and is thus quite easy, as when legislators and staffers in the late 1980s persistently pushed the Japanese on behalf of American chocolate manufacturers who wanted help in reducing excessive Japanese tariffs that inhibited the export of their candies (and who provided legislative offices with unending supplies of one-pound bags of M&Ms as a constant reminder of the issue).[53]

The Changing Foreign Policy Elite

As part of this effort to identify the principal centrifugal forces pulling more and more members of Congress into the foreign policy realm and diversifying the community of policy activists on Capitol Hill, one other development deserves brief reference. That development is the transformation that has occurred in the community of policy watchers and policy intellectuals that is considered the "foreign policy elite." Once consisting chiefly of the old establishmentarians who generally shared a centrist and bipartisan (or nonpartisan) worldview and who managed to articulate (and through their connections, help implement) their vision of U.S. global economic and military leadership, the community of foreign policy idea brokers in Washington today is much different: it is larger, more heterogeneous, more publicity conscious, and more ideological than it was in the past. Indeed, the old and the new foreign policy establishments (if such a word can be used) are so different as to make a drawing of contrasts almost superfluous.

The home base of the most prominent of the new foreign policy professionals is the think tank. There are scores of these quasi-academic institutions scattered across the Washington landscape. The Brookings Institution, the American Enterprise Institute for Public Policy Research, the Center for Strategic and International Studies, the Heritage Foundation, the Institute for Policy Studies, and the Carnegie Endowment may constitute the core institutions, but one count suggests there are some sixty such organizations in Washington dealing with matters of foreign policy and national security.[54] Whereas the old foreign policy elite tended to be drawn primarily from the East Coast, educated at Ivy League schools, and apprenticed in law and investment banking firms to which they would return after any period of public service (or, at least, that is the image), the new foreign policy professionals are more inclined to see government service as part of a career move, perhaps a step up. Their objective is to enhance their visibility and their reputations.

These incentives impel foreign policy professionals toward behavior that their predecessors might find a bit gauche: seeking publicity and visibility. One of the principal things that distinguishes the old from the new elite is that the former, like early White House staffers, had more of a "passion for anonymity." Not so the new breed. They are constantly presenting testimony before congressional committees, churning out signed op-ed pieces, and trying to nail down television appearances, because these things are so crucial to career advancement.

In the main, the policy impact of most of these experts is indirect, but that is not to say that it is negligible. Some of their influence comes in the form of giving members of Congress ideas to run with—intellectual capital for their entrepreneurial appetites. In much the same vein, the experts are able to supply members of Congress with defensible policy rationales for actions the members want to take (but may not know how to explain). But some of their influence is even more indirect: since many of them frequently appear on television and radio and in print, they are able to influence public opinion about American foreign policy.

Some of the recent literature on American politics and public policy has suggested that in domestic policy there has been a gradual supplementing, if not a complete supplanting, of the "politics of interests" with a new "politics of ideas," in which the power of a good idea or a symbolically appealing cause can be just as persuasive or forceful in the policy-making process as resourceful importuning by a vested interest group.[55] It would probably be an overstatement to say that the same thing has happened in foreign policy, but there is little doubt that the changing foreign policy elite is one element in this reconstituted politics of American foreign policy. Indeed, as at least one observer sees it, the disappearance of the old establishment has merely paved the way for organized interests to have more influence, and the consequences are less than salubrious:

> Governments have invariably relied on informal networks of private citizens, political organizations, and elite groupings like the Council on Foreign Relations to fill the interstices between individual will and public power.
>
> For three decades, the old Establishment occupied this area, holding study groups, publishing papers, and providing the officials that filled the upper echelons of government. But as it has disintegrated, narrow lobbies and pressure groups have filled the vacuum. Worse still, these lobbies and pressure groups represent no underlying consensus but only their own separate interests. American foreign policy, once the realm of the gods, has become the domain of mere influence peddlers.[56]

CONGRESS IN THE NEW FOREIGN POLICY SYSTEM

The new policy community for foreign affairs and national security is incredibly varied, fragmented, and porous. Issue experts move in and out of the limelight, trading it off with issue amateurs; ideas and proposals gain and lose political currency; elected officials on the Hill variously bluster and reason their way from one topic to another. Because ideas and opportunities for participation and power are all so widely dispersed, and because so many different people want to influence policy outcomes, the legislative environment for foreign and defense policy making often seems prone either to incoherence or paralysis. Just as in domestic politics, policy coherence sometimes appears to get bargained away and compromised beyond the point of reasonable dilution. And just as in domestic arenas, U.S. foreign policies are increasingly displaying the same sorts of policy dysfunctions and the same pattern of systemic paralysis long evident in domestic policy. Thus, for example, Congress has managed to pass only two foreign assistance authorization bills since 1980 (in 1981 and 1985).

But the foreign policy landscape long was spared this sort of breakdown not only because it was primarily executive driven but because of a prevailing consensus borne both of objective international conditions and of shared beliefs at home about the desirability of a bipartisan, united front. As that consensus has eroded over the past quarter-century, and as the choir of voices heard in Washington has come to include more self-appointed soloists on the Hill and in the think tanks and newsrooms, the politics of American foreign policy has become more like U.S. domestic policy—fractious and ideologically riven—and it yields policy outcomes riddled with multiple and contrary objectives, when it yields legislative outcomes at all.

Some of this exaggerated conflict and increased contentiousness is the fault of legislators who seem to care most of all about posturing and positioning themselves before the camera lenses and grabbing a moment's attention on the evening news. They know that one way to attract that attention is to hop onto one hobbyhorse or another and tilt away at assorted windmills (or, at least, at other windbags). This tendency is supported by the oversupply of idea merchants in Washington—the members of the new foreign policy elite populating the capital's scores of think tanks—many of whom know that the quick way to enhance their personal and professional visibility is to make as bold a splash as possible by attacking the currently ruling paradigms, identifying the latest threat to American national interests, or huffing with righteous indignation about the motives or interests of their competitors.

But in its contemporary incarnation Congress at least performs one of its functions quite well. Responsible for representing society's interests, opinions, and preferences, Congress is institutionally well designed for purposes of representation. As the most profound institutional expression of American representative democracy, Congress is pervasively open to virtually any organized interest that wishes to present its views. With its lengthy and complex procedures and its specialized committees and subcommittees that serve to decentralize power, Congress offers multiple points of access for those wishing to express a view. Moreover, by encouraging the expression of multiple and diverse viewpoints, Congress counters the narrowness of executive branch decision making. These are considerable virtues.

Congress has many flaws as an institutional actor in foreign policy making, and, as much of the recent literature on the subject shows, it is too easy to get carried away in identifying them.[57] Political scientists—like journalists—tend to get carried away, as Lawrence Brown has noted, by "a public-spirited desire to be the first authority on the block to identify the latest threat to American democracy." In the process, we may be led to "decry each new development or change as a grave threat to the roles and functions of time-tested institutions" and to be overly impressed with the fragility of what are in fact very durable processes and institutions.[58] There are, after all, beneficial consequences to the cacophonous increase in the number and range of voices heard on Capitol Hill on foreign policy issues. As a consequence of more widespread involvement, foreign policy decisions are somewhat more likely to be tested in the political process and shaped and tempered through consensus-forming procedures, all of which makes for policies that take into account competing values and interests and that are more likely to enjoy popular legitimacy.

Notes

1. Roger Davidson and Walter Oleszek, *Congress and Its Members,* 3d ed. (Washington, D.C.: CQ Press, 1990), p. 4.

2. See Burdett Loomis, *The New American Politician: Ambition, Entrepreneurship, and the Changing Face of Political Life* (New York: Basic Books, 1988); David E. Price, *The Congressional Experience: A View from the Hill* (Boulder, Colo.: Westview Press, 1992).

3. Rochelle Stanfield, "Floating Power Centers," *National Journal,* December 1, 1990, pp. 2916, 2918. See also Michael Barone and Grant Ujifusa, *The Almanac of American Politics, 1992* (Washington, D.C.: National Journal, Inc., 1991), p. 594. Barone and Ujifusa note that in May 1990 the House passed a Moakley amendment to cut military aid to El Salvador by 50 percent. And although the larger bill later was voted down, Moakley managed to get similar language into other legislation that was passed, the 1991 foreign aid bill.

4. David Forsythe, *Human Rights and World Politics,* 2d ed. (Lincoln, Nebr.: University of Nebraska Press, 1989), p. 143.

5. House rules include strict germaneness requirements, but most bills were considered in those years under ground rules that allowed all germane amendments. Consequently, any member could force a vote on a pet policy proposal if he or she could find a bill to which it was germane, and usually such a bill would come to the floor at some time during the session, giving those members wanting to make law an opportunity to try. Today, access to the floor—and to the amendment process—is much tighter.

6. Robert A. Pastor, "Congress and U.S. Foreign Policy: Comparative Advantage or Disadvantage?" *Washington Quarterly* 14 (Autumn 1991): 105.

7. Christopher Deering, "National Security Policy and Congress," in Christopher Deering, ed., *Congressional Politics* (Chicago: Dorsey Press, 1989), pp. 287–88.

8. Sam Nunn, quoted in Mackubin Thomas Owens, "Micromanaging the Defense Budget," *Public Interest,* no. 100 (Summer 1990): 134.

9. Robert J. Art, "Congress and the Defense Budget: Enhancing Policy Oversight," *Political Science Quarterly* 100 (Summer 1985): 227. Much the same problem prevails in congressional handling of foreign aid programs, where (as we shall see in more detail later) the political incentives push members of Congress to earmark funds for specific countries, particular projects, or pet causes. In the process, legislators have crafted a badly muddled accumulation of foreign assistance programs that are riddled with conflicting objectives. The programs may meet the demands of legislators catering to special interests, but, according to critics, they do little to promote U.S. foreign policy objectives.

10. For the true story about the allegedly wasteful expenditure on hammers, see James Fairhall, "The Case for the $435 Hammer," *Washington Monthly,* January 1987, pp. 47–52. See also the interesting explanation of the constraints faced by military program managers and civilian contract officers in the whole process of weapons procurement in James Q. Wilson, *Bureaucracy: What Government Agencies Do and Why They Do It* (New York: Basic Books, 1989), pp. 317–25.

11. G. Calvin Mackenzie, *The Politics of Presidential Appointments* (New York: Free Press, 1981), pp. 174–75.

12. James M. Lindsay and Randall B. Ripley, "How Does Congress Matter in Foreign and Defense Policy Making," in Randall B. Ripley and James M. Lindsay, eds., *Congress Resurgent: Foreign and Defense Policy on Capitol Hill* (Ann Arbor: University of Michigan Press, forthcoming 1993).

13. Ibid., ms. p. 24.

14. These points are drawn from the useful discussion of reports in Arthur Maass, *Congress and the Common Good* (New York: Basic Books, 1983), pp. 218–21.

15. Owens, "Micromanaging the Defense Budget," p. 138.

16. Robert Gates, quoted in L. Gordon Cravitz, "Micromanaging Foreign Policy," *The Public Interest,* no. 100 (Summer 1990): 112.

17. James M. Lindsay, "Congress and Diplomacy," in Ripley and Lindsay, *Congress Resurgent,* ms. p. 3.

18. Bruce Jentleson, "American Diplomacy: Around the World and along Pennsylvania Avenue," in Thomas E. Mann, ed., *A Question of Balance: The President, the Congress, and Foreign Policy* (Washington, D.C.: Brookings Institution, 1990), p. 155; Lindsay, "Congress and Diplomacy."

19. Lindsay, "Congress and Diplomacy," ms. p. 7.

20. The term "framing" comes from Lindsay, "Congress and Diplomacy"; Lindsay and Ripley, "How Does Congress Matter in Foreign and Defense Policy Making."

21. See Benjamin I. Page, Robert Y. Shapiro, and Glenn R. Dempsey, "What Moves Public Opinion," *American Political Science Review* 81 (March 1987): 23–44; Stephen Hess, *The Ultimate Insiders: U.S. Senators in the National Media* (Washington, D.C.: Brookings Institution, 1986), chap. 7.

22. Loomis, *New American Politician,* p. 96.

23. Hedrick Smith, *The Power Game* (New York: Random House, 1988), p. 139. On Stephen Solarz's relentless pursuit of publicity for his foreign policy preferences, see Loomis, *New American Politician,* pp. 96–100.

24. James McCormick, "Decisionmaking in the Foreign Affairs and Foreign Relations Committees," in Ripley and Lindsay, *Congress Resurgent,* ms. pp. 28–31. Some observers argue that the Foreign Relations Committee's rate of decline steepened in 1987 when Claiborne Pell (D-R.I.) became chair and Jesse Helms (R-N.C.) rose to the panel's ranking Republican slot. See, for example, Stanfield, "Floating Power Centers," p. 2916; Christopher Madison, "Awaiting a Wake-up," *National Journal,* March 28, 1992, p. 750.

25. McCormick, "Decisionmaking in Committees," ms. pp. 11–16.

26. Ibid., ms. pp. 38–41.

27. See David C. Morrison, "Sharing Command," *National Journal,* June 13, 1992, pp. 1394–98.

28. McCormick, "Decisionmaking in Committees," ms. p. 4. McCormick notes that Representative David Obey (D-Wis.), chair of the House Subcommittee on Foreign Operations, has been characterized as "the most powerful foreign affairs spokesman in the House." Stanfield, "Floating Power Centers," p. 2916.

29. Pietro Nivola, "Trade Policy: Refereeing the Playing Field," in Mann, ed., *Question of Balance,* p. 233.

30. Christopher Matthews, *Hardball* (New York: Harper and Row, 1988), p. 44.

31. Walter J. Oleszek, "House-Senate Relations: A Perspective on Bicameralism," in Roger H. Davidson, ed., *The Postreform Congress* (New York: St. Martin's Press, 1992), p. 203.

32. Ibid.

33. Jentleson, "American Diplomacy," p. 158.

34. Thomas E. Mann, "Making Foreign Policy: President and Congress," in Mann, ed., *Question of Balance,* pp. 11–13.

35. On the changing role of interest groups in foreign policy, see John T. Tierney, "Interest Group Involvement in Congressional Foreign and Defense Policy," in Ripley and Lindsay, *Congress Resurgent,* chap. 5.

36. See, for example, Mohammed E. Ahrari, ed., *Ethnic Groups and U.S. Foreign Policy* (New York: Greenwood Press, 1987); Mitchell Bard, *The Water's Edge and Beyond: Defining the Limits to Domestic Influence on United States Middle East Policy* (New Brunswick, N.J.: Transaction Publishers, 1991); Stephen A. Garrett, "Eastern European Ethnic Groups and American Foreign Policy," *Political Science Quarterly* 93 (Summer 1978): 301–23; David H. Goldberg, *Foreign Policy and Ethnic Interest Groups* (Westport, Conn.: Greenwood Press, 1990); F. Chidozie Ogene, *Interest Groups and the Shaping of Foreign Policy: Four Case Studies of United States African Policy* (New York: St. Martin's Press, 1983); Abdul Aziz Said, ed., *Ethnicity and U.S. Foreign Policy* (New York: Praeger, 1977); Edward Tivnan, *The Lobby: Jewish Political Power and American Foreign Policy* (New York: Simon and Schuster, 1988); Eric Uslaner, "A Tower of Babel on Foreign Policy?" in Allan J. Cigler and Burdett A. Loomis, eds., *Interest Group Politics,* 2d ed. (Washington, D.C.: CQ Press, 1991), pp 236–257; Paul Y. Watanabe, *Ethnic Groups, Congress, and American Foreign Policy* (Westport, Conn.: Greenwood Press, 1984).

37. Lloyd Grove, "Israel's Force in Washington: The Power of AIPAC Is Respected and Resented," *Washington Post National Weekly Edition,* June 24–30, 1991, p. 8.

38. On the reasons for the Jewish lobby's failure in the loan guarantee episode, and its subsequent effects on relations between Israel and the United States, see Christopher Madison, "Strained Friendship," *National Journal,* April 18, 1992, pp. 924–28.

39. Indeed, an imposed 7:10 ratio of Greek to Turkish military aid still persists, which is, from a policy standpoint, quite bizarre (even if understandable politically as a reflection of

Greek-American political power) since, by any measure, Turkey has been a far better security partner to the United States than has Greece, both before the Cyprus incident and after.

40. See, for example, Irving Louis Horowitz, "Ethnic Politics and U.S. Foreign Policy," in Said, ed., *Ethnicity and U.S. Foreign Policy,* pp. 175–80.

41. See Elizabeth A. Palmer, "Exiles Talk of PACs and Power, Not Another Bay of Pigs," *Congressional Quarterly Weekly Report,* June 23, 1990, p. 1992–93.

42. See Larry Rohter, "A Rising Cuban-American Leader: Statesman to Some, Bully to Others," *New York Times,* October 29, 1992, p. A18.

43. Rochelle Stanfield, "Ethnic Politicking," *National Journal,* December 30, 1989, pp. 3096–99.

44. See Hawley Traux, "Coming to Terms with Trade," *Environmental Action* 24 (Summer 1992): 31–34.

45. Forsythe, *Human Rights and World Politics,* p. 4.

46. It is illegal for foreign nationals to make campaign contributions in the United States, but the Federal Election Commission has ruled that a U.S. subsidiary of a foreign corporation may sponsor a political action committee (PAC) so long as U.S. citizens provide all the money and decide how it is spent. A Congressional Research Service study in 1989 found that such PACs made contributions totaling $2.8 million in the 1987–88 election cycle. *National Journal,* January 13, 1990, p. 93.

47. See Pat Choate, *Agents of Influence* (New York: Alfred A. Knopf, 1990).

48. See the discussion in Pamela Fessler, "Do Lobbying Dollars Shape the U.S. Trade Debate?" *Congressional Quarterly Weekly Report,* March 31, 1990, pp. 972–75.

49. I. M. Destler, *American Trade Politics* (New York: Institute for International Economics, 1986), pp. 158–59.

50. Ibid., p. 163.

51. See Andrew J. Stritch, "State Autonomy and Societal Pressure: The Steel Industry and U.S. Import Policy," *Administration and Society* 23 (November 1991): 288–309.

52. See Douglas Arnold, *The Logic of Congressional Action* (New Haven: Yale University Press, 1990), p. 240.

53. Mark Bisnow, *In the Shadow of the Dome: Chronicles of a Capitol Hill Aide* (New York: Morrow, 1990), p. 308.

54. See James Allen Smith, *The Idea Brokers* (New York: Free Press, 1991), p. 213.

55. See, for example, Martha Derthick and Paul J. Quirk, *The Politics of Deregulation* (Washington, D.C.: Brookings Institution, 1985); Paul Schulman, "The Politics of 'Ideational Policy', " *Journal of Politics* 50 (1988): 263–91; Timothy Conlan, Margaret Wrightson, and David Beam, *Taxing Choices: The Politics of Tax Reform* (Washington, D.C.: CQ Press, 1990).

56. John B. Judis, "Twilight of the Gods," *Washington Quarterly* 14 (Autumn 1991): 55.

57. See, for example, Dick Cheney, "Congressional Overreaching in Foreign Policy," in Robert A. Goldwin and Robert A. Licht, eds., *Foreign Policy and the Constitution* (Washingtion, D.C.: American Enterprise Institute, 1991), 101–122; L. Gordon Cravitz and Jeremy A. Rabkin, eds., *The Fettered Presidency: Legal Constraints on the Executive Branch* (Washington, D.C.: American Enterprise Institute, 1989); Aaron L. Friedberg, "Is the United States Capable of Acting Strategically," *Washington Quarterly* 14 (Winter 1991): 5–23; Peter W. Rodman, "The Imperial Congress," *National Interest* 1 (1985): 26–35; George Szamuely, "The Imperial Congress," *Commentary,* September 1987, pp. 27–32; Jay Winik, "The Quest for Bipartisanship: A New Beginning for a New World Order," *Washington Quarterly* 14 (Winter 1991): 115–30.

58. Lawrence D. Brown, *New Policies, New Politics: Government's Response to Government's Growth* (Washington, D.C.: Brookings Institution, 1983), p. 5.

Part IV / The Executive and Congress: Cross-Cutting

6 / A Government Divided: The Security Complex and the Economic Complex

I. M. DESTLER

INTRODUCTION

U.S. government institutions for foreign policy making are divided into two groupings. One—the *security complex*—deals with traditional diplomatic and military issues, giving priority to *foreign* policy goals and relationships. The other, the *economic complex,* addresses trade and money and finance, with emphasis on their *domestic* impact. When the two complexes come into conflict, the security grouping generally prevails. But what is striking is how self-contained the two are, how separate from each other, in their day-to-day operations.

Over the past three decades, this intragovernmental division has deepened, notwithstanding regular calls for greater integration. One reason is that a separated foreign economic policy-making structure enabled U.S. economic interests affected by growing interdependence to have greater impact on trade and financial decisions, even as the cold war continued and—for a time—intensified. But the costs of this separation have been significant for the United States.

With the end of the cold war and rising concerns about domestic welfare and global competitiveness, Washington is being driven to give economic concerns at least equal priority. To this end, the Clinton administration has created a National Economic Council parallel to the forty-five-year-old National Security Council. This organizational reform could actually deepen the security-economics divide, however, unless compensatory steps are undertaken.

A CAUTIONARY TALE

An American president is closing out his third year in office. As maker of international security policy, he has been unusually successful: leading Europe into a post–cold war political structure; organizing a global economic and military coalition first to sanction and contain, and then to reverse, a conquest by a dangerous dictator in a strategic area of the world. He is aided in addressing these thorny challenges by his close relationships with a particularly seasoned and congenial team of national security advisers.

But suddenly, as his year of hoped-for reelection approaches, the president becomes embattled on another front. The economy is mired in recession; his domestic adversaries blame him for this, even turning his foreign successes against him with the slogan, "It's time to take care of our own." When his party suffers a stunning loss in a special Senate election, the president responds by canceling a planned trip to Japan and then rescheduling and recasting it as an economic mission. The goal will now be "jobs, jobs, jobs" for Americans, to be obtained through Japanese trade concessions.

In designing and executing this strategy, however, the president has no team for foreign economic policy making comparable to that which has served him so well on international security. He has talented individual advisers, to be sure, including a strong and resourceful U.S. trade representative. But he is accustomed to *delegating* responsibility for these issues, becoming engaged when aides seek his personal involvement but not initiating action himself. So when he suddenly alters that pattern, he consults the wrong people, asks the wrong questions, and makes the wrong decisions. The resulting mission is widely perceived as a fiasco, undercutting both its policy goals and the president's reelection prospects.

The president in question is, of course, George Bush. His January 1992 visit to Tokyo is best remembered for his acute stomach upset (captured by Japanese television cameras) at a formal state banquet. A more serious effect of the trip was to portray the president as ineffectual in managing America's most important trading relationship. In its aftermath, high-level invective crossed the Pacific in both directions, eating into support for the U.S.-Japan alliance and open U.S. trade policies, both of which Bush strongly favored. It illustrated, in especially dramatic form, the damage that can flow from a policy-making system that treats *security* and *economic* policy separately.

George Bush was extreme, to be sure, in the degree to which he separated foreign from economic policy. The *Washington Post* columnist Mary McGrory even suggested in 1990 that there were "two George Bushes": a "cool, measured, tough, coping" man in charge of foreign affairs, and a "strident, petulant, self-pitying" man on economic policy and the budget. In contrast to his smooth-functioning national security team, a contemporary press exposé found that "President Bush's top economic advisers . . . have been divided by personal animosity and turf fights that are fierce even by Washington standards."[1]

And Bush's successor, Bill Clinton, won election on a very different platform. The famous sign at his Little Rock, Arkansas, campaign headquarters declared the dominant issue to be "THE ECONOMY, STUPID." Not only did Clinton declare this his priority, he also demonstrated considerable awareness of the economy's impact on American power and performance in the world. Yet his first major organizational decision, one month after his election, was to establish within his White House a new, cabinet-level National Economic Council separate from and equal to the National Security Council. This action raised the prospect that, in his administration also, security and economic decisions would be debated in isolation from one another.

If so, Clinton will perpetuate the "government divided" he inherited from Bush. But Bush inherited the structural divide as well. It has its roots in policy crises and institutional responses that date from the 1930s and 1940s. Across several administrations, particularly in the 1970s and 1980s, it has led to major decisions being taken in one sphere without sensitivity to their implications for the other. The following discussion will show this separation first by *illustration,* through brief description of how a few key U.S. government decisions were actually made. This will be followed by political-institutional *history,* describing where the separate security and economic complexes came from and how the division has deepened. Then, after some analysis of how each complex has operated, the *conclusions* will discuss why this division seems particularly inappropriate in the post–cold war era and what might be done to overcome it.

ONE-SIDED DECISIONS

- In December 1979, troops from the Soviet Union invaded Afghanistan. Within a month, the Carter administration had imposed a grain embargo in response. Though economics was a critical factor, the decision was taken within the tight circle of national security and senior political advisers. Economic officials were brought fully into the process only when the question became how to implement the decision already taken.[2]
- In 1986–88, the Reagan Pentagon sought to strengthen defense cooperation with Japan by negotiating an agreement to co-develop an advanced fighter aircraft. In the process, it repeatedly rebuffed concerns of officials in the Office of the U.S. Trade Representative and the Department of Commerce that the deal be examined for its impact on future competition in civilian industries. It was only when critics took advantage of the Reagan-Bush transition to reopen the issue that the question was raised to the presidential level.[3]
- After Iraq invaded Kuwait in August 1990, Bush regularly summoned key officials to debate the details of the response. This group became known as "the big eight." But though the central choice was between initiating military action and continued reliance on economic sanctions, not one of the eight held an economic policy portfolio.[4]

In these three instances, economic officials were shut out of the dialogue, notwithstanding the obvious relevance of what they had to say. But recent history also contains instances of critical decisions being taken in the economic policy context with neglect of their serious and durable foreign relations impact.

- On August 15, 1971, Richard M. Nixon announced a historic decision: contrary to a policy pursued since World War II, the United States would no longer support a fixed international value for the dollar. This step shocked our European and Japanese allies and led—within two years—to the unravelling of the postwar system of exchange rate management. But foreign policy officials were excluded from the crisis conference that led to this action.[5] As emphasized in a contemporary study of government organization, "The group [at Camp David that weekend] contained neither the Secretary of State nor the Assistant for National Security Affairs, nor any senior subordinate of either."[6]
- Ten years later, national security officials stayed out of President Ronald Reagan's 1981 decision to seek major tax cuts, as well as most subsequent efforts to reverse them. They did so even though these cuts, predictably, ballooned future budget deficits and hence reduced the funds available for defense and international affairs.[7]

Why does the U.S. government behave this way? Why does it repeatedly make major decisions while excluding important officials and perspectives that are undeniably relevant to those decisions? These episodes occurred over a span of twenty years, in four out of five presidential administrations. Each episode had its particular features, of course, and these may have driven presidents and their aides to keep participation narrower than its substance would seem to demand. But all five were important policy episodes, and most were historic episodes. The fact that all exhibited this separation between security and economics suggests that they reflect not just the idiosyncrasies of one or two groups of leaders but a recurrent tendency in Washington decision making. To this we now turn.

TWO HISTORIC PREOCCUPATIONS

For more than half a century, U.S. presidents have spent their policy time mainly on two sets of issues. Their goals are neatly summarized in the old campaign slogan of President Dwight D. Eisenhower, "Peace and Prosperity," or in two provisions of the preamble to the Constitution: "provide for the common defense" and "promote the general welfare." Forcing attention to these goals were the searing national experiences that marked the Presidency of Franklin D. Roosevelt. One of these—burned into George Bush's generation—was World War II and the cost of weakness, symbolized by the "lesson" of appeasing Adolf Hitler at Munich. The second was the Great Depression of the 1930s.

To pursue the first goal, presidents and their top aides have looked outward, at the threatening world. They are expected to protect the United States in and from that world, as Jimmy Carter discovered in 1980, when Americans were being held hostage in Iran, and as Michael Dukakis learned in 1988, when voters found him insufficiently tough on national security matters. Protecting the nation has involved defense—armaments and alliances. It has required negotiations with allies and adversaries. Throughout the postwar world, foreign policy has been dominated by these traditional interstate diplomacy and political-military concerns. They were what the public thought of, what the press spotlighted, and what U.S. foreign policy institutions have mainly addressed.

The commitment to prosperity, however, has focused presidential (and congressional) attention at home. And the post-Depression tradition has joined with politicians' self-interest to force attention to the near term. They have needed to get the economy moving as close as possible to capacity, as soon as possible, but with price stability, too, and above all as election day approaches. The indicators that matter in pursuit of these goals are *domestic* rates: those measuring unemployment, inflation, and economic growth. Presidents (and legislators) feel that these are what voters hold them accountable for; scholars generally conclude that the politicians are right. President Gerald Ford paid the penalty in 1976 for economic troubles that began on Nixon's watch. Jimmy Carter fell in 1980 to "stagflation," a combination of economic slowdown and sharp price increases. And while prosperity propelled Republicans to victory in 1984 and 1988, George Bush was pulled down in 1992 when the Reagan boom ended during *his* administration.

From the end of World War II, the security and economic tasks were typically viewed as separate.[8] So they were given to separate sub-governments.

THE SECURITY COMPLEX

The formal U.S. system for managing political-military issues dates from the National Security Act of 1947. Enacted in a year of incipient cold war, it created not just the National Security Council (NSC)—originally envisaged as a "supercabinet" to advise the president—but also the Central Intelligence Agency (CIA) and a coordinated military under the secretary of defense and the Joint Chiefs of Staff. Together with the Department of State, these quickly became the preeminent institutions of postwar American foreign policy making.

Every postwar president has gotten the preponderance of his national security advice from the heads of these agencies—as a group, as well as one-on-one.[9] The National Security Council, however, changed its primary role in the 1960s from serving as a forum for such advice, as originally intended, to serving as the institutional umbrella for a specialized White House staff serving the president directly. The head of this staff, the assistant to the president for national security affairs, has come to rival (and sometimes supplant) the secretary of state as the president's principal foreign policy aide. The NSC staff has risen commensurately in influence.

Presidents deal with political-military and diplomatic issues through this national security channel, in a process that typically engages the responsible agencies but is managed by the assistant and tailored to each president's personal predilections.[10] The president also handles certain operational matters directly: communication with foreign leaders in person or by phone, for example. Secretaries of state find their power rising and falling according to their relationship with the president. This dependence—plus State Department structure and tradition—pulls them toward stress on political-military-diplomatic issues and away from economic issues, particularly in periods when presidents have been preoccupied with security matters.

Congress possesses no institution like the NSC, but its committees—Foreign Relations and Foreign Affairs, Armed Services, Intelligence—are counterparts to the State and Defense Departments and the CIA and thus share their priorities. Committee and subcommittee chairs guard their own jurisdictions as well. In the early postwar years, they did so with minimal staffs and much deference to presidents, at least concerning the actual or potential use of force. The Vietnam experience, together with a general rise in policy activism, made legislators more skeptical of presidential wisdom and more likely to contest administration priorities. To do so, they felt they needed staff help. So there was a multiplication in the number of foreign and defense policy aides on Capitol Hill—on personal and committee staffs and in support institutions like the Congressional Research Service, the Congressional Budget Office, and the Office of Technology Assessment.

THE ECONOMIC COMPLEX

The United States emerged from World War II fearing a relapse into economic depression. The Employment Act of 1946 responded to this concern by making prosperity an explicit governmental goal and creating institutions to help achieve it: a Council of Economic Advisers (CEA) to advise the president, and the Joint Economic Committee (JEC) of the Congress to review the annual Economic Report, which the president and the CEA were required to submit. The CEA gave particular emphasis to fiscal policy: the balance between taxes and spending and how the resulting deficit (or surplus) affected the overall level of demand in the economy. But it played no operational role. Instead, its chair worked closely with the director of the Bureau of the Budget (brought into the Executive Office of the President by Franklin Roosevelt in 1940 and responsible for overall spending) and the secretary of the treasury (whose jurisdiction included taxes). Together, they were given the informal label of "the troika," expanded to "the quadriad" when they were joined by the chair of the Federal Reserve Board. This independent board, known as "the Fed," controlled the prime instruments of monetary policy: setting rates at which banks could borrow money, buying and selling securities on the open market. Thus the Fed had more impact on monetary conditions—interest rates, the overall money supply—than any agency under the president's direct control.

The secretaries of commerce, labor, and agriculture formed the second tier of economic officials—important on micro and product issues within their fiefdoms but lacking consistent engagement in broader economic matters. On these, presidents generally heard from—and worked with—the troika. There was, however, no statutory White House staff institution paralleling the NSC.[11] The CEA advised the president and made its analytic contribution to interagency debates but did not manage the decision process or coordinate implementation.

As in the security complex, congressional economic organization replicated the executive branch. The JEC played an advisory role parallel to CEA, and there were separate committees for spending (Appropriations) and revenue (House Ways and Means, Senate Finance), as well as for agriculture, commerce, and labor.

THE SHIFTING LOCUS OF "FOREIGN" ECONOMIC POLICY

In the immediate postwar years, the chief economic advisers saw their policy sphere as overridingly domestic. The U.S. economy was largely self-sufficient: imports were less than 6 percent of U.S. goods production in 1960. So international influences and effects were not important factors in promoting prosperity at home. This situation meant that *foreign* economic policy could be handled primarily by the security complex.[12] In turn, this responsibility encouraged the security people to take a relatively broad view of their domain. "In the first two decades of the postwar era," notes Ellen L. Frost, "American definitions of national security encompassed economic and industrial strength as well as military power."[13] Congress even passed a National Defense Highway Act.

During this period, the security complex used foreign economic policy to advance U.S. political-strategic interests. The State Department employed multilateral trade talks to strengthen ties with key European (and Japanese) allies. In the second Eisenhower administration, the State Department's number two official, Under Secretary C. Douglas Dillon, was effectively the czar of foreign economic policy.[14] And in the Kennedy and Johnson administrations, the deputy assistant to the president for national security affairs (Carl Kaysen, then Francis Bator) was typically a professional economist responsible for both international economics and U.S.-European relations.[15]

But by this time, a major shift had begun in the locus of responsibility for foreign economic policy. In 1962 leadership in the Kennedy Round of trade negotiations was moved from the State Department to a newly created Office of the Special Representative for Trade Negotiations (STR). By the late 1970s, STR (soon to become the Office of the United States Trade Representative, or USTR) had gained executive branch primacy over trade issues generally and was no longer typically aligned with the State Department in interagency debates. Meanwhile, bilateral foreign assistance, which had been Douglas Dillon's other primary policy lever at the State Department, was growing progressively less important within the U.S. policy arsenal. The issues where the Treasury Department led, by contrast, were becoming more central: international money and debt, for example, and the Bretton Woods institutions—the World Bank and International Monetary Fund.

From the 1970s onward, it was the economic complex that had primacy in foreign economic policy. When coordination of this sphere was imperfectly achieved in Washington, it would now be through Treasury Department dominance (Secretary John Connally under Richard Nixon; Secretary James Baker under Ronald Reagan) or through White House–based institutions like Gerald Ford's Economic Policy Board.

Behind this shift were broader economic forces, specifically the "internationalization" of the American economy. Trade became much more important, with imports rising from under 6 percent of U.S. goods production in 1960 to 9 percent in 1970 and more than 20 percent from 1980 onward. As the impact of international transactions on the U.S. economy rose, domestic interests grew more concerned about them. They insisted that foreign economic policy not be visibly under the dominance of persons who would subordinate home economic interests to diplomatic concerns. Their congressional champions pressed this cause, and those with trade policy jurisdiction were in the forefront. Wilbur Mills (D-Ark.), chair of the House Ways and Means Committee, made creation of STR a condition for enactment of the Trade Expansion Act of 1962. And the campaign to strengthen USTR was pressed by Russell Long (D-La.), chair of the Senate Finance Committee in the 1970s and, in the 1980s, by Senators John Danforth (R-Mo.) and Lloyd Bentsen (D-Tex.), and House Ways and Means chair Dan Rostenkowski (D-Ill.). In fact, USTR's exceptional ties to these committees made successful international negotiations possible in a sphere where the Constitution gave Congress primary jurisdiction.[16]

Sometimes these committees have been pressed by competitors seeking policy even more responsive to domestic interests, as when the House Energy and Commerce Committee reported out "local content" legislation for automobiles in 1982 and 1983. Or occasionally a reorganization proposal would seek even clearer separation from foreign policy institutions—as when Senator William V. Roth, Jr., (D-Del.) persistently proposed creation of a Department of Trade. What almost no legislators pressed for was a return to the old State Department role and the reintegration of trade with foreign policy.

SEPARATE, DIFFERENT, AND UNEQUAL

The security complex and the economic complex tend to operate autonomously, with much interplay within each but relatively little between them. There are, of course, officials whose job is to link the two. These include desk officers (labeled "country directors") in the State Department, economic aides on the White House national security staff, and the president himself. And certain policy choices do engage both from time to time—how much to control exports to adversary nations of materials that might be militarily useful;[17] whether the president should follow a recommendation by the U.S. International Trade Commission to protect a specific industry; how hard to press the Japanese prime minister for trade concessions on a forthcoming Washington visit. But what is striking is their day-to-day separation on most issues, most of the time.[18]

Both complexes have ties to Congress. Particularly important on the security side are the Armed Services and Defense appropriations panels and (since the 1970s) the Intelligence Committees. But the economic grouping is closer to the Hill. USTR is unique in its ongoing relationship to the Finance and Ways and Means Committees, and the Treasury Department works closely with them on taxes—as do the Treasury Department and the Federal Reserve Board with the banking committees. The Office of Management and Budget, established as successor to the Bureau of the Budget in 1970, has a love-hate relationship with the Senate and House Budget Committees and the Congressional Budget Office, established by statute in 1974.

Neither complex is perfectly coordinated, but the security complex is typically much more centralized. The engagement of the president pulls it together—at least on the issues he cares about. And leeway vis-à-vis Congress allows for relatively closed policy processes—from Richard Nixon's two-man show with Henry Kissinger to George Bush's close ties with Secretary of State James Baker and National Security Assistant Brent Scowcroft—that would simply not be tolerated in the economic sphere. In fact, the economic complex is quite fragmented, driven by institutional and policy subgroups covering trade, finance, macroeconomics, and so forth. Occasionally a strong secretary of the treasury can pull things together. Sometimes a cabinet-level economic policy committee can act as a forum for coordination. And annually, presidents host or travel to international economic summits with their counterparts in the Group of Seven (G-7) advanced industrial countries. Yet these have not brought durable centralization of foreign economic policy making.[19]

The two complexes also draw upon distinct and oft-incompatible intellectual traditions: strategy and economics. Even though economists such as Thomas Schelling played a major creative role in strategic analysis, what has evolved are two professional groupings that speak different languages, focus on different variables, and have difficulty communicating with one another.[20]

In both complexes, short-term coping has taken precedence over long-term strategy. This failing is normal to government, but it has been particularly true of the economic complex. For macroeconomic policy, the primary focus has been on getting the economy operating close to full capacity, in the tradition of John Maynard Keynes.[21] On trade, the emphasis was on balancing pressures among trade policy interests and seizing opportunities to negotiate reductions of specific trade barriers. In foreign currency markets, officials became skilled at staving off financial crises. Top-level security policy making has also given precedence to short-term management of crises and relationships, but there has been a larger element of forward planning in the military and arms control arenas.

The security-economic division is reinforced by the press, which has given heavy priority to security complex issues and has separate reporters covering the two spheres. International economic issues are seldom pressed in presidential debates, for example: the trade deficit and unprecedented U.S. international borrowing went virtually unmentioned in the presidential debates in 1988.[22]

Reflecting the gap in attention and prominence is the outcome when the two spheres do come into conflict. As William Hyland has noted, with only modest exaggeration, during the cold war period "in almost every instance where there was a clash in priorities between economic policy and national security, the latter prevailed."[23] The advocates of the former knew this, which is why they labored so long and so hard to escape the clutches of the security side.

POLICY MAKING FOR THE POST–COLD WAR WORLD

The cold war is now over. The Soviet Union has come apart. But the relative economic position of the United States has eroded significantly and risks further erosion. Both these developments suggest that Americans should give greater priority to the long-term economic roots of national strength. This shift would require softening of the separation between security and economic institutions and the elimination or reversal of the security-over-economics hierarchy.

There is certainly greater consciousness of a need. Even before the communist collapse, the historian Paul Kennedy was warning of the threat of economic decline, and the former official Clyde Prestowitz was arguing that it had already happened.[24] By the early 1990s, distinguished veterans of the security complex were arguing for a shift of priorities to the home front: Leslie H. Gelb of the *New York Times*, William Hyland of *Foreign Affairs*, and Steven Rosenfeld of the *Washington Post.* When the American Assembly brought foreign policy experts together in a 1992 study entitled *Rethinking America's Security,* they concluded that "the most urgent challenge for America is to get its own house in order."[25] A national commission sponsored by the Carnegie Endowment for International Peace declared later that same year that "America's first foreign policy priority is to strengthen our domestic economic performance."[26]

It has become conventional wisdom that future U.S. prosperity and global influence will depend on steps to strengthen the American economy. The best description of the complex relationships the United States has with today's other major powers—the advanced industrial nations—is captured by the oxymoron, "competitive interdependence."[27] To extract its share of the benefits, a higher priority to economic policy and closer integration of economic and security policy now seem essential.

These new imperatives became clear early in George Bush's presidential term. But they brought no immediate shift in priorities. One reason was personal predilection: the national security arena was where Bush felt most competent and comfortable and where he had assembled his most congenial group of advisers—indeed, the best *team* assembled by any president since Harry Truman. Another reason was the world's demands. The Warsaw Pact began coming apart in 1989, but there was a need to manage the transition to German unity in 1990, and then to mobilize a world coalition to challenge and reverse Saddam Hussein's conquest of Kuwait.

The president responded impressively to these national security challenges.[28] And the public responded impressively to his performance, particularly on the Persian Gulf War. For months after Hussein was driven from Kuwait, Bush's standing in the polls was higher than Reagan's ever was. In the June 1991 poll of the *New York Times*/CBS News, for example, 75 percent of the public approved of the way Bush was "handling his job as President." This was down from his March peak of 88 percent. But Ronald Reagan's best had been the high sixties.

Moreover, Bush had warning, fairly early in his administration, that economic policy engagement might not be similarly popular. The only time in his first thirty-three months when Bush's approval rating dropped below 60 percent was in late October 1990, during the final turmoil surrounding the budget agreement with Congress. It was in the course of that summer and fall that Mary McGrory discovered not merely two policy processes but "two George Bushes," with the "strident, petulant, self-pitying" man on economic policy struggling to disassociate himself from the agreement on taxes and spending that his agents were negotiating at his behest. It is hardly surprising that Bush returned immediately to the more congenial world of the international telephone.

Yet if Bush prospered in pursuit of security priorities for almost three years, he suffered enormously thereafter. The fiasco of his January 1992 Japan trip, described at the outset of this chapter, proved to be part of a larger personal-institutional failure on economic policy, as the sluggish economy dragged on, month after month, and met an even more sluggish Bush response. From the July 1992 Democratic convention onward, the president found himself trailing Democratic challenger Bill Clinton in the polls, with public approval ratings down below 40 percent. Finally, he was reduced to promising a shift in priorities as an election ploy, announcing in the first 1992 presidential debate that he would invite James Baker to "do in domestic affairs what you've done in foreign affairs," to serve as "the economic coordinator."[29]

Clinton had long since made the economy his prime campaign issue and his "economic plan" the centerpiece of his reform program. Charging his opponent with the worst presidential economic record since Herbert Hoover,[30] the Arkansas governor stressed both the current economic slump and the longer-term problem of sluggish growth in productivity and incomes. Meanwhile, independent candidate Ross Perot was making the budget deficit his single, overriding issue.

When the votes were in, the result was a sharp repudiation of the president, and a clear signal of the primacy of the economy. Clinton beat Bush decisively, 43 to 38 percent (to Perot's 19 percent) in the popular vote, and 370 to 168 in the electoral totals. Clinton's presidency would clearly depend on whether he could get economic growth moving again and renew Americans' hope in a brighter future. The problems, and the solutions, cut across the domestic-international divide. How might they be addressed institutionally? And how might bridges be built between economic and security policy institutions?

One option would be to broaden the National Security Council to cover foreign economic issues. As noted by a sophisticated presidential transition report of 1988, the president could "expand [the NSC's] working membership . . . to include the Treasury Department, the U.S. Trade Representative, and other agencies." But as this report also notes—and this essay underscores—such an approach does not square with the reality of international economic policy, because it ignores "the major role played by domestic as well as international factors."[31] It is such growing pressures from the "domestic" side that have pulled issues like trade and international debt out of the security orbit.

Of course, one might try to go further, to transform the NSC into something truly comprehensive, a National *Policy* Council with jurisdiction over *all* U.S. economic and security policy. Such a move would solve, at least formally, the problem of the governmental divide. The danger is that, in practice, the council might behave very much like the old NSC. For it would retain the White House Situation Room, the communications system, the intelligence links, and other capabilities and routines that help the current NSC staff stay on top of security crises. They would continue to carry the aura of urgency; it is hard to see why they would not prevail.

Such was the overwhelming conclusion of the Commission on Government Renewal, a body of senior experts created by two Washington think tanks to advise the incoming administration on how it should reorganize to deal with the changing world.[32] Interestingly, several of the members with strong security policy credentials were among the most vehement in asserting the impracticality of broadening the NSC. So the commission recommended creation of a separate Economic Council, parallel to the NSC, to address both domestic and international issues. Managed by an assistant to the president for economic affairs, it "would be [the President's] instrument for assuring that economic policy gets attention equal to traditional national security."[33] Such a creation would respond very well to the question of priority, and was in fact foreshadowed by Clinton's campaign pledge to establish an "Economic Security Council." Amid criticism that this name had too strong a flavor of international economic conflict, the president-elect changed the label. But when he announced his economic team, his first group of major appointments, Clinton included a new "National Economic Council" and a senior presidential assistant to head it.

Given the situation, and the alternatives, this step was desirable, even necessary. Among its virtues is to join responsibility for domestic and international economic policy, a connection that makes both substantive and political sense. The big drawback of such a reform is that it risks institutionalizing the security-economics divide. Clinton's action, therefore, creates a need for compensatory steps to bridge it.

To some degree, the president can do that himself. If Clinton fulfills his promise and becomes more deeply engaged in economic policy than any of his recent predecessors, he will personally be able to guard against the worst forms of compartmentalized decision making. But people beneath him need to be organized to do so as well. He could enlarge the number of top officials who make security-economics connections by meeting regularly with a small group of senior officials—an executive cabinet[34] perhaps—which would advise him on broad policy for security, economics, and domestic issues. Another step would be to assign to the White House chief of staff the task of coordinating the coordinators, of assuring input from both complexes on issues that affect both significantly. The White House chief of staff would need to strike a careful balance: not preempting either the national security or the economic assistant, nor blocking their regular access to the president, but assuring that neither cuts the other out of cross-cutting issues. In addition, the chief of staff might himself act as the coordinator on issues for which neither of the two councils has a clear claim to priority.

An additional measure would be to build modest substantive overlap into the two staffs—a few economic policy experts serving the NSC, and some specialists in national security economics working for the NEC. These aides would participate in joint staffing of cross-cutting issues, and they would be on the watch for pending decisions and actions that, though properly within one council's jurisdiction, needed input from the other.[35] And issues like national security export controls or major foreign relationships—with Japan and Germany, for example—might be managed by working groups drawing members from both NSC and NEC.

The end of the cold war did not herald the end of serious security challenges. Iraq's invasion of Kuwait made this clear in 1990. Further perils surely lie ahead, in, for example, the volatile brew of nuclear capacity and political-economic incapacity in the states of the former Soviet Union. But challenges to the U.S. economy, at home and abroad, are unlikely to take second place. Ask Bill Clinton. Or ask George Bush.

Notes

1. Mary McCorory, "Two Contrasting Bushes," *Washington Post,* August 16, 1992; Bob Woodward's "The President's Key Men: Splintered Trio, Splintered Policy," *Washington Post,* October 7, 1992.

2. According to Robert L. Paarlberg, National Security Assistant Zbigniew Brzezinski asked the Central Intelligence Agency—not the Department of Agriculture—for an estimate of the potential impact. The Agriculture Department, urged perhaps by White House political advisers opposed to an embargo, provided one anyway, estimating far less damage to Soviet meat production. The Agriculture Department estimate proved the more accurate. See *Food Trade and Foreign Policy* (Ithaca, N.Y.: Cornell University Press, 1985), pp. 174–75. Brzezinski remembers matters a bit differently in *Power and Principle* (New York: Farrar, Straus, Giroux, 1983), pp. 430–31.

3. At this point, George Bush was forced into an eleventh-hour compromise that angered the Japanese by changing the deal previously negotiated but did not satisfy critics concerned that technology transferred through the deal could put Japanese firms in a position to compete with Boeing in building civilian aircraft.

4. The eight were the president, the vice-president, the secretaries of state and defense, the White House chief of staff, the assistant and deputy assistant for national security affairs, and the chairman of the Joint Chiefs of Staff. The index to Bob Woodward's *The Commanders* (New York: Simon and Schuster, 1991) contains two references to Treasury Secretary Nicholas Brady and one to Energy Secretary James Watkins. There are eighty-two references to National Security Assistant Brent Scowcroft.

5. See William Safire, *Before the Fall: An Insider's View of the Pre-Watergate White House* (Garden City, N.Y.: Doubleday, 1975).

6. Graham Allison and Peter Szanton, *Remaking Foreign Policy: The Organizational Connection* (New York: Basic Books, 1976), p. 145.

7. James Schlesinger, one of a handful to have played senior security *and* economic policy roles, wrote a prescient op-ed piece arguing that Ronald Reagan was gutting the revenue base for foreign policy. But national security officials inside the administration stayed out of that issue in the critical early years and made only token efforts thereafter—even after House Appropriations Subcommittee Chair David Obey (D-Wis.) told Secretary of State George Shultz that his international affairs budget would continue to be a casualty. Defense Secretary Caspar Weinberger protected his own turf in 1981 by getting the Pentagon budget excluded from Office of Management and Budget Director David Stockman's budget cuts. But by his third year, Congress was cutting the Defense Department budget back sharply, with the burgeoning budget deficit an oft-cited reason.

8. During that war, the two goals were intertwined: the United States had to outproduce its enemies in order to outfight them, and "economic mobilization" was considered an integral part of the conflict with Germany and Japan. But with postwar decontrol of the economy, this connection was severed.

9. The group has typically been leavened by one or more senior political aides or confidants. Among the names that come to mind are Robert Kennedy and Theodore Sorenson in the Kennedy administration, John Mitchell under Nixon, Walter Mondale and Hamilton Jordan for Carter, Edwin Meese for Reagan, and John Sununu for Bush.

10. Since much ink has been spent on this topic—by this author, among others—the presentation here is suitably cryptic. For an extended, fairly recent treatment of the topic, see I. M.

Destler, Leslie H. Gelb, and Anthony Lake, *Our Own Worst Enemy: The Unmaking of American Foreign Policy* (New York: Simon and Schuster, 1984 and 1985), esp. chaps. 4, 5. For a wide-ranging selection of essays and documents, see Karl F. Inderfurth and Loch K. Johnson, eds., *Decisions of the Highest Order: Perspectives on the National Security Council* (Pacific Grove, Calif.: Brooks/Cole, 1988).

11. Several presidents established nonstatutory economic policy committees at the cabinet level. The only one that was consistently effective was Gerald Ford's Economic Policy Board, which met daily and served as the primary forum for a broad range of issues.

12. There was concern over international programs that could have major impact on the domestic economy. For this reason, Congress shifted responsibility for administering the Marshall Plan to an agency independent of the State Department.

13. Ellen L. Frost, "The New Interaction of Economics and National Security in U.S. Foreign Policy: Notes from the Muddy Mainstream," draft paper, 1991, p. 2.

14. C. Douglas Dillon's leadership was not uncontested by the Treasury Department, and as he prepared to depart after the 1960 elections he is said to have drafted (for his successors) a strong memorandum calling for abolition of a Treasury Department–chaired statutory committee created by the Bretton Woods legislation, one that he found a significant thorn in his side. But he was suddenly invited down to Palm Beach, where President-elect John F. Kennedy offered him the position of secretary of the treasury. Dillon reportedly telephoned the home office with a brief command: "Burn that memo!"

15. On experience through the 1970s, see U.S. Congress, House of Representatives, Committee on Foreign Affairs, *U.S. Foreign Economic Policy: Implications for the Organization of the Executive Branch,* Hearings before the Subcommittee on Foreign Economic Policy, 92 Cong., 2d sess. (Washington, D.C.: Government Printing Office, 1972); I. M. Destler, *Making Foreign Economic Policy* (Washington, D.C.: Brookings Institution, 1980). Concerning the legislative role over this period, see Robert A. Pastor, *Congress and the Politics of U.S. Foreign Economic Policy, 1929–1976* (Berkeley: University of California Press, 1980).

16. For extended discussion, see I. M. Destler, *American Trade Politics,* 2d ed. (Washington, D.C., and New York: Institute for International Economics and Twentieth Century Fund, 1992), chap. 5.

17. The ongoing struggle over routine national security export controls inevitably engages both complexes, with the security side usually the winner. See William J. Long, *U.S. Export Control Policy: Executive Autonomy vs. Congressional Reform* (New York: Columbia University Press, 1989).

18. The Carter NSC aide handling Persian Gulf issues had his office on the third floor of the Old Executive Office building, adjacent to the White House. So did the staff of the Council of Economic Advisers. This aide could recall no occasion during his four-year tenure when he met with any of the CEA people. Personal conversation.

19. During the Carter administration, Henry Owen was able to build a limited but useful White House policy coordinating role with the title (and role) of president's special representative for economic summits.

20. In the Nixon administration, Henry Kissinger was known to have an "economic allergy." In the Reagan administration, the economist George Shultz had difficulty coming to grips with the analytics of arms control during his service as secretary of state.

21. By the 1980s, economists—chastened by inflation—were less inclined to see Keynesian fiscal policy as a useful instrument of short-term adjustment and more inclined to emphasize monetary policy.

22. In the 1988 vice-presidential debates, Lloyd Bentsen forced the issue to prominence by attributing Reaganite prosperity to writing $100 billion or so a year in bad checks. And in 1992, as discussed below, the relative international economic position of the United States has been a prominent issue.

23. William Hyland, "America's New Course," *Foreign Affairs* 69 (Spring 1990): 7–8.

24. See Paul Kennedy, *The Rise and Fall of the Great Powers* (New York: Vintage, 1987); Clyde Prestowitz, *Trading Places: How We Allowed Japan to Take the Lead* (New York: Basic Books, 1988).

25. "Final Report of the Seventy-Ninth American Assembly," in *Rethinking America's Security: Beyond Cold War to New World Order,* ed. Graham Allison and Gregory F. Treverton (W.W. Norton and Co., for the American Assembly, 1992), p. 448.

26. *Changing Our Ways,* Report of the Carnegie Endowment National Commission on America and the New World (Washington, D.C.: Carnegie Endowment for International Peace. 1992), p. 18.

27. C. Fred Bergsten, *America in the World Economy: A Strategy for the 1990s* (Washington, D.C.: Institute for International Economics, 1988) chaps. 1, 8.

28. This statement holds notwithstanding the administration's lamentable failure to toughen policy toward Iraq before August 1990.

29. Transcript of presidential debate of October 11, 1992, *Washington Post,* October 12, 1992, p. A17.

30. Actually, the recessions of 1973–75 and 1980–82 had been deeper than the one on Bush's watch. But if one looked at average growth, adjusted for inflation, across a president's entire tenure, Bush's record was the worst—an average of .91 percent/year over 3.5 years. Harry Truman's *first* term averaged negative real growth, due to postwar readjustment, but the Korean War boom during his second term brought his overall average up to .925 percent.

31. "Presidential Transition in International Affairs: An Organizational Manual," prepared by the Presidential Transition Project of the Center for Strategic and International Studies, Washington, D.C., November 10, 1988, p. 17.

32. The commission, a joint project of the Carnegie Endowment for International Peace and the Institute for International Economics, was chaired by Richard C. Holbrooke, President Jimmy Carter's assistant secretary of state for East Asia. Serving as co-chair was former Representative William Frenzel (R-Minn.). Commission members totaled thirty in number, with balance between parties, between security and economic expertise, and across administrations. The report, released in early November 1992, was entitled "Memorandum to the President-Elect: Subject—Harnessing Process to Purpose." The author of this chapter was a member of the commission.

33. Chaired by the president, "the members of the Economic Council might include: the Vice President, the Secretary of the Treasury, the Director of the Office of Management and Budget, the Secretary of State, the United States Trade Representative, the Secretary of Commerce, the Secretary of Labor, the Secretary of Agriculture, the Secretary of Energy, the Chairman of the Council of Economic Advisers, and Secretary of the Environment (a new post we recommend elsewhere in this Memorandum)." A Domestic Council was also proposed, with even larger membership. Actual meetings of the councils, the commission suggested, would "rarely include every one of their members" but would rather "consist of those members with a concern for the issue under discussion." Ibid., pp. 8–9.

34. This reform was proposed by Allison and Szanton in *Remaking Foreign Policy.*

35. The Clinton administration has addressed this problem a slightly different way, by making international economic staff aides report jointly to the NEC and the NSC.

7 / Politics, the Constitution, and the President's War Power

ROBERT SCIGLIANO

THE UNDERSTANDING

A new understanding of the president's war power has emerged since World War II: that the president may engage the country in war without the consent of Congress. This understanding appeared most recently, as of this writing, in the events that preceded the attack by an American-led coalition of nations against Iraq in January 1991 for the purpose of compelling that nation to withdraw from Kuwait, which it had occupied the previous August. Iraq's action had been condemned by the United Nations Security Council in several resolutions, the most important of which, adopted in November 1990, called on member nations to use "all necessary means" to compel its withdrawal if it did not do so by January 15, 1991. Repeatedly during this period, President George Bush and other administration officials claimed authority for the president to take action against Iraq. In early December, the secretary of defense told a congressional committee that he did not think the president needed "additional authorization" from Congress—that is, any authorization, as he had received none as of that time—before committing American forces assembled in the Persian Gulf region to action.[1] Even when Congress authorized hostilities on January 12, just four days before the President launched them, he reiterated his view, in signing the legislation, that he already possessed authority.[2]

What is the basis of this understanding of the war power? I shall attempt to answer this question by considering both the Constitution and the practice of government over the course of American history, and, with respect to government practice, I shall place special emphasis on certain actions taken by President Franklin D. Roosevelt prior to the United States' entry into World War II.

This question is important, concerning the constitutional balance between the presidency and Congress in relation to the conduct of war and, more generally, foreign policy. For example, the committee that reported the War Powers Resolution of 1973 to the Senate floor claimed that "the residual authority over the entire domain of foreign policy—not just the war power—was placed in Congress by the Constitution."[3] The question's importance can be seen in the debate that preceded Congress's decision to give President Bush authority to act against Iraq. A prominent senator warned the president that he would precipitate a "constitutional crisis" if he went to war against Iraq without it, and there was some murmuring in Congress of possible impeachment.[4] This was not the first time that a claim of presidential warmaking power had been made by, or on behalf of, a president, and it is not likely to be the last. Thus it deserves examination.

THE CONSTITUTION

I need not spend much time in considering whether the Constitution allows the president to engage in war. The document is clear on this point. It states that "the Congress shall have power . . . To declare war, grant letters of marque and reprisal, and make rules concerning captures on land and water," and that "the President shall be commander-in-chief of the army and navy of the United States, and of the militia of the several states, when called into the actual service of the United States." Why does it give Congress authority to declare war, subject to the president's veto as in the case of other legislative acts, if the president may engage in war on his own authority? As Alexander Hamilton stated: "it is the peculiar and exclusive province of Congress, *when the nation is at peace,* to change that state into a state of war."[5] The president's authority, to borrow again from Hamilton, "would amount to no more than the supreme command and direction of the military and naval forces, as first General and Admiral of the Confederacy; while that of the British king extends to the *declaring* of war."[6]

To my knowledge, no early American ever suggested that the Constitution invested the president with war-making power. Nor have modern advocates of this view, such as those persons I have quoted, ever supported their claim with arguments drawn from the document. So far as the Constitution is concerned, the claim rests on assertion.

THE PRACTICE OF GOVERNMENT

Instead of relying on arguments drawn from the Constitution, supporters of recent presidential initiatives in war making have cited precedents. For example, in the debate over President Bush's authority to go to war against Iraq, the president's supporters, relying on a report prepared by the Congressional Research Service in 1989, referred to over 200 instances of presidential decisions to use armed forces abroad "in situations of conflict or potential conflict."[7] This list descended more or less lineally from ones used in earlier contests over the president's war powers. In 1971, for instance, Senator Barry Goldwater (R-Ariz.) produced an enumeration of "153 Military Actions Taken by the United States Abroad without a Declaration of War" (some, he conceded, were "arguably" based on legislation or treaty), to bolster his opposition to enactment of the War Powers Resolution.[8] In 1966 the State

Department supported the claim made by President Lyndon B. Johnson of authority to go to the aid of South Vietnam a few years earlier with a list of "at least 125 instances" in which other presidents had "ordered the armed forces to take action or maintain positions abroad without prior congressional authorization."[9] Still earlier, in 1950, the State Department defended President Harry Truman's action in defending South Korea from invasion by North Korea with a list of "eighty-five instances of the use of American armed forces without a declaration of war."[10] On several of these occasions, supporters of the president have punctuated their argument with the statement that in all American history Congress has declared war only five times.

Past practice may properly be invoked to justify current actions, no matter what we may think the Constitution intended, especially if that practice has been of long duration and generally accepted. Precedents do go a good way to conferring legitimacy. As an opinion of the Supreme Court put the matter, "a systematic, unbroken, executive practice, long pursued to the knowledge of the Congress and never before questioned . . . making as it were such exercise of power part of the structure of our government, may be treated as a gloss on 'executive power' vested in the president" by the Constitution.[11] So, too, in the specific case of the president's war-making power. The argument that presidents have used the armed forces so often "in situations of conflict or potential conflict" or in "military hostilities" or "without a declaration of war" over the course of American history, with Congress having declared war so seldom appears to make an impressive case for the legitimacy of that power.

In saying that Congress has declared war in only five instances, supporters of a presidential war power imply that if the Constitution and not precedents had been followed, war would have been declared in all or most of them. Rarely do they indicate what these instances covered, and never do they do so in any detail. It is my purpose here to examine the actions behind the numbers to see whether they do, in fact, support the claim that a long history of precedents supports the position that the president has an independent power to make war.

To begin, it is beside the point and misleading to speak only of the number of times Congress has declared war, for Congress may authorize war by means other than a declaration. The difference between a war that is declared and one that is authorized by ordinary acts of legislation corresponds to the difference, well understood by early Americans, between general war and limited war, between war that "destroys the national peace and tranquillity" and war that "interrupts it only in some particulars."[12] According to the 1989 Congressional Research Service list, Congress declared war against Great Britain in 1812, against Mexico in 1846, against Spain in 1898, against Germany in 1917, and against Germany, Italy, and Japan in 1941.[13] But the list fails to mention that Congress authorized war by ordinary statutes on more than a dozen occasions, most of which occurred in the first few decades of the nation's history and in those following World War II.

In the 1790s, when the Indians along the frontiers were formidable enough to command respect, in part because they had ties with European powers, Congress authorized expeditions against the Miami and related tribes in the Ohio territory (and declined to authorize any against tribes to the south of the Ohio River). The so-called undeclared war with France, which was fought between 1798 and 1800 and which has frequently been singled out in recent decades by advocates of presidential power, was authorized by several congressional acts. Other early limited wars authorized by Congress include those with Algiers in 1794 (not fought), with Tripoli in 1802, with any Barbary power that committed hostilities against the United States in 1806 (nor was this fought), with any country that attempted to possess Spanish Florida, by a secret law in 1811 (Andrew Jackson drove the British out in 1814), and again with Algiers in 1815. The only authorized war acknowledged by the 1989 list is the second war with Algiers.

Since World War II, Congress has authorized hostilities on several occasions. It empowered President Dwight Eisenhower to protect Formosa and countries in the Middle East from external aggression in 1955 and 1957. It also empowered John F. Kennedy to use armed force against Cuba in 1962 and Lyndon B. Johnson to assist nations in Southeast Asia to defend their freedom in 1964, although neither thought he needed legislative permission. More will be said about these congressional authorizations a little later. In 1983, to give a more recent example, Congress authorized Ronald Reagan to maintain in Lebanon the marines he had sent there a year earlier as a peacekeeping force, having decided they were now exposed to hostilities. Like Kennedy and Johnson, Reagan denied that he needed approval. And in January 1991, Congress authorized George Bush to go to war against Iraq.

Actually, the United States has never *declared war* against anyone. In its wars with Great Britain, Spain, and Germany and its allies, Congress declared that war existed, implying that the United States had been drawn into it by the actions of others. Indeed, it declared both times against Germany that war had been "thrust upon" the United States. Such expressions befit a nation whose founding principles allow it to fight only in defense of its rights. And in the case of Mexico, Congress did not declare anything but rather, at the request of President James K. Polk, merely "recognize[d]" the existence of war, because, or so Polk claimed, Mexico had started the war by attacking an American army detachment on American soil, in Texas.

Congress's action in the Mexican War reflects the view that Congress's power to declare or otherwise authorize war is that of deciding whether the United States should go from peace to war, as I quoted Hamilton earlier. But, to continue with Hamilton, "when a foreign nation declares or openly and avowedly makes war upon the United States, they [the United States] are then by the very fact already at war, and any declaration on the part of Congress is nugatory; it is at least unnecessary."[14] Abraham Lincoln acted on this understanding when he "call[ed] out the war power of the government" after South Carolina attacked Fort Sumter, a federal fortress in

Charleston Harbor, in April 1861. The Supreme Court agreed with him and upheld the legality of the blockade that he imposed on southern ports immediately thereafter.[15] Congress may, as Hamilton indicated, still act—it may declare that war exists or recognize its existence, or, as it did in July 1861, it may approve and make legal (in case there was any question) all things done by the president prior to its own action.

It is pertinent to ask here, What did early Americans understand to constitute war? I take my bearing from an early Supreme Court opinion, which defines foreign war as "a contention by force between two nations, in external matters, under the authority of their respective governments."[16]

By this definition of war, most of the incidents on the lists cannot be considered precedents for what presidents in recent decades have done or claimed the power to do. Indeed, some of them did not involve the use of force against anyone. Take the small party of soldiers, under Lt. Zebulon M. Pike, that was sent west in 1806 to explore the domain that the United States had recently acquired from France. Even if the party intentionally entered into land that was in dispute between the United States and Spain—Jefferson said it had strayed there—its entry hardly constituted an act of war. And yet the 1989 list says that Pike and his men "invaded Spanish territory." Or take the employment of American troops in 1922 to supervise Panamanian elections, at the request of local political parties. How could this incident be counted as a precedent for later presidential war making?

Nor do all uses of force constitute acts of war. Most often prior to World War II, in the period we are considering, the United States used force abroad, or was prepared to use it, in limited ways in order "to secure adequate safety and protection for its citizens and their property." This is how a State Department memorandum issued in 1912 and updated in 1934 described it in enumerating eighty-five such incidents between 1812 and 1933. This appears to have been the first of the lists of presidential actions taken abroad.[17] It cannot be said that the United States was engaged in war when its navy chased pirates and other adventurers on the high seas or punished inhabitants of Pacific islands for threats or attacks against American citizens and their property. Generally such incidents involved action by single naval vessels. Nor was it war when marines or other forces were sent onto the soil of other countries for the purpose of protecting Americans during disturbances at the request or with the acquiescence of the authorities there. Such incidents were rather common at one time, especially in Latin American and Far Eastern port cities, as the lists testify, and they still happen, as when American helicopters flew into Cyprus in 1974 to evacuate Americans and other foreigners during fighting between Turkish and Greek Cypriot forces.

Constitutionally speaking, does it matter whether American forces enter another nation's territory with or against the permission of its government, or whether their efforts are directed against unruly mobs or the government itself, if the latter cannot effectively oppose the American interventions?

Does a "contest by force" take place in any meaningful sense in such situations? A congressional committee gave this answer in 1824 in considering the question of the pursuit of pirates by American naval parties onto Spanish territory: "Powerful nations never permit feeble neighbors to enter their territory for this purpose but enter without scruple in pursuit of their enemies the territory of such neighbors. . . . Practically, the question is one not of right but of relative power."[18]

It is not always easy to draw a line between those uses of military force that fall short of war and come within the president's authority and those that cross the line and require the consent of Congress. This is true also of line drawing in other areas. Consider, for example, the related area of treaty making. Which international arrangements made by the president must be submitted to the Senate, and which, falling short of treaties, may be made by the president alone as executive agreements? By the early twentieth century an accumulation of precedents had made the distinctions tolerably clear in both areas. William Howard Taft offered his successors in the presidency a rule of thumb for deciding on which side of the line a contemplated military action fell: It depends on whether you can make do with using (shipboard) marines or whether you must send in the army.[19] And Hamilton, always prudent in matters of statecraft, offers this cautionary advice: In a "delicate" case, "one which involves so important a consequence as that of war—my opinion is that no doubtful authority ought to be exercised by the President."[20]

I do not suggest that no president before WWII engaged in acts of war when there was time to go to Congress or, where emergency action was necessary, declined to go there after taking it. Lincoln thought that President Polk had engaged in war unconstitutionally against Mexico by sending American troops onto Mexican soil or, at least, to territory to which Mexico had a strong claim. (However, if Mexico had commenced hostilities, it would have been a different matter.) Taft thought that his successor in the presidency, Woodrow Wilson, had committed an act of war against Mexico in 1914 by ordering the American navy to attack and the army to occupy the Mexican city of Vera Cruz because Mexican authorities refused to make what Wilson deemed a proper apology for the brief arrest of some crew members from an American naval vessel. Perhaps we could add President William McKinley's dispatch of about 3,000 American troops to northern China in 1900 to assist forces sent by other nations in protecting foreign nationals from attacks during the Boxer Rebellion, and President Theodore Roosevelt's naval threat off Panama that assured Panama's independence from Columbia and enabled the United States to fulfill its ambition to construct a canal connecting the Atlantic and Pacific oceans.

Whatever their precise number, the instances in which presidents prior to World War II engaged in war making independently of Congress were few. They do not provide precedential support for war making by the executive that has been claimed for them. In particular, they do not make the Constitution's commander-in-chief clause an alternative means of going to war.[21]

THE ROOSEVELT PRECEDENT

In the debate in Congress over going to war with Iraq, one House member said that the United States was in the fortieth year of a constitutional crisis, begun when Harry Truman sent troops to Korea without the authority of Congress. He might have looked a little farther back, to actions that Franklin Roosevelt sanctioned during the months of 1941 preceding formal American entry into war against the Axis powers. In the spring of that year, Roosevelt ordered the occupation of Greenland after Denmark, which exercised sovereignty over the territory, was overrun by Nazi Germany. Roosevelt wished to prevent the Germans from establishing a military garrison there. In July, an American force was sent to Iceland and Roosevelt announced that the American navy would escort convoys of merchant ships between Iceland and the United States. Encounters between American destroyers and German submarines occurred, and American merchant ships came under German attack. According to what Prime Minister Winston Churchill revealed to his cabinet at the time, Roosevelt told him when the two met off Newfoundland in August 1941 that he would "wage war" but he would "not declare it."[22] The next month Roosevelt announced that American vessels and aircraft would henceforth "shoot on sight" German and Italian submarines and war vessels anywhere in American defensive waters, which were defined to extend most of the way across the Atlantic. In October he announced that the "shooting" had begun.

An important debate occurred in the Senate following President Roosevelt's decision to send forces to Iceland and to provide convoy escorts. During the debate, Tom Connally, the administration's main spokesman in Congress on matters of foreign policy, defended the president's actions in the face of charges that they would lead inevitably to war and that the president had, consequently, usurped Congress's authority. Connally produced a list of eighty-five incidents occurring between 1812 and 1933 to justify what the president had done, interpreting them to mean that a president could send the armed forces wherever he believed "the interests of the nation" required.[23]

Connally's list of incidents was taken in its entirety from the State Department memorandum of 1912, updated to 1933. That memorandum, as I have shown, was concerned with measures taken by presidents to protect Americans and their property abroad—that is, nearly always with minor uses of force against pirates or island natives or lawless mobs or weak governments. What Connally did was to make these incidents into more than they were in order to claim sustained government practice for what was, in fact, a clear break with the past. The list that the State Department issued in 1950 to justify President Truman's going to war in Korea was identical to Connally's list. But whereas Connally avowed in Senate debate that only Congress could make war, Dean Acheson, who was Truman's secretary of state, cited the 1950 list to support his belief that Truman had "constitutional authority to do what he did."[24] And yet war, at first limited and then full scale, resulted from Roosevelt's actions, inevitably as some predicted at the time. The broad terms in which Connally defended the president's right to act in foreign affairs—wherever the national interests were concerned, I should add—were used later by Roosevelt's successors and their supporters to support greater claims of presidential power. President Bush, for example, referred to "the president's constitutional authority to defend vital U.S. interests" in discounting what Congress had given him in authorizing the use of force against Iraq.[25]

The modern understanding of the president's war power has its origin in the presidency of Franklin Roosevelt. In this as in other important ways, Roosevelt cast his influence upon the future. To sustain the president's power to use the armed forces as he deemed necessary, including their use in war, became an article of belief among New Dealers. And then, gradually, as with other innovations wrought by the New Deal, its influence spread into the ranks of conservative Democrats and liberal Republicans, and then further. Indeed, eventually the Republicans became the bearers of Roosevelt's legacy regarding the president's use of military power, at a time when Democrats were reconsidering their position and rejecting that legacy.

Harry Truman was the first of Roosevelt's successors to view his powers through the prism provided by his great predecessor. He did not go to Congress for authority to engage American forces in support of South Korea against invasion from the north in June 1950, after he had taken his initial emergency measures. Truman could have had Congress's approval from many Republicans as well as from his own party. Senator Robert A. Taft (R-Ohio), the most influential Republican in the Senate at the time, stated on the floor of that body about a week after Truman acted that he would have voted for legislation approving of what the president had done and providing supplies for the war, if Truman had requested it.[26] According to Acheson, Truman did not go to Congress because he wished to avoid "the possibility of endless criticism" (ironic, in view of the criticism that dogged the rest of his presidency for his failure to do so) and because he was determined "to pass [his office] on unimpaired by the slightest loss of power or prestige."[27]

Eisenhower had no such constitutional qualms. In asking for authority to act in defense of Formosa against invasion from China in 1955, he stated that he might be able to do some things if that happened—he referred to "emergency action [that] might be forced upon us"—but that suitable legislation "would clearly and publicly" establish his authority to employ the armed forces if it became necessary.[28] In explaining why he wanted authority to act in defense of nations of the Middle East, the president was not as clear, perhaps because he was sensitive to the need for Democratic support, and the Democrats, as he later noted, were criticizing him for seeking Congress's approval for the use of the armed forces.[29] He referred both to the greater effect that would result from "a consensus of executive and legislative opinions" and to his "regard for constitutional procedures."[30]

In the Eisenhower years, support for presidential war-making power came mainly from liberals, that is, from Democrats, and Congress was under their control in 1955 and 1957. In the House of Representatives, John W. McCormack (D-Mass.), the floor leader for the Democrats, stated in the Formosa debate that "the general feeling is that the President has the power inherent in the authority of the commander in chief to issue the necessary orders to take such action as might follow the passage of the pending joint resolution." In the Senate, Mike Mansfield (D-Mont.), the assistant Democratic floor leader and a leading spokesman for his party on matters of foreign affairs, made a similar declamation of the president's power to act without Congress. In committee consideration of the Middle East resolution, for example, he asked Eisenhower's secretary of state, "Mr. Secretary, why does the President want to limit his power, thereby abrogate the Constitution to an extent, and create a precedent which may become recognized as a part of the Constitution through custom, especially in this hydrogen age?"[31] This view was even more strongly held by constitutional scholars at the time. It was observed in Senate discussion of the Middle East resolution that "practically all" such persons held that Eisenhower already possessed the powers he had asked for.[32]

A Republican senator expressed the predominant view held by conservative members of Congress, especially by those in his own party, on the matter of presidential war making when he stated during debate on the Formosa resolution that passage of the resolution would "restore a long line of precedents which, over the past ten years, have been disregarded, unconstitutionally in my judgment. . . . What I cannot understand is why some members of the Congress are so eager to deny that we have any right to participate in this momentous decision."[33]

President Eisenhower got his way in regard to Formosa and the Middle East. Congress embodied its approval of what he might undertake in joint resolutions—acts submitted to his approval that had the force of law—and not, for example, in concurrent resolutions, which are approved only by the two houses and are typically employed as expressions of opinion. If Congress wished merely to register its support for actions the president might undertake on his own authority, the latter method was the way to do it. Those in Congress who believed that the president needed authority to engage in acts of war, mostly Republicans, won out. But they were in a minority, because the majority of Democrats were unwilling to go against a popular president. Their victory was more clearly marked in the Formosa resolution, for it "authorized" the president to employ the armed forces in combat, whereas the Middle East resolution was intentionally evasive—a compromise of contending views providing that, "If the President determines the necessity thereof, the United States is prepared" to employ the forces in this way.[34] Those who believed in a presidential war power might have been reconciled to voting for joint resolutions by the argument made to them that foreigners would be more impressed by an action that showed the unity of the two branches of government. Besides, some of them probably asked themselves what difference the form by which Congress acted made if, as they believed, Congress was not giving the president power that he did not already have.[35]

Roosevelt's precedent was reestablished in the White House in the presidencies of John F. Kennedy and Lyndon B. Johnson. Kennedy and Johnson had as members of Congress sustained Truman's action in Korea and questioned Eisenhower's turn to the legislative branch in the Formosa and Middle East situations, and they acted on their earlier views when they became president. Kennedy did not ask Congress for its consent to take military action against Cuba and possibly its ally, the Soviet Union, in September, 1962, as the Cuban Missile Crisis began to develop, but before there was incontrovertible evidence that the Soviet Union was arming Cuba with missiles. When members of Congress began to press for a joint resolution to authorize him to take such action, he sought to divert their efforts. "As commander in chief," he said, "I have full authority now to take such action" as might be necessary against Cuba; however, it would be "useful" for members of Congress, "if they desire," to "express their view" on the situation.[36] Nor did Lyndon B. Johnson ask for consent to use the armed forces in defense of South Vietnam prior to sending them there in May, 1965. Indeed, when Johnson's undersecretary of state testified before a congressional committee in August 1967 that the Southeast Asia resolution, adopted three years earlier, was "a functional equivalent" of a declaration of war, the president repudiated the statement, insisting that in conducting hostilities in Vietnam "we did not think the resolution was necessary to do what we did and what we are doing. We think we are well within the grounds of our constitutional authority."[37]

However, Congress acted in both cases by joint resolution. Kennedy accepted one that cloaked the question of authority in compromise language similar to that used in the Middle East resolution. It said, "The United States is determined, by whatever means may be necessary, including the use of arms" to prevent Cuba from carrying out revolutionary activities elsewhere in the hemisphere or from permitting on its territory a foreign (that is, Soviet) military capability dangerous to the United States. Johnson actually proposed a joint resolution to Congress: "It could well be based upon similar resolutions enacted by Congress in the past," though he thought of it as a means for "expressing the support of the Congress for all necessary action" that he might take to protect American forces in Vietnam and to help nations in the region to defend themselves. As enacted, the resolution declared that "the United States is prepared, as the President determines, to take all necessary steps, including the use of armed force" to help those nations.[38]

In discussing the Southeast Asia resolution, members of Congress tended to regard it as President Johnson did, as a way of registering their support for the president. This view helps explain why some members of Congress complained later that they did not realize what they were doing at this time or, more often, that they had not authorized Johnson to do what he was doing in Vietnam and hence that the war in Vietnam was unconstitutional. But it would be more correct to say that they were confused as to the constitutional issue. J. William Fulbright (D-Ark.) who guided the measure through the Senate, explained the resolution to his colleagues: "[The resolution] would authorize whatever the commander in chief feels is necessary," including the landing of large American armies in Vietnam or China. Yet it "would [not] in any way be a deterrent, a prohibition, a limitation or an expansion of the President's power to use the armed forces in a different way or more extensively than he is now using them." And, "We are not giving the President any powers he does not have under the Constitution as commander in chief." Finally, when Fulbright was asked by another senator if Congress was giving the president advance authority to do whatever he thought necessary in the defense of South Vietnam, his answer was, "I think that is correct."[39]

Many who were influenced by the precedent that Franklin Roosevelt created by his actions in the North Atlantic believed there were limits to how far a president might go on his own, but they were not very specific about them. Perhaps typical of their views was the remark made by a Democratic senator in the Formosa debate: "Of course, if he [the President] undertook to engage in a major war, he would be expected to ask Congress for a declaration of war."[40] And yet belief in the importance of Congress's power to declare war had undergone erosion. This erosion can be seen in the remark that Senator Fulbright—an early subscriber to the Roosevelt precedent—made on the Senate floor back in 1945. He did not see why the war power should be retained in Congress, he said, when it had never been "important."[41]

Democrats in and out of Congress were induced to rethink their view of the president's war power when they moved from support of the Vietnam War to opposition. The process was undoubtedly helped by the accession of a Republican, Richard M. Nixon, to the presidency in 1969. They now became the "party of the Constitution," some of them with the fervor of converts as they discovered the document's grant of war-making power to Congress to be clear and relevant, where once they had found it ambiguous and obsolescent. The Republicans became defenders of presidential war making, speaking of the precedents that sustained it and for the need for flexibility in interpreting the Constitution. In short, the Republicans, most of them, became heirs of the Roosevelt legacy and praised him and also Truman and Kennedy, and sometimes Johnson, for their leadership in foreign affairs.

Thus the Democrats were mainly behind the enactment of the War Powers Resolution of 1973, hoping that it would make it more difficult for presidents to get the country into future wars and easier for Congress to get it out of them. The resolution states that the president may introduce the United States armed forces into hostilities or a hostile situation on his own authority only when the country or its armed forces have been attacked. The president's right to engage in defensive war is recognized but not hostile actions short of war, such as the rescue of U.S. citizens from foreign danger. But supporters of the Resolution in Congress were unsure as to how far they should go in controlling presidential war making; and they did not give the resolution's definition of the president's independent war-power legal effect. The result was that the law offers its opinion on the president's constitutional authority and allows action based on personal opinion. Instead of trying to control the president before the fact, the Resolution seeks to do so afterward, mainly by requiring the withdrawal of armed forces from hostilities or a hostile situation within sixty days (in some instances, ninety days), unless Congress has declared war or extended the duration of their use.

As of 1993, the War Powers Resolution has not had much influence on presidential conduct. No president has accepted the resolution's definition of war-making authority (Bill Clinton has not yet made his views known) or its command that he obtain congressional consent in order to maintain them in hostile situations longer than sixty (or ninety) days. Indeed, no president has conceded that the resolution was applicable to situations in which he has used the armed forces in hostile actions or situations. It is true that Jimmy Carter, unlike Republicans in the White House, has spoken well of the law, saying that it made an appropriate reduction in the authority that the executive had prior to Vietnam. Also, his State Department's legal adviser assured a Senate committee that the law would be observed by the new administration, and yet he would not say that it was legal "in all respects." And when President Carter ordered military aircraft to enter Iran in 1980, in an effort to rescue Americans being held hostage there, he declared the Resolution to be inapplicable to the effort.[42] Republican presidents have been more insistent on their assumed rights in the face of the War Powers Resolution. In 1982, Congress authorized President Reagan to maintain an American peacekeeping force in Lebanon for eighteen months, and, in response, he asserted his independent right to do so while signing the legislation. And in September, 1990, with the build-up of American armed forces in Saudi Arabia for use against Iraq underway, President Bush's secretary of state told a committee of Congress that the administration considered unconstitutional the War Powers Resolution's provision requiring the president to obtain Congress's consent in order to maintain them in a hostile "situation" for longer than sixty days.[43]

The new political division regarding the president's power to engage in hostilities can be seen in the vote that took place in the House of Representatives on January 12, 1991. Congress had shortly before rejected legislation that would have continued sanctions imposed earlier on Iraq by the United States and other nations and it approved hostilities against that country. The House of Representatives voted on a concurrent—nonbinding—resolution that contained two propositions: that the Constitution vests the power to declare war in Congress and that the president must obtain Congress's approval before undertaking military action against Iraq. (The resolution did not come to a vote in the Senate.) The resolution was adopted by a vote of 302 to 131, with 260 Democrats and 1 Socialist (associated with them) voting in favor and 5 Democrats voting in opposition and with Republicans opposing by a margin of 126 to 41 members.

THE NEW UNDERSTANDING OF DEFENSIVE WAR

It has generally been accepted that the Constitution allows the president to meet war with war when "a foreign nation declares or openly and avowedly makes war upon the United States," to recall Hamilton. Even those, like Madison, who would not go this far have acknowledged a right in the president to engage in measures of self-defense until Congress could decide what to do. Jefferson would have had it both ways. When he sent a naval squadron to the Mediterranean in 1801, he instructed it (through his secretary of the navy) "to chastise" Algiers or Tripoli "in case of their declaring war or committing hostilities" on American shipping. But when a vessel of that squadron attacked a Tripoli warship, after it was learned that Tripoli had, in fact, declared war, Jefferson misstated the vessel's action as strictly defensive and left it to Congress to authorize offensive hostilities against that Barbary state.[44]

During World War II, the idea of self-defense began to assume a broader meaning, to include the defense of other nations. This meaning was incorporated into the Charter of the United Nations, whose Article 51 recognizes the right of nations to engage in collective as well as individual self-defense in case of armed attack. The defense treaties into which the United States entered following the war, which in the main were directed at the Soviet Union and communism generally, all embraced the idea of collective defense. Typical of the others, the North Atlantic Treaty, ratified by the United States in 1949, stated as follows: "The parties agree that an armed attack against one or more of them in Europe or North America shall be considered an attack against them all."[45]

Just as several of the joint resolutions we have considered avoided specifying just who would decide on going to war, so, too, did the mutual defense treaties now under review. In the Southeast Asia Collective Defense Treaty, for example, it was agreed that in the event of an attack upon a state covered by the treaty, each party would "act to meet the common danger in accordance with its constitutional processes."[46] It was easy for many Americans to think of the president as the one who would decide for this country, inasmuch as he was the one to whom the people looked when the nation itself was attacked. In the new thinking, an attack upon an ally or, by extension, any country whose security was important to that of the United States (the United States had no treaty with South Korea in 1950) was considered to be tantamount to an attack upon the United States. The circumstances of the cold war played a large part in this thinking. For example, it was explained in the Senate in 1950, when Truman decided to go to the assistance of South Korea, that "the basic reason" for his decision was that if the communists had been successful there they would have carried out similar invasions elsewhere and "our own ultimate safety would then have been threatened."[47] In a similar vein, the committee that reported the Formosa resolution to the House saw in Communist China's threat to the island not only the fate of Formosa to be at stake but that of "the non-communist peoples of Asia, and of our own defensive line."[48] So long as the war that the United States fought in support of any ally was defensive, there was no need to have recourse to Congress. In explaining why the president had not requested a declaration of war to fight in Vietnam, the State Department gave as one reason the circumstance that the struggle had to be won primarily in South Vietnam, and it "is in that context a defensive military effort."[49] Similarly, President Bush's supporters in Congress argued in early 1991 that the United States would not be engaging in "offensive action"—would not be going to war—against Iraq, inasmuch as Iraq started the war when it occupied Kuwait.[50]

The connection between the security of the United States and that of other nations lost much of its compelling force when the long confrontation between the United States and the Soviet Union dissolved in the latter's collapse at the beginning of the 1990s. The right continues to be asserted by American politicians but in a different way. President Bush and other administration officials cited the United Nations Charter and Security Council resolutions calling for Iraq to withdraw from Kuwait and then for other nations to compel its withdrawal in support of the position they took with regard to going to war against Iraq. When President Truman went to war in Korea, his action was justified by the citation of Article 51 of the United Nations Charter and U.N. Security Council resolutions calling on North Korea to withdraw from South Korea and asking other nations to help bring about that result. But the connection of Iraq's action to American security, especially in the absence of the cold war, was not very apparent. Indeed, the president emphasized an emerging new world order in which aggression would be shown not to pay in defending his right to force Iraq out of Kuwait. Congress challenged the president's right to act for the nation in the Iraq case, whereas it had not at the time of Korea; it, too, cited the U.N. Charter and Security Council resolutions as if seeking to wrest from the president any legitimacy they might confer.

Thus the new idea of self- or collective defense seems to have a broader dimension that I had earlier recognized: the action of all peace-loving nations against aggressors. This aspiration, too, exists in the United Nations Charter. When the charter was before the Senate for ratification in 1945, one of that body's leading advocates, Warren R. Austin (R-Vt.), expressed it in this way: "We recognize [in the charter] that a breach of the peace anywhere on earth which threatens the security and peace of the earth is an attack upon us."[51] This was Franklin Roosevelt's vision of the United Nations. In it, the United States would play a major role in policing a new world order, helping the cause of human rights everywhere.[52] Long submerged in the more pressing concern for security and survival during the period of the cold war, this idea was never absent even then. It is not without significance that Senator Austin's remarks, quoted above, were reproduced by the State Department in the memorandum that justified President Truman's going to the aid of South Korea. The idea seems to be gaining strength in the world at present. As to who will decide on the disposition of American forces in this brave new world, I cannot predict. Senator Austin, however, was sure that he knew back in 1945. "I am bound to say that I feel that the president is the officer under our Constitution in whom there is exclusively vested the responsibility for maintenance of peace."[53]

Notes

1. U.S. Congress, Senate, Committee on Armed Services, *Crisis in the Persian Gulf Region: U.S. Policy Options and Implications,* Hearings, September 11, 13, November 27, 28, 29, 30, December 3, 1990, 101st Cong., 1st sess. (Washington, D.C.: Government Printing Office, 1990), testimony of Secretary of Defense Dick Cheney, December 3, 1990, pp. 701–2.

2. George Bush, "Statement on Signing the Resolution Authorizing the Use of Military Force against Iraq," January 14, 1991, *Weekly Compilation of Presidential Documents, January–March, 1991* 27, no. 3 (January 21, 1991): 48–49.

3. U.S. Congress, Senate, Committee on Foreign Relations, *War Powers,* report no. 220, June 14, 1973, 93d Cong., 1st sess., 1973, p. 13.

4. Edward M. Kennedy (D-Mass.), January 3, 1991, *Congressional Record,* Senate, 102d Cong., 1st sess., 1991, vol. 137, no. 1, p. S13.

5. Alexander Hamilton, "The Examination, No. I," December 17, 1801, in Henry C. Syrett et al., eds., *The Papers of Alexander Hamilton,* (New York: Columbia University Press, 1977), 25: 455.

6. Alexander Hamilton, *The Federalist No. 69,* in Alexander Hamilton, James Madison, and John Jay, *The Federalist: A Commentary on the Constitution of the United States,* introd. Edward M. Earle (New York: Modern Library, 1937), p. 448, emphasis in the original.

7. "Instances of Use of United States Armed Forces Abroad, 1798–1989," CRS Report for Congress, December 4, 1989, in U.S. Congress, January 10, 1991, *Congressional Record,* Senate, 102d Cong., 1st sess., 1991, vol. 137, pt. 8, pp. S130–35. See, e.g., Representative Bruce A. Morrison (R-Wash.), House of Representatives, ibid., January 12, 1991, p. H404; Senator John W. Warner (R-Va), in U.S. Congress, Senate, Committee on Armed Services, *Crisis in the Persian Gulf Region,* Hearings, November 27, 1990, pp. 112–13; Secretary of Defense Dick Cheney, ibid., December 3, 1990, p. 702, also pp. 730–31.

8. U.S. Congress, Senate, Committee on Foreign Relations, *War Powers Legislation,* Hearings, March 8, 9, 24, 25, April 23, 26, May 19, July 26, 27, October 6, 1971, 92d Cong., 1st sess. (Washington, D.C.: Government Printing Office, 1972), pp. 359–79. This and earlier enumerations are taken up in Francis D. Wormuth and Edward B. Firmage, *To Chain the Dog of War: The War Power of Congress in History and Law* (Champaign: University of Illinois Press, 1986), pp. 133–49.

9. U.S. Department of State, Office of the Legal Adviser, "The Legality of United States Participation in the Defense of Viet Nam," March 4, 1966, p. 36.

10. "Authority of the President to Repel the Attack in Korea," Department of State Memorandum, July 3, 1950, *Department of State Bulletin,* July 31, 1950, p. 174.

11. *Youngstown Sheet and Tube Co. v. Sawyer,* 343 U.S. 579 (1952), Justice Frankfurter concurring, at 610–11. My attention was drawn to this statement by Harold H. Koh, *The*

National Security Constitution: Sharing Power After the Iran-Contra Affair (New Haven, Conn.: Yale University Press, 1990), p. 71.

12. *Case of the Resolution,* 2 Dall. 12, 20 (1781).

13. The 1989 list did not concern itself with declarations of war against Germany's lesser allies in the two world wars, nor will I notice them.

14. Alexander Hamilton, "The Examination, No. I," p. 456.

15. Abraham Lincoln, "Special Session Message," July 4, 1861, in James D. Richardson, ed., *A Compilation of the Messages and Papers of the Presidents, 1789–1897* (Washington, D.C.: Government Printing Office, 1897), 6:23; *The Prize Cases,* 2 Bl. 635 (1863).

16. Bushrod Washington, *Bas v. Tingy,* 4 Dall. 37 (1800), at 40.

17. Department of State, *Right to Protect Citizens in Foreign Countries by Landing Forces,* memorandum of the solicitor [Reuben J. Clark, Jr.], October 5, 1912, 3d rev. ed., with supplementary appendix up to 1933 (Washington, D.C.: Government Printing Office, 1934), p. 33.

18. U.S. Congress, House of Representatives, Committee on Foreign Relations, "Piracy and Outrage on Commerce of the United States by Spanish Privateers," report no. 398, 18th Cong., 2d sess., in *American State Papers: Foreign Relations,* 2d ser., vol. 5 (Washington, D.C.: Gales and Seaton, 1858), p. 586.

19. William Howard Taft, *Our Chief Magistrate and His Powers* (New York: Columbia University Press, 1916), p. 95.

20. Alexander Hamilton to James McHenry, [May, 17, 1798], in Syrett et al., eds., *Hamilton Papers* (1974), 21:462.

21. Using the incidents reported in the State Department's 1966 list as their basis, Wormuth and Firmage conclude in *To Chain the Dog of War* that American presidents have unconstitutionally engaged American forces in war or acts of war (they distinguish between the two) probably between a dozen and two dozen times, a conclusion they think disproves the claims made for such lists in recent decades. But their understanding of the president's constitutional power to use force is much narrower than the one I and those statespeople I have cited in this essay (e.g., Hamilton, Lincoln, William Howard Taft, and Robert A. Taft) have used. These two authors do not believe the Constitution allows the president ever to intrude the public force onto foreign territory. See pp. 147–49.

22. British war cabinet minutes, cited in Thomas A. Bailey and Paul B. Ryan, *Hitler vs. Roosevelt: The Undeclared Naval War* (New York: Free Press, 1979), p. 166.

23. Tom Connally, July 10, 1941, *Congressional Record,* Senate, 77th Cong., 1st sess., 1941, vol. 87, pt. 6, pp. 5927–930. Connally was unable to put his hand on a longer list that, apparently by including actions prior to 1812, would have brought the total to more than one hundred incidents.

24. Ibid., p. 5927; Dean Acheson, *Present at the Creation: My Years at the State Department* (New York: W. W. Norton, 1969), p. 414.

25. Bush, "Statement on Signing the Resolution," p. 48.

26. Robert A. Taft, June 27, 1950, *Congressional Record,* Senate, 81st Cong., 2d Sess., 1950, vol. 96, pt. 7, p. 9320.

27. Acheson, *Present at the Creation,* p. 415. President Reagan's secretary of state gave a similar reason for his president's refusal to acknowledge Congress's grant of authority to him in 1983 to keep marines in Lebanon: The President had "no intention of turning over to Congress his constitutional responsibilities as commander in chief." Secretary of State George Shultz, cited in "A Reluctant Congress Adopts Lebanon Policy," *Congressional Quarterly Almanac, 1983* (Washington, D.C.: Congressional Quarterly, Inc., 1984), p. 117.

28. Dwight D. Eisenhower, "Message to Congress," January 24, 1955, *Congressional Record,* Senate, 84th Cong., 1st sess., 1955, vol. 101, pt. 1, p. 601.

29. Dwight D. Eisenhower, *The White House Years: Waging Peace, 1957–1961* (Garden City, N.Y.: Doubleday, 1965), p. 182.

30. Ibid, pp. 179, 273.

31. John W. McCormack, January 24, 1955, *Congressional Record,* House of Representatives, 84th Cong., 1st sess., 1955, vol. 101, pt. 1, p. 659; Mike Mansfield, January 24, 1955, Senate, ibid., p. 622; U.S. Congress, Committees on Foreign Relations and Armed Services, *The President's Proposal on the Middle East,* Hearings, pt. 1, 85th Cong., 1st sess., (Washington, D.C.: Government Printing Office, 1957), p. 118.

32. John F. Kennedy, March 1, 1957, *Congressional Record,* Senate, 85th Cong., 1st sess., 1957, vol. 103, pt. 3, p. 2878.

33. John W. Bricker, January 28, 1955, *Congressional Record,* Senate, 84th Cong., 1st sess., 1955, vol. 101, pt. 1, pp. 953, 954.

34. The Formosa Resolution, January 29, 1955, in U.S. Congress, Senate, Committee on Foreign Affairs, Subcommittee on National Security Policy and Scientific Development, *Background Information on the Use of United States Armed Forces in Foreign Countries,* 1970 rev., Committee Print, 91st Cong., 2d sess. (Washington, D.C.: Government Printing Office, 1970), pp. 59–60. The Middle East, Cuba, and Southeast Asia (Tonkin Gulf) resolutions are also reproduced here, pp. 58–62.

35. See, e.g., Senate, Committee on Foreign Relations, report no. 13, to accompany S.J. Res. 28, January 26, 1955, 84th Cong., 1st sess., p. 9; "Prepared Statement and Testimony of Secretary of State John Foster Dulles," Senate, Committee on Foreign Relations and Committee on Armed Services, *President's Proposal,* pt. 1, pp. 9–10, 21.

36. John F. Kennedy, "Presidential Press Conference," September 13, 1962, *Congressional Record,* Senate, September 14, 1962, 87th Cong., 2d sess., 1962, vol. 108, pt. 14, pp. 19537, 19539.

37. Lyndon B. Johnson, "Presidential Press Conference," August 21, 1967, *Congressional Record,* 90th Cong., 1st sess., 1967, vol. 113, pt. 17, p. 29393. Perhaps reflecting some confusion or uncertainty on his part, President Johnson stated later in his remarks both that the resolution "authorized the president and expressed the Congress's willingness to go along with the president in doing whatever was necessary to deter aggression." *Ibid.*

38. Lyndon B. Johnson, "Message to Congress," August 5, 1964, *Congressional Record,* Senate, 88th Cong., 2d sess., 1964, vol. 110, pt. 14 p. 18132.

39. J. William Fulbright, (D-Ark.) August 6, 1964, *Congressional Record,* Senate, 88th Cong., 2d sess., 1964, vol. 110, pt 14, pp. 18403, 18407, 18409.

40. John J. Sparkman (D-Ala.), January 28, 1955, *Congressional Record,* Senate, 84th Cong., 1st sess., 1955, vol. 101, pt. 1, p. 933.

41. J. William Fulbright, December 4, 1945, *Congressional Record,* Senate, 79th Cong., 1st sess., 1945, vol. 91, pt. 8, p. 11396.

42. Jimmy Carter, cited in "Prepared Statement of Honorable Clement J. Zablocki," U.S. Congress, Senate, Committee on Foreign Relations, *War Powers Resolution,* Hearings, July 13, 14, and 15, 1977, 95th Cong., 1st sess. (Washington, D.C.: Government Printing Office, 1977), p. 302; Herbert J. Hansell, ibid., pp. 190, 207.

43. U.S. Congress, House of Representatives, Committee on Foreign Affairs, in *Crisis in the Persian Gulf,* Hearings and Markup, September 4, 27, 1990, 101st Cong., 2d sess. (Washington, D.C.: Government Printing Office, 1990), testimony of Secretary of State James A. Baker III, September 4, 1990, pp. 4, 82.

44. See Acting Secretary of the Navy Samuel Smith to Capt. Richard Dale, May 20, 1801, in U.S. Navy Department, Office of Naval Records and Library, *Naval Documents Related to the United States Wars with the Barbary Powers* (Washington, D.C.: Government Printing Office, 1944), 1:465–68; Lt. Andrew Sterrett to Capt. Dale, Aug. 6, 1801, ibid., p. 537; Thomas Jefferson, "First Annual Message to Congress," December 8, 1801, in Richardson, ed., *Messages and Papers,* 1:326–27.

45. North Atlantic Treaty Between the United States of America and Other Governments, April 4, 1949, in U.S. Congress, House of Representatives, *Collective Defense Treaties,* 91st Cong., 1st sess., (Washington, D.C.: Government Printing Office, 1969), p. 77.

46. Southeast Asia Collective Defense Treaty and Protocol Thereto, September 8, 1954, in U.S. Congress, Senate, Committee on Foreign Relations, *Background Information Relating to Southeast Asia and Vietnam,* 4th rev. ed., Committee Print, 90th Cong., 2d sess. (Washington, D.C.: Government Printing Office, 1968), pp. 100–102.

47. Paul H. Douglas (D-Ill.), July 5, 1950, *Congressional Record,* Senate, 81st Cong., 2d sess., 1950, vol. 96, pt. 7, p. 9649.

48. U.S. Congress, House of Representatives, Committee on Foreign Affairs, Report no. 4, *Authorizing the President to Employ the Armed Forces of the United States for Protecting the Security of Formosa, the Pescadores, and Related Positions and Territories of that Area,* January 24, 1955, 84th Cong., 1st sess., 1954 (Washington, D.C.: Government Printing Office, 1955), p. 4.

49. "The Question of a Formal Declaration of War in Vietnam," State Department Position Paper, November 19, 1965, in Senate, Committee on Foreign Relations, *Background Information,* p. 168.

50. See, e.g., Henry J. Hyde (R-Ill.), January 12, 1991, *Congressional Record,* House of Representatives, 102d Cong., 1st sess., 1991, vol. 137, pt 8, H391.

51. Warren R. Austin, July 26, 1945, *Congressional Record,* Senate, 79th Cong., 1st sess., 1945, vol. 91, pt. 6, p. 8065.

52. Cf. Willard Range, *Franklin D. Roosevelt's World Order* (Athens, Ga.: University of Georgia Press, 1959), esp. chap. 11.

53. Warren R. Austin, July 26, 1945.

Part V / The Force of the Media

8 / The Media and the Foreign Policy Process

W. LANCE BENNETT

INTRODUCTION

It does not take an advanced degree in communications to see the media intruding within the frame of world events. Policy makers have recognized the presence of the press and high-speed global communications at trade negotiations, at peace conferences, and in war zones, with the result that foreign policy has taken on a public relations, or media diplomacy, dimension of substantial proportions. Different players in the policy process try to shape news coverage, either to mobilize domestic support or to communicate directly with foreign leaders and their publics. Elites, interest groups, and foreign governments have taken the task of media management as an increasingly important element of the policy process.

Vivid examples are legion. In 1989 the power of grass-roots actors to use the media was demonstrated as Chinese students in Tiananmen Square used television to play the goddess of liberty and other symbols of Western democracy back to Western audiences. Although those symbols had little significance in the Chinese political context, they engaged the popular imagination in North America and Europe, inspiring grass-roots and human rights groups to complicate, if not fully change, the China policies of various nations, including the United States.[1] During the Persian Gulf crisis of 1990, leading to the war against Iraq in 1991, both Iraqi leader Saddam Hussein and U.S. officials used the Cable News Network (CNN) as something of a diplomatic back-channel, sending complex signals to each other and to their respective publics at the same time.[2] In 1992, television images of starvation in Somalia helped mobilize an international humanitarian intervention led by a large U.S. military presence. Among the most vivid scenes from that operation was one in which startled marines in war paint hit beaches already secured by television news crews to record the landing.

Viewers of the Persian Gulf War against Iraq witnessed the first large-scale application of the Pentagon's post-Vietnam resolve never again to lose a public relations war.[3] Many policy officials in the Defense Department and the State Department became convinced that the U.S. military defeat and eventual withdrawal from Vietnam resulted, in part, from critical media coverage of battlefield activities and sympathetic coverage of domestic opposition to government policies at home.

The invasions of Grenada and Panama during the 1980s went more according to government plan, due to severely restricted news coverage before and during the military actions. The Persian Gulf War demonstrated the ability of an administration to steer a policy course through months of public scrutiny and to use sophisticated news management techniques to turn saturation coverage by hundreds of news organizations into a public relations bonus. One public relations expert remarked:

> The Department of Defense has done an excellent job of managing the news in an almost classic way. There's plenty of access to some things, and at least one visual a day. If you were going to hire a public relations firm to do the media relations for an international event, it couldn't be done any better than this[4]

Indeed, it is hard to overlook the Gulf crisis as a textbook case in the selling of a foreign policy, as persuasive messages with relatively little competition or rebuttal were delivered on a daily basis to the living rooms of the United States through a mix of smoothly run daily press-briefings coordinated between Washington and Dhahran, officially guided press-pool tours to approved news sites, and well-produced daily video handouts. Working the same policy front from another angle, the Kuwaiti government hired top public relations firms to put the right spin on the news reaching both the American people and Congress.[5]

Intense media coverage makes press management a must for modern policy makers, but it is not clear how policy makers and their policies are affected by struggles over political information. Under what conditions are policy options critically examined? Are informed publics often created? To what extent are policy makers held accountable for their actions? All in all, it is easier to talk about how news coverage of world events has changed than it is to decide how changes in the news have affected the policies that drive world events.

HOW PRESS COVERAGE OF FOREIGN POLICY HAS CHANGED

The press has often played a part in foreign policy. Lively policy debates over issues like alliances with France and England were common in the era of partisan newspapers during the early decades after the nation's founding. By the late 1800s, however, the news had become a mass-market, commercial commodity. Dramatic tales of foreign intrigue sold papers, simultaneously advancing the political interests of newspaper owners. At the turn of the century, cartoons and news-adventure tales in the "yellow press" (named for a popular New York paper's front page cartoon character, the "yellow kid") told of the need for U.S. involvement to bring order, sound economics, and civilization to a savage world. Many newspaper owners and business elites shared an interest in expansionist policies during this heyday of a growing

U.S. empire abroad. Shortly before the outbreak of the Spanish-American War, for example, the story goes that newspaper mogul William Randolph Hearst wired photographer Frederick Remington in Cuba that if Remington would send the pictures, Hearst would furnish the war.

During the 1920s, the rise of a more professional journalism aimed at a middle-class audience tempered the political cheerleading of the press. However, the adoption of a norm of objective reporting translated into a tendency for leading papers like the *New York Times* to support most foreign policies (both in editorial writing and in the selection of sources in news reports) unless there was substantial elite debate and opposition to them in Washington, a tendency that is still evident among today's mainstream news organizations.

Throughout much of the period from the 1920s until well into the Vietnam War (when the press played up growing opposition to the war after 1967), the press was often a sideline player and occasional cheerleader in the policy process simply because the process, itself, was anything but open to public view. Power over key decisions was frequently held by a relative handful of elites—a so-called foreign policy "establishment"—who controlled information about their activities and worked hard to maintain a consensus on policy goals to avoid losing control of the policy process. As the journalist Walter Lippmann observed as early as the 1920s, images that the public received from news about world affairs were designed to create public support for policies, not to bring the public into the decision process itself. Lippmann lamented that the realities behind the scenes in trade negotiations, treaty settlements, and military interventions often had little connection to the information that elites fed journalists for reporting as news.[6]

Following the 1920s, public relations techniques similar to those developed to sell consumer products and create images for celebrities were used to create public reactions to foreign policy situations. Among the many public relations triumphs in the foreign policy field was the media campaign designed to convince the public that the democratically elected government of Guatemala in the 1950s was really a communist front and that the Central Intelligence Agency–sponsored military coup that overthrew that government in 1954 was really a popular uprising in support of progressive new leadership.[7] The leading news organizations reported the Guatemala story (along with many other world events during this period) according to the scripts written by policy elites.

Many observers have noted a change in news coverage patterns after the late 1960s (the late Vietnam era), attributed mainly to the crumbling elite consensus about the world interests of the United States.[8] As most of the authors in this volume suggest, with the subsequent changes in the global economy and power alignments, the expansion of domestic political forces into the foreign policy arena has been considerable. There is more available information on the public record for the press to cover, while more people in the news audience may see the domestic relevance of a broader range of foreign policy issues, from trade and the environment to human rights and war. In addition, the outbreak of more frequent elite debates and congressional opposition to presidential foreign policy initiatives provides journalists with a more regular supply of safe, reportable opposition views to put in the news. Finally, the emergence of global electronic communication networks makes instantaneous and dramatic coverage of events possible from almost anywhere on the planet.

All of this means that policy makers cannot ignore news management as a key part of the policy process, as they once may have had the luxury of doing. The journalist Marvin Kalb argues that at least since the Iranian hostage crisis of 1979–80—in which President Jimmy Carter was held hostage in the White House by the press for 444 days—decision makers have been forced to engage in a new dimension of political activity. Kalb calls this new dimension of the policy process *press politics,* reflecting the inseparability of foreign policy from its management in the news.[9]

What is less clear, as noted above, is whether changes in the quality of policy have resulted from the changes in the political presence of journalists. Is policy more subject to open, democratic debates because of media scrutiny, or do politicians simply apply more sophisticated media management and public relations techniques than ever before, continuing to turn the news into a public relations tool? In this chapter, I argue that there probably is not a simple either-or response to this question. Rather, the news has the capacity to shape public communication about foreign policy in a whole variety of ways, depending on how a number of (primarily domestic) political factors operate in a given policy situation. The next section reviews various claims about how news coverage can affect the foreign policy process. After that, we will see when these various potential effects are, and are not, likely to occur.

WHAT DIFFERENCE CAN PRESS COVERAGE MAKE?

According to many observers, we have entered a new era of political communications. Even foreign policy, once the private domain of pinstripe bureaucrats and business elites, that gray world of threats, promises, wars, espionage, and diplomacy, may have become transformed by a combination of new communications technologies and global media systems.[10] The problem is that different observers see different and often contradictory implications of this mass-mediated foreign policy.

A common impression is that the speed and portability of communications equipment, combined with a public fascination for live, saturation events coverage, force officials to make calculations based on the publicity surrounding their actions. Such calculations might result in policies that are hasty, ill conceived, damaging to future options, or tempered by domestic opinion rather than long-term state interests. In the words of one former official, the media have not only put foreign policy "on deadline," but changed its substance to fit what may be misleading yet convincing television images.[11] These kinds of claims need to be examined carefully to see if they have any theoretical standing or are merely the emotional, if understandable, reactions of harried officials wishfully seeking an easier time of their daily encounters with the press.

A related concern also worth careful investigation is what happens when the unblinking eye of television brings the world into the living rooms of what William Schneider has called "a vast inadvertent audience."[12] If, as has been suggested, this mass audience also turns out to be skeptical of foreign intrigues since Vietnam, the result may be a change in the methods of promoting domestic consensus by forcing officials to sell almost every policy initiative to the general public.[13] If the mass marketing of foreign policy is a new wrinkle, it then becomes important to decide whether those sales efforts—which also rely on media imagery—create more demagoguery than democracy in foreign affairs.

As noted above, the intrusions of domestic politics, again magnified by the media, have increased the degree of instability in the once highly managed foreign policy process. It is tempting to think that domestic news angles, from jobs to jingoism, affect how international stories are covered.[14] In addition, a more familiar cast of domestic players on the foreign policy scene may prompt more familiar scripting or framing of stories, creating more interest in foreign policy among the general public. Such changes may complicate policy debates in areas as diverse as trade, the environment, arms control, defense projections, and the use of force.[15]

To this already vibrant news picture, add the growing sophistication of public relations and marketing techniques that enable domestic interests and foreign governments alike to intervene directly in shaping both the news and public opinion. Going directly to the public may increase the leverage of various players when lobbying Congress and the bureaucracy and, in the case of foreign governments, enable them to bypass or supplement traditional diplomatic channels.[16] Now, top all of this off with the decline of the cold war ideological consensus and the absence of policy doctrines like containment, and there is reason to suspect that the foreign policy process involves more players, domestic and foreign, and more complex institutional and interest group politics than in previous eras, as pointed out by John T. Tierney in Chapter 5 of this volume. All of this leads many observers to presume that mass media news organizations sit more squarely than ever as mediators of more complex policy situations that defy the control of single actors, factions, or institutions.

These presumptions may be true, but there is still the troublesome question of whether these factors affect media policy debates and policy results in any systematic way. Some observers argue that for all the complexity of modern-day domestic politics, the press continues to display an underlying pattern of deference to foreign policy elites that results in largely unchallenged reporting of elite definitions of political situations. Thus the public continues to hear mainly what elites choose to make public, but the range of policy debate expands when the elites, themselves, are in more open disagreement or conflict, as they have been for various reasons since Vietnam. Those who support this view offer evidence spanning a broad historical range of policy situations from Vietnam, to the Central American conflicts of the 1980s, to the Persian Gulf War and other episodes from the 1990s.[17]

In the case of the Persian Gulf War, for example, the quality of the actual public policy debate may have been affected relatively little by all the high-tech communications, the saturation coverage, the slick Pentagon public relations, the overt press censorship, or the press's putting policy makers "on deadline." In many respects, the reasons offered by the Bush administration for defending Kuwait against invasion by Iraq were not particularly artful by public relations standards, but perhaps they did not need to be. As William Dorman puts it, the early stages of the policy debate concerning what to do about Iraq were distinguished by the absence of much opposition from elites in Washington, leading the press to report uncritically on the way the Bush administration framed the issues:

> By accepting the frames offered by Bush, the press helped to limit debate at the time when it would have done the most good—before we got in so deeply that the war became inevitable. Perhaps the most important question of this discussion has to do with why the press behaves as it does. There are a number of reasons, some more significant than others. First, Congress was on vacation, and American journalism is very closely indexed to structured institutional debate. The press didn't know who to turn to for critical reaction to the President's initial policy. In other words, when it comes to foreign policy, if a member of Congress doesn't say it, it isn't as likely to be covered.[18]

In an interesting comparison of the U.S. invasions of the Dominican Republic in 1965 and Grenada in 1983, Brigitte Lebens Nacos suggests that this deference may have grown over the decades as presidents have stepped more into the news spotlight to sell the policies of their administrations. Given the visibility of the president as a news maker and the sophisticated techniques for managing the news, the press may be all the more vulnerable to manipulation despite the capacity to cover world events more quickly and more intensively than ever before.[19] In addition, the growing ranks of media critics and ideologically oriented experts in foreign policy think tanks may open the press to charges of being unpatriotic or having a leftist bias when criticisms of foreign policy do not come directly from the mouths of public officials. All of this may reinforce the tendency to frame policy debates in the news mainly through the exchanges among members of Congress, present or former executive branch officials, and representatives of respected interest groups.

This is not to downplay the importance of the press. The news is and has been an important variable in the foreign policy equation. It is worth asking, however, whether modern-day news organizations are any more or less susceptible to being captured by policy makers and their ideologies than when Walter Lippmann warned citizens more than three-quarters of a century ago about the dangers of taking world news seriously.[20] We might even ask whether the growing importance and sophistication of public relations and media management techniques constitute such a radical break with long-standing government practices. Since the dawn of mass advertising and scientific public relations with the pioneering work of Edward L. Bernays and others, both government and private interests have a long tradition of selling images of foreign adventures wholesale to the press and other information brokers, who repackage them for retail distribution to the general public. It is not clear that selling policies to larger, less ideological (so-called inadvertent) audiences presents any greater challenge than the traditional task of selling them to opinion leaders and informed publics. To pick a provocative case, it is not clear that the public debate phase of the policy process leading up to the war against Iraq in 1991 established any greater levels of coherence, popular understanding of goals or options, serious challenge to administration control over the definition of the situation, or greater accountability for the results than, say, the similar phase leading up to the U.S.-sponsored Guatemalan coup in 1954.[21] Nor is it clear that the policy process was, on any of these dimensions, any worse. Although there were higher levels of public protests during the Persian Gulf crisis than in the Guatemala case, suggesting a tendency toward broader grass-roots interest in foreign policy, the media tend to downplay or marginalize grass-roots opinion in foreign policy coverage for reasons explained in a number of studies.[22]

The moral of these stories is simply that we need to be cautious about jumping to conclusions. It is an important and still unresolved question whether the *qualities* of public policy debates about the world role of the United States have changed fundamentally due to the rise of a global media, new information technologies, the emergence of CNN as a video wire service, or shifts in world and domestic politics.[23] The way to resolve questions about what is going on with the media and policy, and what may or may not be changing, is to think about how the news is constructed, how news content shapes public opinion, and how news coverage and opinion may constrain policy makers and their activities.

THE NEWS, PUBLIC OPINION, AND POLICY

The goal of the remainder of this chapter is to explain what goes on in three domains of the mass-mediated policy process: (1) the production of news images by journalists and political actors; (2) the effects of those news images on patterns of public opinion and participation; and (3) the policy effects resulting directly from news coverage and indirectly from the impact of news on opinion and participation. In addition, an important aim of this discussion is simply to bring the news media more formally into our thinking about how foreign policy is made.

Traditional theories of foreign policy have focused heavily on behind-the-scenes politics, putting the spotlight on things like bureaucratic bargaining, elite perceptions of enemies and crises, the workings of the security community, and levels of consensus within the so-called policy establishment.[24] Attempts to look at more public aspects of the foreign policy process have focused primarily on public opinion and its linkages to elites. These studies have looked at opinion linkages in different ways, including direct linkages between public opinion and policy;[25] game theory analyses that view opinion as another piece of information known to both friendly and enemy players in the policy game;[26] and institutional analyses that see opinion as a power lever along with lobbying and elections.[27]

While all of these approaches suggest the importance of a public phase of the policy process, there is no general model that provides an account of what this public phase looks like and how it connects the various pressure points of lobbying, polls, elections, or protests. The approach here offers a broad view of how the news record of a policy debate is constructed, how that record in turn affects opinion formations, and how those formations may enter different institutional arenas and the calculations of policy officials. To begin with, consider a range of imaginable differences across policy situations.

At one extreme, foreign policy debates in the news can be relatively rich, with more (and more diverse) voices participating, more views and options being introduced, and continuing for more extended periods of time, with noticeable effects on the course of policy itself. In this scenario of an *open public debate,* public opinion formations can be expected to change in response to information changes. For example, opinion rallies may rise and fall as opposition voices do or do not enter the fray or new information is brought to bear by the media. In some cases, guidelines for evaluating policies may be introduced into the news record so that politicians are called into account for actions of the past. Perhaps the historical context is built up so that it becomes difficult for officials to change course, as often happens in foreign policy, by simply reinventing history. Under these conditions, both public opinion in response to news images and elite concerns about future public reactions to the way a story is developing (or might develop) can affect the course of policy.

At the other extreme are cases involving shorter debate periods, with fewer and less varied voices and views making it into the news. In this type of relatively *closed public debate,* there is little ground established for evaluating policies or holding officials accountable, and the lack of much historical reference within the debate leaves officials free to reinvent history and make untested claims about policies.

Different cases can be located along this continuum. For example, the media debates surrounding the invasions of Grenada and Panama probably fall closer to the closed end of the continuum, while the debate pattern from the post-1967 Vietnam era would probably fall toward the open end. The Persian Gulf War case to be discussed in more detail below falls somewhere in between, although more toward the closed end. Recent debates over trade with other countries have been relatively open, reflecting differences between key political players in Washington and congressional attention to the pressures of voters concerned about jobs.

The point of the continuum, of course, and the reason for the tentativeness at this point are to avoid blanket judgments that news coverage of all foreign policy situations is the same. Different political conditions may create quite different kinds of news coverage of different situations—which brings us back to understanding the relations among journalists, publics, and the political actors at the core of the decision process.

First, we need to understand how regular patterns of press-government relations affect news content—content that can be regarded as the public record of a policy debate. The goal here is to identify the organized ways in which journalists choose sources and screen and weigh information. Equally important is to understand the news management strategies that politicians use to turn the predictable behavior of journalists to their political advantage. Sometimes officials are successful in structuring information so that journalists are effectively trapped or limited by their own operating procedures. At other times, those political strategies are poorly implemented, or journalists find other sources to help them get around official definitions of policy situations. Hanging in the balance of this interaction are things like the duration and the information diversity of news coverage. By fleshing out the reporting habits of journalists and showing how they interact with the information strategies used by politicians, we may be able to account for the following kinds of differences in the reporting of foreign policy situations:

- the length of public debate periods
- the diversity of actors who make the news
- the ratios of official, nonofficial, expert, elite, and nonelite voices in the news, on editorial pages, and on public affairs programs
- the dominant themes, images, and arguments in news coverage
- the diversity of policy options reported in the news and the weight given to those options
- the degree of historical background established in news coverage
- the degree of accountability established for decision makers, both short term (reports of opinion polls, editorial positions, and experts' judgments) and long term (development of standards for measuring the success or failure of policies themselves).

Second, and following from this understanding, we must account for how these characteristics of the news information record affect public opinion and create opportunities for various forms of popular participation. Of concern here are the relations between the above information characteristics in the news and patterns of public reaction, opinion formation, and change. Among the obvious concerns here are:

- how reporting what officials and experts say shapes opinion among publics with varying degrees of interest and information about an issue
- how reporting political opposition in the news contributes to the rise and fall of opinion rallies supporting presidents and their policies
- how the framing of news accounts (e.g., as narratives around human interest angles, as ideological choices, or in more abstract economic or political terms) affects popular reasoning and information processing about a policy situation.

Third, we must understand the impact of both the news information record and public reactions on the actual policy options chosen, the ones not chosen, and the record of accountability established for the policy itself. In this third public domain of the foreign policy process, officials may find themselves either limited or licensed to take particular actions, depending on factors like these:

- when prominent and favorable news coverage is given to opinion polls, demonstrations, or protest movements, officials may be more limited in their pursuit of particular policy options
- reporting past policies, along with indications of success or failure, may affect the credibility of official claims about new situations
- patterns of historical reporting and public awareness of historical conditions may affect how officials can define the current situation.

The key factors operating in each of these three domains of the policy process are specified more fully in the sections that follow.

Press-Government Relations and News Production

The place to begin understanding how reporters and officials interact to produce news is with one of the most established findings in media research: Reporters overwhelmingly turn to officials as sources for stories and for framing the content of stories.[28] In some of the pioneering research in this field, Bernard Cohen concluded that journalists and political officials were engaged in a process of symbiosis or mutual dependence, in which each side used the other to promote particular organizational (press or government) goals.[29] This finding has been confirmed and expanded in a number of subsequent studies with important implications for the production of foreign policy media debates: The dominance of official, and particularly executive branch sources has been found to be more pronounced in national security stories than for the news as a whole.[30]

This basic finding has been expanded in subsequent research into several important points. To begin with, the most illuminating explanations for the overwhelming reliance of reporters on officials arise from studies of how news organizations work. U.S. news gathering is organized around beats. In foreign policy, the key beats are the "golden triangle" of the White House, the Pentagon, and the State Department, with Congress playing a role in many stories as well. The beat system has obvious bureaucratic and economic advantages for organizations seeking to schedule reporting assignments, "cover" a large amount of news territory, and produce a steady, predictable supply of stories to fill the daily news hole in a cost-effective way. In addition, there is an ideological rationale that in a democracy the news should tell the people what their elected officials say and do, not make independent judgments about where to find a more representative set of democratic actors for each story covered.

All of these may seem like perfectly good reasons to cover the world from the official viewpoints of Washington, but to a sociologist, they look like the makings of a normative order. Any normative order favors some social values, ideas, and actors over others and creates certain tendencies for deception, manipulation, and conflict. As Timothy Cook has noted in his study of bureaucratic routines and reporting in the Persian Gulf War, U.S. news about the war looked a good deal different from, for example, French news, largely because of differences in the structures of news organizations themselves. In the United States, the beat system has become so routinized that it provides reporters with the ability to anticipate official reactions during crises and breaking events, even when they have not been able to interview their usual sources before filing a story.

> For [the] French news broadcast, the world was first globally constructed, then ideologically constructed. By contrast, for [the] American broadcast, the world was first domestically constructed, then institutionally constructed. Even though [the American reporters] presented themselves as impartial observers at their newsbeats, it was clear that none of them had the opportunity to discuss the breaking news with any sources; their reactions to [the anchor's] questions reflected their understanding of the institution and the individuals in charge. In effect, [they] were almost as much spokespersons for the newsbeats as were the party spokespersons who appeared on French television.[31]

Needless to say, the bureaucratic interdependence of reporters and officials places some obvious boundaries around policy debates in general, making them heavily structured by elite cues, official information, policy options considered viable by insiders, and all of this weighted by the prominence of information sources within the Washington power hierarchy (at least as it is perceived by the insiders in the national press corps). What these boundaries mean, among other things, is that even in cases where national debates spill beyond the efforts of officials to define the policy situation, resulting in protests, opinion divisions in society, and lively media debates, the final word in the news is usually reserved for journalists' well cultivated official sources, led by the president, top administration officials, and key members of Congress.

All of this leads to the second important theoretical element that explains many of the differences in patterns of news coverage. Obviously, even if most news is elite driven, there can be important differences from one policy situation to another in things like the duration of coverage, the play given to different sides of the story, and other information patterns of great consequence to the formation of public opinion and the legitimation of policy. These differences suggest that while journalists may be operating with one norm referring them to official sources, they also operate with an equally important professional norm that discourages taking sides by looking to report different sides of a debate. So strong is this oppositional norm that journalists have been described as going into something of a "feeding frenzy" when conflict and controversy break out within official circles.[32]

The operative phrase here remains "within official circles." When a relatively open policy debate occurs along our idealized news debate continuum, it most often develops because different officials along the beat system have broken ranks and decided to "go public" with their policy differences, thus providing journalists with reportable opposition voices and viewpoints. Conversely, when for any reason elite differences are resolved, or attempts at resolution lead officials to retreat behind closed doors, the reportable official conflict disappears, and the debate story dries up. U.S. policy toward El Salvador turned from one of the most widely reported and hotly contested stories of the early 1980s to one of the most neglected, at least in the view of a national panel of media critics who put it by mid-decade at the top of the list of ten most (self-) censored stories by the media. Although the U.S.-funded war escalated, and El Salvador became the third leading U.S. aid recipient in the world during the period of declining news coverage, journalists complained that Washington officialdom had closed ranks around the administration war policy, leaving the press without much of a policy story—never mind the presence of an easily accessible war with many domestic and foreign opponents willing to offer themselves as news sources.[33]

Differences in the structure of news debates (e.g., diversity of sources, range of opinions, and policy options) can be explained in large part by this tendency to "index" news coverage to the intensity and duration of official conflicts.[34] Indexing has at least two important effects on the information structure of the resulting news. First, journalistic perceptions of division among officials open the news gates to contending elite views of policy options, rationales, historical arguments (if there are any), and standards of accountability for evaluating policy successes and failures. Second, when official conflict is sustained, the news gates also tend to open to grass-roots groups, interest groups, opinion polls, and broader social participation in media policy debates because the ongoing story offers news organizations opportunities to follow different angles—opportunities that may turn into imperatives as journalists look for new ways of reporting the same old story.

Grass-roots voices and interest groups do, of course, make the news on the strength of their own public relations strategies, but studies suggest that unless their issues find a place on the institutional or electoral agenda and stimulate official conflict, there is little chance that policy debates will be sustained from below.[35] In general, however, the views of activist groups are not taken as seriously in news about foreign policy as they are on many domestic issues. Studies of policy debates as different in time and substance as the antiwar movement during Vietnam and the nuclear freeze movement in the early 1980s find that compared to government officials, grass-roots groups and their spokespeople were portrayed in more negative terms, while being less able to get preferred versions of their messages into the media debate.[36] The bottom line remains that the news gates tend to open or close depending on the levels of conflict among powerful players in the policy situation on Capitol Hill, the White House, the State Department, the Defense Department, and other relevant institutions.

Studies of different historical periods and different policy issues confirm the importance of indexing in constructing the information values of the news product. Daniel Hallin attributes much of the media debate during Vietnam to the gradual opening of sustained opposition to administration policies within Congress and even within the executive branch itself. With the expansion of what Hallin calls the "sphere of legitimate controversy," many features of the news policy record began to change, including the frequency of opposition views in the policy debate and the entry of nonofficial (i.e., nongovernment official) voices into that debate.[37]

A much more restricted version of this pattern emerges in studies of the media debates during the Persian Gulf War, with the news becoming more open to opposition views and policy alternatives during the relatively brief periods of congressional debate in November 1990 and January 1991 and closing down almost completely after Congress passed a resolution supporting the administration. Also characteristic of indexing, the narrow margin of passage for the congressional resolution was not as important as the fact that after its passage, official debate all but ended, and so did debate in the leading news media.[38]

Adding to the constraints on news production are the policy officials working both behind the scenes and in front of the cameras to shut off institutional opposition or at least to influence its course. These efforts to frame news stories advantageously also shape the information patterns in the media debate. While journalists are looking for signs of official conflicts on their beats, officials are trying to feed reporters the daily news line most advantageous to their policy preferences. In short, the same officials who make up the journalist's news index are, themselves, active players in a press management game, applying various techniques of strategic communications to elevate the volume of their own messages and reduce the credibility of their opponents.[39]

While journalists are aware of and may try to resist their dependency on officially packaged news, the growing sophistication of press management techniques holds the constant possibility that some officials will be more successful in engineering favorable media framings for their positions than others. The result is that the news record of any policy debate is liable to be inflated in favor of one faction or another, depending on the success of press management techniques in keeping the preferred spin on the story. The textbook example, of course, was the news management strategy of the early Reagan administration, which has by now become legendary. A host of techniques (controlling leaks, putting out a story of the day, setting up computer files scripting the official line of the day for different administration officials) often added up to favorable policy coverage even in many situations where opposition in Congress and in opinion polls might warrant more critical media treatment.[40]

To summarize the discussion so far, two important dynamics affect how the mass media construct the news record of policy debates. First, reporters and editors tend to index the voices and viewpoints in stories to the range of official conflict available to reporters on the newsbeats of decision-making institutions of government. Second, government officials try to influence how their views are played in news debates by implementing public relations and news management strategies. The interaction of these two sides of press-government relations affects the content, duration, and intensity of media policy debates, along with the degree to which grass-roots or social voices are included in those debates. The information in these policy debates, in turn, has important effects on the formation of public opinion.

How Media Debates Affect Opinion Formation

There is considerable evidence that public opinion is cued heavily by the media framing of issues and reported patterns of elite conflict in news about those issues.[41] The contributions by Thomas W. Graham and by Robert Y. Shapiro and Benjamin I. Page (Chapters 9, 10) in this volume suggest that there is an overall stability to opinion about many foreign policy issues. However, it is also clear that considerable energy is devoted to educating the public on important issues and to forging popular support for dominant official positions. It is important to understand how these opinion formations are shored up or shifted in specific policy situations. Shanto Iyengar's work makes it clear that the way an evolving story is framed (for example, around human interest vs. more analytical themes) affects how people following the debate think about policy options and preferred outcomes.[42] In the case of the Persian Gulf War, market research showed that the most effective framing for the policy problem that needed to be solved was to turn Saddam Hussein into the personal enemy of the American people and bring George Bush into that frame to do battle.

The important question, of course, is whether this framing permitted adequate consideration of other options, or even invited careful scrutiny of the war policy, since there was ample evidence that the Bush administration had been sending Saddam fairly positive diplomatic signals on the eve of the crisis. As a study by Robert M. Entman and Benjamin I. Page shows, the Saddam framing, magnified by the compressed time frame of the congressional debate, made it possible to dismiss the leading policy alternative of economic sanctions against Iraq long before there was any empirical basis for doing so (i.e., long before it was reasonable to determine whether sanctions were working).[43] Hence, media framing downplayed the leading policy alternative to war, suggesting that framing can go beyond opinion formation to have direct effects on policy choices.

The next theoretical question becomes, Within the news frame, what information cues help people to structure their thinking? The enduring theoretical proposition here is one that fits well with the earlier account of news production. As people become more informed about the policy issues in question, their opinions become more responsive to cues from the elites who appear in media debates on those issues. As John Zaller describes it:

> Evidence from a half a century of polling in the United States firmly supports the proposition that the more citizens know about politics and public affairs, the more closely they are wedded to elite and media perspectives on foreign policy issues. When elites are united in support of a foreign policy, highly informed Americans support that policy more strongly than any other segment of the public. When elites divide along partisan or ideological lines, better informed citizens are more likely than less informed ones to align their opinions with that segment of the elite which shares their party or ideology. And when elite opinion changes, the best informed are quickest to respond to the new view.[44]

This iron law of opinion and foreign policy offers one reason why the indexing of news stories (with its potential for inflated indexes due to news management) becomes so important in a foreign policy debate. The media index in a given situation becomes the best predictor of opinion, particularly among informed publics. Further limiting the impact of opinion, polls are reported far less often (roughly half as often) in foreign policy stories compared to domestic policy stories, while the qualifications of the general public to hold opinions on foreign policy issues are more often questioned by reporters (often through interviews with experts).[45] Such screening and information discounting do not go on at anything approximating the same rate in news coverage of domestic issues.

There are, of course, significant moments when public opinion can dominate the media debate record. The most obvious case is the so-called opinion rally in support of a government (usually presidential) response to a crisis.

Of equal importance are those moments when rallies come to a sudden end with a crash in support for administration policy. Not only can strong opinion rallies become big news, sustaining the imbalance of voices in an ongoing media debate, but when rallies crash at important moments in a delicate policy situation, openings for renewed debate and opposition may appear in the news.

Research by Richard A. Brody suggests that, like other opinion formations, rallies are cued (on the downward cycle) by media reports of elite opposition. This finding conforms to the general indexing dynamic that moves the whole news debate theory and marks an important departure from the traditional view that rallying publics were solidly bound together by patriotism, following leaders blindly through crises.[46] Brody and Catherine R. Shapiro suggest that while crisis and uncertainty may trigger a rally, particularly in the absence of initial elite opposition, the subsequent outbreak of elite criticism and more general opposition in the media will almost certainly end it.[47] In a case study of the Persian Gulf War, Brody (with Hollis Robbins) elaborates on the rally phenomenon with this explanation:

> Political elites frame the public's response to uncertain international events. If and when opinion leaders publicly interpret a crisis as a result of policy failure a rally will not take place. If the elite is silent or openly supportive of the administration's position, the public will respond to the administration's generally positive one-sided view of the events—to the administration's "spin"—and rally behind the president. A rally will last as long as the president's tacit or explicit support coalition persists.[48]

Two implications of this finding seem important for understanding media dynamics in foreign policy situations. First, the timing and management of institutional conflict (timing administration requests to Congress, framing those requests, sequencing policy actions and initiatives, scheduling legislative debates) may have important effects on the structuring of media debates and, in turn, on the maintenance or collapse of opinion rallies. Second, the management of news surrounding a crisis becomes important both for actors benefiting from a rally and for those who would seek to undermine it. All of this points to the effects both news images and the opinion formations that flow from them have on the policy options actually selected in a situation.

The Policy Domain

Many early observers concluded that public opinion did not have much direct influence on elite policy preferences, largely because opinion on foreign policy tended to be uninformed and subject to volatile mood swings (see Graham's opening discussion in Chapter 9 in this volume). In the traditional view, there is at most a tendency, which Bernard C. Cohen called elite "responsiveness," in which decision makers try to anticipate public concerns and present their policies with those hopes and fears in mind.[49] There is growing evidence, however, that opinion can enter more directly into policy calculations, as noted by Graham and by Shapiro and Page in Chapters 9 and 10 of this volume. At the very least, impressive evidence now suggests that there is substantial convergence between public opinion trends and foreign policy trends in a broad range of policy areas.[50]

The tricky question, of course, is whether elite control of information makes public opinion largely a dependent variable or an output in the policy process, or whether opinion is more independent of propaganda and elite influence, making publics more the initiators of policy.[51] Whether leaders prepare publics to accept already developed policies or policies are shaped by public opinion may depend on the quality of news debates and the degree to which leaders are held accountable to public opinion. The direction in which opinion flows (i.e., as input or output) in the policy process may depend a great deal on the interactions of journalists and officials, resulting in relatively more open or closed policy debate in the news. In other words, open opposition among Washington elites leads to more sustained news coverage of different viewpoints, creating the conditions necessary for greater public arousal and participation in the policy process.[52]

Patrick O'Heffernan adds that surrounding these core relations between reporters and officials today are more complex communications and political environments.[53] These relations mean that media management has become more complex and the potential for media damage to policies may have become greater due to the interworkings of foreign and domestic political players, global media systems, public relations campaigns, and experts, among other things that complicate the foreign policy picture. Officials can become trapped by the news record of their decisions and the resulting opinion dynamics that make those decisions appear popular or unpopular. Interview studies of decision makers show that news coverage has become integral to their decision processes. For example, O'Heffernan's studies of national security officers show that these elites monitor the media, particularly television, to follow emerging news angles in event coverage.[54]

It is worth considering in the case of the Persian Gulf War that an earlier debate in Congress might have created a different set of news and opinion dynamics and changed the framing of the policy options. Whether this earlier debate would have affected the eventual decision to use force is open to question, but it might well have introduced greater levels of accountability about the reasons for going to war, the reasons for deciding when to stop the war, and later evaluations of the success of the policy.

We are now in a better position to see how the news can have different effects on the course of foreign policies. For example, we might predict that when a news debate is relatively intense, elites are more likely to lose control of policies, publics are more likely to have input into those policies, and more decision makers and decision points enter the process, for better or worse. On the other hand, when news debate dies, the number of domestic political players and their options are reduced, smaller circles of elites take control, and public accountability is relatively weak.

CONCLUSION

How can we understand the role of today's global, high speed communications media in the foreign policy process? From trade talks to war zones, the television cameras are more ever-present, the speed of communication is faster, and the risk of arousing public opinion with vivid, attention-getting political drama may be greater than ever before. These factors mean that the most important change in the public world of foreign policy is that the news media simply cannot be ignored, relegated to the sidelines, or counted on to endorse government policies as elites may have done more often before Vietnam. Politicians must develop ever more sophisticated strategies for playing what Marvin Kalb calls "press politics." There is near universal agreement by government officials about the importance of public relations, strategic communications, and other efforts to shape the reporting of policy decisions.

These changes are, however, swirling around a solid core of press-government relations and public opinion effects that have been observed over a long period of time. In other words, long before the advent of instant global television coverage, the news media were capable of very different reporting styles on foreign policy, from relatively open to closed debates, and relatively critical to uncritical coverage, depending largely on how key government officials lined up politically. Equally enduring are the patterns of public opinion in response to these different information patterns in the news. If there have been more lively policy debates over trade, treaties, or military roles since Vietnam than before, the reasons may have to do with the breakdown of elite political consensus and the rise of more domestic angles in foreign policy news. As the case of the Persian Gulf War suggests, however, the techniques of effective news management can compensate for the decline of the old foreign policy establishment and the rise of domestic political forces.

Given the potential of electronic and satellite communications to create more democratic participation in, and critical surveillance of, the foreign policy process, the important question is whether the news media should take a more active role in structuring open foreign policy debates. In deciding this issue, it may help to recognize that the news is neither a sacred nor a fixed commodity but something that has continually evolved over the last two centuries in response to social, economic, and political forces. Perhaps the greatest challenge for the future is for journalists themselves to decide if the interests of democracy are best served by allowing the information gates to be opened and closed largely by the political relations between the White

House and Capitol Hill. Perhaps more debate is needed within the profession concerning how journalists should handle an information flow often managed by public relations techniques from partisan players in the foreign policy game.

Should news organizations create more independent agendas of policy issues and challenge current policies even when government opponents do not? Should public opinion be taken more seriously more often in establishing a base line for political debate in foreign policy news? If the authors of the next two chapters in this volume are right, there are reasons to be supportive of greater popular participation in the foreign policy process, and the news media, after all, hold considerable sway over whose voices are heard.

Notes

The author would like to thank the Social Science Research Council for providing opportunities to think about these issues. The comments of Brigitte Nacos and Ellen Hume on an earlier version of this chapter were very helpful. Thanks, in addition, to Regina Lawrence for the valuable research assistance on this project.

1. See Donald R. Shanor, "The 'Hundred Flowers' of Tiananmen," *Gannett Center Journal* 3 (Fall 1989): 128–36. See also, in the same issue, Leonard Pratt, "The Circuitry of Protest," pp. 105–15.
2. Lewis A. Friedland, "Democracy, Diversity, and Cable: The Case of CNN," University of Wisconsin, 1991, unpublished p. 14.
3. See Bob Woodward, *The Commanders* (New York: Simon and Schuster, 1991).
4. Michael Deaver, as quoted in the *New York Times,* February 15, 1991, p. A9.
5. See Jarol Manheim, "All for a Good Cause: Managing Kuwait's Image during the Gulf Conflict," paper presented at the Social Science Research Council Workshop on the Media and Foreign Policy, University of Washington, Seattle, September 26–29, 1991.
6. Walter Lippmann, *Public Opinion* (New York: Free Press, 1922).
7. See Richard Immerman, *The CIA in Guatemala* (Austin: University of Texas Press, 1982).
8. See, for example, Daniel C. Hallin, *The Uncensored War: The Media and Vietnam* (Berkeley: University of California Press, 1986).
9. Marvin Kalb, director of the Joan Shorenstein Barone Center on the Press, Politics, and Policy at the Kennedy School, Harvard University, personal communication.
10. See, among others, Simon Serfaty, ed., *The Media and Foreign Policy* (New York: St. Martin's Press, 1991); Lloyd Cutler, "Foreign Policy on Deadline," *Foreign Policy* 56 (Fall 1984): 113–28; Patrick O'Heffernan, "Mass Media and U.S. Foreign Policy: An Inside-Outside Model of Media Influence in U.S. Foreign Policy," paper presented at the Annual Meeting of the American Political Science Association, San Francisco, August 30–September 2, 1990; James F. Larson, "Global Television and Foreign Policy," *Headline Series,* no. 283 (New York: Foreign Policy Association, 1988); James F. Larson, "Quiet Diplomacy in a Television Era: The Media and U.S. Policy toward the Republic of Korea," *Political Communication and Persuasion* 7 (1990): 73–95; George H. Quester, *The International Politics of Television* (Lexington, Mass.: D. C. Heath, 1990).
11. Cutler, "Foreign Policy on Deadline."
12. William Schneider, "Public Opinion," in Joseph Nye, Jr., ed., *The Making of America's Soviet Policy* (New Haven: Yale University Press, 1984), p. 19.
13. Richard A. Melanson, *Reconstructing Consensus: American Foreign Policy since the Vietnam War* (New York: St. Martin's Press, 1991).
14. John B. Judis, "Twilight of the Gods," *Wilson Quarterly* 15, no. 4 (Autumn 1991): 43–56.
15. For various perspectives on this possibility, see the collection of articles in the special

issue on "Democracy and Foreign Policy: Community and Constraint," *Journal of Conflict Resolution* 35, no. 2 (June 1991).

16. See Jarol B. Manheim, and Robert B. Albritton, "Changing National Images: International Public Relations and Media Agenda Setting," *American Political Science Review* 78 (1984): 641–54. See also, Manheim, "All for a Good Cause."

17. See, for example, Hallin, *Uncensored War;* Edward S. Herman, "Diversity of News: Marginalizing the Opposition," *Journal of Communication* 35, no. 3 (Summer 1985): 135–46. See also, W. Lance Bennett, "Toward a Theory of Press-State Relations in the United States," *Journal of Communication* 40 (1990): 103–25.

18. William Dorman, address to the conference, in *The Media and the Gulf: A Closer Look,* Proceedings of the Conference held on May 3–4, 1991, at the Graduate School of Journalism, University of California, Berkeley, pp. 15–16.

19. Brigitte Lebens Nacos, *The Press, Presidents, and Crises* (New York: Columbia University Press, 1990).

20. Lippmann, *Public Opinion.*

21. See Immerman, *CIA in Guatemala.*

22. See, among others, Todd Gitlin and Dan Hallin, "Cultural Themes and News about War," work in progress, Social Science Research Council Workshop series on the Media and Foreign Policy; Todd Gitlin, *The Whole World Is Watching: Mass Media in the Making and Unmaking of the New Left* (Berkeley: University of California Press, 1980). This generalization even applies to the press coverage of the nuclear freeze movement of the early 1980s, as shown by Robert M. Entman and Andrew Rojecki, "Freezing Out the Public: Elite and Media Framing of the U.S. Anti-Nuclear Movement," paper presented at the Annual Meeting of the American Political Science Association, Chicago, September 4–6, 1992.

23. For what it means to talk about the rise of a global media, see Anthony Smith, *The Age of Behemoths: The Globalization of Mass Media Firms* (New York: Priority Press, 1991). See also Roger Wallis and Stanley Baran, *The Known World of Broadcast News: International News and the Electronic Media* (London: Routledge, 1990).

24. For a review of this literature, see Charles F. Herman, "Changing Course: When Governments Choose to Redirect Foreign Policy," *International Studies Quarterly* 34 (1990): 3–21.

25. See, for example, Robert Y. Shapiro and Benjamin I. Page, "Foreign Policy and the Rational Public," *Journal of Conflict Resolution* 32 (June 1988): 211–47.

26. See Bruce Bueno de Mesquita and David Laiman, "Domestic Opposition and Foreign War," *American Political Science Review* 84 (September 1990): 747–65.

27. Thomas Risse-Kappen, "Public Opinion, Domestic Structure, and Foreign Policy in Liberal Democracies," *World Politics* 43 (July 1991): 479–512; Bruce Russett, "Doves, Hawks, and U.S. Public Opinion," *Political Science Quarterly* 105 (Winter 1990–91): 516–38; Miroslav Nincic and Barbara Hinckley, "Foreign Policy and the Evaluation of Presidential Candidates," *Journal of Conflict Resolution* 35 (June 1991): 333–55.

28. This tendency has been reported widely in various studies, including Bernard C. Cohen, *The Press and Foreign Policy* (Princeton, N.J.: Princeton University Press, 1963); Gaye Tuchman, *Making News: A Study in the Construction of Reality* (New York: Free Press, 1978); Herbert Gans, *Deciding What's News* (New York: Pantheon, 1979). The classic study remains Leon Sigal, *Reporters and Officials* (Lexington, Mass.: D. C. Heath, 1973).

29. Cohen, *Press and Foreign Policy.*

30. Based on a study of national security news in seven major national papers reported by Daniel C. Hallin, Robert Karl Manoff, and Judy K. Weddle, "Sourcing Patterns of National Security Reporters," paper presented at the Annual Meeting of the American Political Science Association, San Francisco, August 30–September 2, 1990.

31. Timothy E. Cook, "Domesticating a Crisis: Washington Newsbeats, Human Interest Stories, and International News in the Persian Gulf War," paper presented at the Social Science Research Council Workshop on the Media and Foreign Policy, University of Washington, Seattle, September 26–29, 1991.

32. Larry Sabato, *Feeding Frenzy: How Attack Journalism Has Transformed American Politics* (New York: Free Press, 1991).

33. See W. Lance Bennett, *News: The Politics of Illusion,* 2d ed. (New York: Longman, 1988), pp. 46–51, 98–99, 137–38.

34. For a more complete discussion of "indexing," see Bennett, "Toward a Theory of Press-State Relations in the United States."

35. This generalization holds true even for domestic policy issues, as shown in Edie Goldenberg, *Making the Papers* (Lexington, Mass.: Heath-Lexington Books, 1975).

36. On Vietnam, see Gitlin, *Whole World Is Watching;* on the freeze movement, see Entman and Rojecki, "Freezing Out the Public."

37. Hallin, *Uncensored War.*

38. See William Dorman and Steven Livingston, "Historical Content and the News: Policy Consequences for the 1990–1991 Persian Gulf Crisis," paper presented at the Social Science Research Council Workshop on the Media and Foreign Policy, University of Washington, Seattle, September 26–29, 1991. See also Robert M. Entman and Benjamin I. Page, "The News before the Storm: The Limits to Media Autonomy in Covering Foreign Policy," also presented at the workshop.

39. For a more general discussion of strategic communications, see Jarol B. Manheim, *All of the People, All the Time: Strategic Communication and American Politics* (Armonk, N.Y.: M. E. Sharp, 1991).

40. See Mark Hertsgaard, *On Bended Knee: The Press and the Reagan Presidency* (New York: Farrar, Straus, Giroux, 1988).

41. See, for example, Shanto Iyengar, *Is Anyone Responsible? How Television Frames Political Issues* (Chicago: University of Chicago Press, 1992); John Zaller, "Political Awareness and Susceptibility to Elite Influence on Foreign Policy Issues." Paper presented at the Social Science Research Council Workshop on the Media and Foreign Policy, University of Washington, Seattle, September 26–29, 1991. See also Richard A. Brody, with Hollis Robbins, "Crisis, War, and Public Opinion: The Media and Support for the President in Two Phases of the Confrontation in the Persian Gulf," also presented at the workshop.

42. Iyengar, *Is Anyone Responsible?*

43. Entman and Page, "News before the Storm."

44. Zaller, "Political Awareness and Susceptibility," p. 1.

45. See W. Lance Bennett, "The Psychology of Mass-Mediated Publics," paper presented at the annual meeting of the International Society for the Study of Political Psychology, Washington, D.C., July 11–14, 1990.

46. The traditional view of rallies can be found in John Mueller, *War, Presidents and Public Opinion* (New York: Wiley, 1973). See also Samuel Kernell, "Explaining Presidential Popularity," *American Political Science Review* 72 (1978): 506–22.

47. Richard A. Brody and Catherine R. Shapiro, "A Reconsideration of the Rally Phenomenon in Public Opinion," in Samuel Long, ed., *Political Behavior Annual,* vol. 2 (Boulder, Colo.: Westview Press, 1989). See also Richard A. Brody and Catherine R. Shapiro, "Policy Failure and Public Support: The Iran-Contra Affair and Public Assessments of President Reagan," *Political Behavior* 11 (1989): 353–69.

48. Brody, with Robbins, "Crisis, War, and Public Opinion," p. 6.

49. Bernard C. Cohen, *The Public's Impact on Foreign Policy* (Boston: Little, Brown, 1973).

50. The historical evidence for elite-public policy convergence is demonstrated amply by Shapiro and Page, "Foreign Policy and the Rational Public."

51. For a case suggesting elite responsiveness to public opinion, see Larry M. Bartells, "Constituency Opinion and Congressional Policy Making: The Reagan Defense Buildup," *American Political Science Review* 85 (June 1991): 457–74. For a case suggesting public opinion is often marginalized in the policy process, see W. Lance Bennett, "Marginalizing the Majority: Conditioning Public Opinion to Accept Managerial Democracy," in Michael Margolis and Gary A. Mauser, eds. *Manipulating Public Opinion* (Belmont, CA: Brooks/Cole, 1989), pp. 321–361.

52. See, for example, Bennett, "Toward a Theory of Press-State Relations."

53. O'Heffernan, "Mass Media and U.S. Foreign Policy."

54. Patrick O'Heffernan, "What the Government Thinks? Post War Policy Makers' Perceptions of Media and Foreign Policy," paper presented at the Social Science Research Council Workshop on the Media and Foreign Policy, University of Washington, Seattle, September 26–29, 1991.

Part VI / Public Opinion

9 / Public Opinion and U.S. Foreign Policy Decision Making

THOMAS W. GRAHAM

INTRODUCTION

A revolution is taking place in the academic fields of international relations and American foreign policy. The still-pervasive, but now discredited, elitist paradigm in the field argued that public opinion is volatile or moody, unstructured and poorly informed, changed through a top-down process, and not particularly significant to decision making. Contemporary research, which is both more extensive and methodologically rigorous than work completed through the mid-1980s, challenges these assertions.

This chapter briefly reviews three schools of thought on the relationship between public opinion and foreign policy and emphasizes recent contributions made by scholars associated with an emerging new paradigm. Based on a review of data from more than five hundred national surveys and an examination of primary source documents spanning seven presidential administrations, this chapter develops a model that specifies the conditions under which public opinion influences U.S. foreign policy and national security decision making.

The discussion demonstrates that American public opinion can have a direct impact on foreign policy decision making and implementation. Linkages are complex, and the degree of influence is conditioned on four factors: the level of overall public support or opposition, the stage in the policy process being influenced, the degree to which competing elites and policy makers are aware of the multiple dimensions of mass opinion, and the degree to which political actors develop political communication strategies that use a vocabulary in tune with preexisting mass attitudes.

THREE SCHOOLS OF THOUGHT

Over the past seventy years, scholars who have made important contributions to the study of U.S. public opinion on international issues can be divided into three broad schools of thought.[1] The dominant *elitist paradigm* argues that public opinion is volatile or moody, public opinion on foreign issues is unstructured and poorly informed, mass attitude change is triggered by the attentive public through a top-down, elite-to-mass transmission process, and public opinion is not particularly significant to foreign policy or national security decision making. Influential scholars such as Gabriel Almond, Frank Klingberg, and Jack Holmes argue that Americans' attitudes on international issues are unstable and characterized by extreme shifts between moods of isolationism and internationalism.[2] Walter Lippmann, Philip Converse, and Martin Kriesberg have developed important elements in a view of public opinion as unstructured and poorly informed. Gabriel Almond and Walter Lippmann are largely responsible for development of the concept of an *attentive public*. While Almond did not attempt to quantify the size of the attentive public, since the publication of his influential work dozens of scholars have attempted to define the attentive public in operational terms, calculate its size, or analyze the character of attitudes held by its members.[3] All public opinion theorists believe an attentive public exists, but estimates of its size vary. (See Table 9-1.)

The lasting theoretical importance of the attentive public concept is its relationship with the top-down model of attitude change. This model was the product of innovative research that utilized a panel study during the 1948 presidential election.[4] Berelson, Lazarsfeld, and McPhee (1954) concluded that ideas often flow from the media to local opinion leaders and then diffuse from these leaders to the less-attentive sectors of the population. Even today this elite-driven image of mass opinion change is the conventional wisdom in the foreign policy establishment and has been used by the State Department for decades to sell ratification of various treaties such as the Panama Canal Treaty, the Strategic Arms Limitation Treaty (SALT II), and the Intermediate Nuclear Forces (INF) Treaty. Without realizing its intellectual origins, many foundations, foreign policy interest groups, and politicians implicitly use the top-down model to organize their (largely unsuccessful) public education campaigns.

The proposition that public opinion plays a limited role in influencing government foreign policy decisions was mentioned by Lippmann (1922) and Almond (1956), but its main proponent is Bernard Cohen (1973). Cohen argued that scholars who concluded public opinion had an impact on decisions such as the founding of the United Nations, recognition of the Soviet Union, creation of the Truman Doctrine, and nonrecognition of Communist China in the 1960s made "nonevidential assertions." Cohen strongly criticized the case study approach on methodological grounds and found fault with all the research completed up through 1973 which asserted that public opinion had an important impact on various foreign policy decisions. Since he discounted case studies, his research involved interviewing a sample of foreign service officers and mid-level foreign policy officials and asking them to describe the decision-making process. His study concluded that public opinion has little direct influence on almost all foreign policy decisions. Research by several historians, who mainly studied the Truman administration, also concluded that public opinion plays a minimal role in foreign policy decision making.[5]

Table 9-1 / Estimated Size of the Attentive Public (percent of the population)

Author	*Date*	*Type of Indicator*	*Size*
Free and Cantril	1967	Knowledge	26
Kriesberg	1949	Knowledge	25
Devine	1970	Behavior, interest, media exposure	25
Genco	1984	Interest, media exposure	22
NSB/NSF	1983	Interest, knowledge, media exposure	20
Cohen	1966	Knowledge	19
Marttila & Kiley	1985	Knowledge	18
Levering	1978	Knowledge	15
V.O. Key	1961	Interest	15
Rosi	1965	Behavior, knowledge	13
Rosenau	1961	Behavior, interest	10
Cohen	1966	Behavior, interest	9
SSRC	1947	Knowledge	8

Important research conducted in the 1970s and early 1980s has added to our understanding of public opinion on foreign affairs but did not revolutionize the field because it was conducted under the intellectual influence of the elitist paradigm. A mainstay of research associated with this second school asserts that the Vietnam War broke a preexisting attitude consensus in the area of foreign affairs that was in place from the late 1940s through the late 1960s.[6] This period of consensus was characterized by an operational agreement both on foreign policy goals and broad tactics among all three groups in society first identified by Almond: the opinion and policy elite, the attentive public, and the general public. During and after the war in Vietnam, this picture changed dramatically.[7] While the precise terms and classifications of post-Vietnam attitude clusters differ slightly among the half-dozen authors who support the breakdown in consensus theory, there is remarkable agreement on the basic idea: foreign policy leaders and the attentive public have been split into two or three mutually exclusive groups. According to William Schneider (1985), elites have fragmented into three groups: liberal interventionists, conservative interventionists and noninterventionists. Using slightly different definitions, Holsti and Rosenau (1984) state that elite attitudes have been split into three belief systems: cold war internationalists, post–cold war internationalists, and neo-isolationists. Russett and Hanson (1975) and Barton (1974–75) conclude that the pre-Vietnam consensus has been destroyed and anti-interventionist attitudes have increased among elite businessmen. While most theorists who support the fragmentation hypothesis concentrate their research and analysis on elite attitudes, William Schneider (1986) has concluded that the general public also has been split into three groups: noninterventionists (40–60 percent), liberal internationalists (~20 percent) and conservative internationalists (~20 percent).

A second area of research associated with the lingering influence school has focused on presidential leadership in changing public opinion. Samuel Kernell (1986) has argued that presidential leadership in the foreign policy area is strongly associated with the president's ability to "go public" to mobilize public opinion and thereby increase the chief executive's political power inside Washington. In the foreign policy area, this type of presidential opinion leadership is often equated with statesmanship. Benjamin Ginsberg (1986) reaches complementary conclusions that the persuasive power of presidents to lead public opinion is an important element that helps maintain the status quo.

An emerging new paradigm is beginning to be recognized in the field.[8] Its conclusions differ substantially from research associated with the earlier two schools.[9] The most basic difference is that public opinion on foreign policy and national security issues has been found to be relatively stable, not volatile or moody. Benjamin Page and Robert Shapiro reviewed thousands of survey items from 1935 through 1990 and concluded that "the American public, as a collective, holds a number of real, stable and sensible opinions about public policy and that these opinions develop and change in a reasonable fashion, responding to changing circumstances and to new information."[10] However, one conclusion by Shapiro and Page (1988) provides an explanation for the conventional wisdom about foreign affairs attitude instability. For the minority of foreign and domestic policy attitudes that changed *quickly* (20 percent of all repeated questions in their data set), foreign policy attitudes changed at three times the *rate* of domestic attitudes. Of these, not surprisingly, attitudes toward war changed the fastest. While no other research has demonstrated the scope and methodological sophistication of Shapiro and Page, other studies completed by Graham (1986, 1989a, 1989b), Kohut (1989), and Bardes and Oldendick (1985) also argue that attitudes on international affairs tend to be relatively stable, not volatile, over long periods of time.

Other research refutes the validity of the mood theory as posited by Klingberg (1952, 1983). Public opinion data show that isolationist attitudes disappeared in 1943 during World War II, not after the war as a result of a cold war mood shift. Except for three years immediately after the war in Vietnam, internationalist attitudes have been quite stable through the end of the cold war and on into 1992 during a presidential election year that was devoted almost entirely to domestic issues. Fifty years of attitude stability, well beyond the twenty–thirty-year cycle posited by mood theorists, has undermined credibility in the mood theory.[11] Recently, despite massive amounts of media attention and dramatic events associated with the Persian Gulf War, John Mueller (1992) has shown that many public attitudes were stable during operations Desert Shield and Desert Storm.[12]

Several authors, including Eugene Wittkopf (1990) and Ronald Hinckley (1992), have challenged the assessment that Americans' foreign policy attitudes are unstructured. Using different research approaches, they have concluded that from the 1970s through the 1990s, Americans' foreign policy attitudes have been structured on two dimensions: whether people prefer unilateral, multilateral, or isolationist policies and whether they have a proclivity to approve the use of military force. These belief systems have remained remarkably stable despite the end of the cold war, the Persian Gulf War, and the neo-isolationist 1992 campaigns of Republican presidential candidate Pat Buchanan and independent H. Ross Perot.

Detailed empirical research also challenges the idea that the public is not sufficiently informed to reach logical or stable attitudes on foreign affairs. A thorough examination of Americans' knowledge of foreign and nuclear issues reveals that while knowledge of specific terms or foreign policy jargon is low, functional knowledge can be relatively high, and awareness of many abstract issues can be extremely high.[13] Other scholars, using different research techniques, have concluded that the public is capable of following foreign affairs in sufficient detail to take it into account when making voting decisions.[14]

An array of scholars have called attention to the limits of the top-down attitude change model. Research completed after the model was refined in the 1950s has shown that the correlation between attitudes held by the attentive public and the general public is quite weak, making a causal linkage unlikely. Recent research on attitude change by several scholars emphasizes the diversified nature of the audience, the limited amount of communication between an attentive public and the general public, and the independent role played in attitude change by the mass media, especially television.[15]

Studies completed by Gamson and Modigliani (1966, 1971) have shown that attitudes among the general public have been relatively stable, but those of the attentive public have been variable. This research indicates that elites, who make up the attentive public, are the ultimate followers: Often their attitudes change only *after* government policy changes. If Gamson and Modigliani are correct, the attentive public does not provide any democratic check on government policy, as hypothesized by Lippmann.

A final criticism of the top-down attitude change model is that empirical evidence demonstrates that when the attentive public changes its attitudes, the general public does not always go along with the "new" orthodoxy. Also, the reverse is often the case: The general public can change its attitudes despite the fact that the elite holds a contrary set of opinions.[16]

A consensus is emerging that real world events, not diffusion of ideas from the attentive public or even from presidents "going public," have the most powerful influence on mass attitude change. Research in the 1970s concluded that mass attitudes changed in relation to real world events such as casualty rates during the wars in Korea and Vietnam.[17] Page, Shapiro, and Dempsey (1987) show that presidential opinion leadership (the ultimate top-down phenomenon) takes place only with "popular" presidents (defined as those with over 50 percent approval in job performance ratings), and its impact is relatively limited, changing opinion only 5–10 percent. Recent research has demonstrated that under "extreme" conditions, in which the United States was in the middle of a very popular war, President George Bush was able to shift opinion by only 20 percent.[18] This example may establish the unique conditions under which the top-down model functions and set an upper limit on the degree of top-down attitude change that can take place.

The strong relationship among real world events, mass attitude change, and the limited influence played by the attentive public can be seen in a historical review of public opinion data concerning the use of nuclear weapons. In this example, the public learned one lesson from real world events while the elites (the attentive public, the opinion and policy elite, and policy makers) learned a very different lesson. Since the mid-1950s, the American public has rejected the idea of using nuclear weapons unless these devices were used first by the Soviet Union. The Korean War experience—in which nuclear weapons were not used—and the growing awareness that the Soviet Union had the ability to retaliate with nuclear weapons convinced the public to become cautious regarding the use of nuclear weapons. At the same time this mass attitude was taking hold, both Republican and Democratic policy makers—supported by the attentive public—were implementing nuclear weapons strategies of massive retaliation and flexible response that placed the threat to use nuclear weapons at the center of U.S. defense and foreign policy. The fact that no top-down learning took place on this important issue over an extended period of time demonstrates the limits of the top-down model.[19]

Even the concept of a cold war attitude consensus is under challenge. While some scholars such as Wittkopf (1990) argue that an attitude consensus existed, other scholars challenge this assessment.[20]

The final component of the emerging new paradigm argues that public opinion has a significant impact on foreign policy and national security decision making. Since the mid-1980s, several scholars have refuted Cohen's no-impact hypothesis.[21] This rich body of research leaves one question unanswered: Has public opinion had an impact on policy making only recently in the 1980s, or has public opinion had an impact over a longer time period? As a new conventional wisdom emerges that public opinion has an impact on policy decisions, some scholars have concluded that this phenomenon has occurred only recently.[22] This chapter argues the contrary point that public opinion has had a significant impact on decision making for several decades, and it can be documented as far back as Franklin D. Roosevelt.[23]

A MODEL OF PUBLIC OPINION'S IMPACT

If four conditions are met, American public opinion will have a substantial direct or indirect influence on both the formation and the implementation of U.S. foreign and national security policy. The first factor that determines the degree of influence relates to the magnitude of public attitudes. Public attitudes held by less than 50 percent of the public rarely influence decision makers. Popular presidents can make foreign policy decisions that fly in the face of majority opinion (50–59 percent), but doing so provides their political opposition with fertile ground from which it can launch a counterattack. At the same time, majority-level public attitudes are strong enough, when combined with decisive presidential policy leadership, to enable a foreign policy decision to be made and implemented. Consensus level public opinion (60–69 percent) successfully influences the policy process even if powerful bureaucratic interests have to be overruled. Preponderant level public opinion (70 to 79 percent) not only "causes" the political system to act according to its dictates but also deters political opposition from challenging a specific decision. Nearly unanimous opinion (80+ percent) sweeps all political opposition away, dominating the entire political system so that decisions appear to be automatic. (See Table 9-2.)

In terms of foreign affairs decision making, we do not live in a world of majority rule. Public opinion can have a powerful and direct impact on decisions, but for this impact to take place, opinions have to be substantially larger than a majority. Why is democratic theory so far off? The reasons relate to inertia, resistance to taking mass opinion seriously by policy professionals, and uneven presidential understanding of or sympathy with the opinion-policy connection. With a few exceptions, relatively few senior foreign policy decision makers track public opinion in sufficient detail to be aware of its complex and multidimensional nature. As a result, primary source documents show that in most cases decision makers are aware of public opinion only when attitudes are at the consensus level or higher. If a substantial percentage of the public feel one way, there is an increased chance that this fact will reach decision makers and will be explicitly discussed when issues are decided at the highest level within an administration.

Bernard Cohen (1973) recorded hostile attitudes of mid-level foreign policy officials toward taking public opinion into account when making foreign policy decisions. While the level of explicit opposition to public opinion has declined over the years in the executive branch, a residue of professional opposition to giving public opinion a primary role in decision making still exists.[24] This inertia helps explain why public opinion must reach at least consensus levels (60 percent and higher) before it begins to have a discernible effect on decision making and acts as an "independent variable."

Historical documents reveal that all presidents since Franklin D. Roosevelt have followed public opinion polls and have used them to varying degrees of success in their decision making on foreign policy and national security issues. In all administrations, public opinion is taken into account more at the White House level than at the senior levels inside agencies such as the State Department or Defense Department.[25] However, few White House staff members make the administration's own public opinion data or analysis widely available to senior professional diplomats, military officers,

Table 9-2 / Levels of Public Opinion and Differentiated Impacts on Decision Making

Categories	*Percent*	*Impact*
Nearly unanimous	<79	Almost automatic
Preponderant	70–79	Substantial, deters opposition
Consensus	60–69	Important, may defeat strong bureaucratic opponents
Majority	50–59	Problematic, impact related to presidential policy leadership
Plurality	>50	Insignificant

or intelligence experts throughout the executive branch. Such information is considered extremely sensitive, from the political point of view, and usually is restricted to a select group of longtime political advisers who rarely interact with the foreign policy community. In the Roosevelt administration, even the existence of such polling was a closely guarded secret; it was discussed openly only three decades later by one of the participants, Hadley Cantril.[26]

Public opinion has been tracked and used by different administrations to widely varying degrees that fit no easy characterization in terms of party, ideology, or time period. Presidents have not gotten "better" at the opinion-policy connection over time, and the party usually associated with the people—the Democrats—has on balance done worse than its Republican competition at understanding and using public opinion in decision making. Table 9-3 summarizes this author's evaluation of the degree to which various administrations collected, analyzed, and successfully used public opinion data with respect to foreign affairs decision making. Two presidents have followed public opinion closely, understood the opinion-policy relationship, and used public opinion data heavily in their decision making. Four presidents have been moderately well informed on public attitudes but used public opinion only occasionally in their decision making and with varying degrees of success. Four presidents failed to understand the opinion-policy relationship, were poorly informed on the nature of mass opinion, and did not use it in their decision making on foreign policy and national security issues. (See Table 9-3.)

The second factor determining the extent of public opinion's influence on decisions is the stage in the foreign policy process: getting on the agenda, negotiation, ratification, or implementation. Public opinion has a direct impact during the first and third stages of the policy process (agenda and ratification). In large part because of controls over information, during the negotiation and implementation stages public opinion tends to have an indirect impact, if it has any at all.[27]

Table 9-3 / Presidential Administrations' Understanding and Use of Public Opinion in Foreign Policy Decision-Making

Administration	*Level of Presidential Understanding*	*Level of Successful Presidential Use*
Ronald Reagan	Extensive	Extensive
Franklin D. Roosevelt	Extensive	Extensive
Dwight Eisenhower	Moderate	Moderate
John F. Kennedy	Moderate	Moderate
George Bush	Moderate	Fair
Gerald Ford	Moderate	Fair
Jimmy Carter	Moderate	Poor
Richard Nixon	Fair	Poor
Lyndon Johnson	Fair	Poor
Harry Truman	Poor	Poor

The third factor in the opinion impact model is the ability of competing political elites and policy makers to comprehend the multidimensional nature of mass opinion that is necessary to discover the most politically relevant aspect of public opinion. Historically this third factor has been the single most important determinant explaining when public opinion has or has not influenced government decisions. This determinant is the hardest element of the model to document, however.

Despite the advent of public opinion polling in the mid-1930s, understanding the importance of the opinion-policy relationship has been difficult for many U.S. administrations.[28] First, it requires maintaining a staff trained in all aspects of survey research and funding regular polling dedicated in part to foreign affairs issues. Even more demanding, it also requires that an administration involve pollsters in the policy process early enough to establish the attitude environment before a presidential decision or world event occurs.

Bureaucratic factors and pressures to limit the "need to know" to a small group of national security policy experts exist both in administrations that mastered the opinion-policy process and in administrations that systematically ignored public opinion. Dramatic policy failures have usually occurred, however, when such understandable pressures have led to the exclusion of informed public opinion considerations from the policy-making process. For example, Jimmy Carter and Ronald Reagan both failed to take public opinion into account when making policy with Iran, and both paid a substantial political price for making this mistake.[29]

Public opinion is difficult to understand and to characterize in part because attitudes are multidimensional on any single policy issue.[30] Political salience is based on the relative degrees of mass support for different aspects of an issue. Making this determination is difficult because senior officials are constantly bombarded with independent polls and pseudoindicators of public opinion (letters, demonstrations, editorial opinions) through the media, contacts with members of the opinion and policy elite, or observations of political events. Ironically, large amounts of polling data can lead to a failure to understand the multidimensional nature of mass opinion if the material is not organized coherently.

Understanding public opinion also requires specialists to accurately evaluate the quality of their data and understand the true error margins inherent in survey research.[31] One can characterize public opinion on any specific dimension of a policy issue as falling into one of the categories detailed in Table 9-2 (majority, consensus, preponderant, nearly unanimous) if all polling from a variety of survey organizations using different question wording is reviewed. Contrary to an academic perspective, which places the highest value on trend data or on data collected by a select few organizations such as the University of Michigan, in politics determining the most salient dimension of public opinion requires an all-inclusive research strategy.

Several times throughout the last fifty years, a presidential administration that was conducting regular polling was able to use its superior knowledge of mass attitudes to organize and implement major foreign policy initiatives against organized opposition. In a competitive political environment, detailed knowledge and evaluation of public opinion has proved to be the key to success.

The final factor determining when public opinion influences policy is the real limitation on elite opinion leadership. Since even popular presidents can mobilize public opinion only on the margins, about 10–15 percent except in exceptional circumstances, effective political communications require using a vocabulary that is congruent with preexisting mass attitudes. This phenomenon puts any president in a difficult position because the expert foreign policy establishment, other politicians, and the media value the perceived ability of a leader to mobilize public opinion. Thus to provide leadership, the president needs to articulate themes that must be interpreted as "leading" public opinion when they usually are just reinforcing or calling attention to preexisting attitudes. Opinion leadership translates into bringing political attention to preexisting attitudes that support a leader's particular policy position. To use a sports analogy, political leaders tend to surf the public opinion wave; they rarely create the wave or significantly increase its magnitude.

Since public opinion is multidimensional on most issues, exerting presidential leadership requires that a White House staff conduct sufficient survey research in advance of quick moving events to give it the ability to determine how to frame foreign policy and national security issues to maximize positive public support. Relying simply on a strategy of "going public" will lead to disaster. While this element of the opinion-policy relationship may sound simple, it, too, has been quite difficult to sustain over the years.

EVIDENCE IN SUPPORT OF THE OPINION IMPACT MODEL

The four-stage model of public opinion's influence on decision making derives from research conducted primarily in the area of arms control and nuclear issues from 1945 through 1980. In most cases, the argument that public opinion was the key factor in a decision is based on primary source documents. On occasion, public opinion's impact is deduced from extensive secondary source scholarship.[32]

Policy Impact Proportionate to Level of Mass Support/Opposition

Substantial evidence has been obtained concerning the differentiated impact public opinion has on the policy process depending on the levels of public opinion. The best example where nearly unanimous opinion (80+ percent) drove a decision almost automatically was President Harry Truman's 1945 decision to drop the atomic bomb on Japan. Eighty-six percent of the public supported Truman's decision.[33] Historians who have analyzed every scrap of information on this issue have concluded that there was really no debate over whether the United States should drop the bomb against Japan despite the momentous nature of this decision.[34] This pattern seems to be typical of decision-making environments in which mass attitudes are so clear-cut. Since policy makers live in the same society in which the dominant attitudes are present, they either tend to share those same attitudes or are aware of them even without relying on public opinion polling. In these instances, the decision makers act almost automatically and rarely explicitly examine the pros and cons of their decision.

A second example that illustrates the policy impact of nearly unanimous opinion comes from President John Kennedy's decision to resume atmospheric nuclear testing after the Soviet Union violated the de facto testing moratorium in 1961. Eighty-nine percent of the public supported Kennedy's decision, and a review of the relevant primary source documents shows that public opinion was the primary reason this decision was made.[35]

Preponderant-level public opinion (70–79 percent) has a powerful but not an automatic impact on decisions. In the early 1980s, 70 percent of the American public thought that the United States should negotiate arms control agreements with the Soviet Union. When the Reagan administration took arms control off the agenda and began talking about fighting nuclear wars, the political systems in western Europe and the United States erupted. The nuclear freeze movement was born, and this grass-roots movement was the major force that pressured the administration to resume arms control negotiations (despite the opposition to this decision by major bureaucratic actors inside the government). Public opinion's impact did not stop there, however, illustrating a second aspect of the model. The effects of the Reagan administration's resuming negotiations with the Soviets, toning down Reagan's rhetoric about nuclear war, and conducting substantial survey research that discovered the absence of strong public support for specific elements of the nuclear freeze proposal, combined with the virtual ignorance of freeze movement leaders of the complex, multidimensional nature of public opinion and the failure of the freeze movement to frame its political communications in terms of preexisting mass attitudes, made it impossible for the movement to achieve its stated goals.

At the beginning of the nuclear age, preponderant-level public opinion was opposed to "sharing the secret" with the Soviet Union, an international atomic agency, or the United Nations. This level of opinion was responsible for the policy reversal associated with the transformation of the utopian Acheson-Lilienthal plan into the almost nonnegotiable Baruch plan.[36]

Rarely does one find that public attitudes are so clear-cut that they can be classified as nearly unanimous or preponderant. One can find many examples concerning foreign affairs when attitudes fall into the consensus range (60–69 percent). At this level, public opinion is sufficiently strong to override strong bureaucratic opposition (provided that the other three parts of the model also hold). The clearest case occurred in the Truman administration. An analysis of relevant public opinion data concerning international control of atomic energy from thirty-five national and four state surveys (131 questions) and a review of primary source documents conclude that public attitudes and diplomatic pressure from the British government forced an extremely reluctant Democratic administration to begin negotiations to control atomic weapons in 1946. With major bureaucratic actors strongly against international control and with secret U.S.-British wartime agreements precluding multilateral international nuclear cooperation, the main question was not why this negotiation ultimately failed but, rather, why and how it got onto the policy agenda. Public opinion, measured and reported to U.S. negotiator Bernard Baruch and other senior policy makers by confidential polling conducted for the State Department, provides an important part of the answer. Another example occurred in the Eisenhower administration when the president overruled his military and atomic energy advisers and pressed for negotiations of a bilateral nuclear test ban agreement with the Soviets.[37]

Franklin D. Roosevelt showed that consensus-level opinion was powerful enough to overcome strong congressional and interest group pressure associated with the isolationists to help him move the United States toward war before Pearl Harbor. By closely monitoring public opinion, Roosevelt was able to obtain support for increasing the military budget, win approval for reinstating the draft, and initiate the lend-lease program against concerted and organized political opposition.[38] This case is important because it reveals that public opinion influences presidential decisions primarily about tactics, timing, and political communications strategy rather than determining the ultimate goals of an administration's foreign policy. Roosevelt was intent on leading the United States into World War II.[39] He used public opinion data to help him implement this decision, not to reach his policy choice.

In the late 1950s, majority-level opinion (50–59 percent) opposed U.S. entry into a nuclear test moratorium with the Soviet Union. President Dwight Eisenhower was able to override this attitude and join the Soviets in a moratorium. Upon becoming president, John F. Kennedy continued this decision. This relatively weak level of public opposition was not sufficiently strong to force either president's hand, but it provided fertile ground for those who wanted the moratorium to be lifted. These critics of any possible movement toward a test ban were making substantial progress in forcing President Kennedy to reverse his decision when the Soviets broke the agreement in 1961. Their actions forced Kennedy to resume not only underground nuclear testing but atmospheric testing as well.[40]

Direct versus Indirect Public Impact

The second element of the model argues that public opinion affects getting on the agenda and ratification of an agreement directly, but negotiations and implementation only indirectly. Public opinion was responsible for getting the issues of international control of atomic energy, a test ban agreement, and the Strategic Arms Limitation Talks on the political agenda despite the opposition of various powerful bureaucratic actors and vested interests. Strong public support for ratification of the Limited Test Ban Treaty in 1963 (preponderant level) was essential to its becoming the first major arms control agreement negotiated between the superpowers.

Once the Limited Test Ban Treaty was signed in Moscow on August 5, 1963, it took the Senate less than two months to ratify the document by a vote of 80–19. Most accounts of the ratification process attribute President Kennedy's success to his ability to persuade elite actors, such as former President Eisenhower and the Joint Chiefs of Staff, to support the treaty. While these factors were important, to be complete this conclusion must account for the role played by public opinion. The preliminary political fight over Senate ratification of a test ban began well before the agreement was signed. Throughout 1963 congressional opponents of a comprehensive test ban (CTB) made sure that President Kennedy was aware that a CTB probably would be defeated in the Senate. On different occasions powerful Republican and Democratic members of the Joint Committee on Atomic Energy and the Senate Armed Services and Senate Foreign Relations Committees—such as Representative Craig Hosmer (D-Ca.), Senator Pastore (D.-R.I.), Senator Dodd (D-CT), Senator Jackson (D-WA), and Senator Gore (D-TN)—held hearings and criticized any attempt to negotiate a comprehensive test ban. In May, Senators Dodd and Humphrey (D-Minn) introduced a resolution recommending a limited test ban, and a private survey of senators showed that a comprehensive test ban would fall short, by ten votes, of that needed for ratification. The White House was aware of this potential vote, and when combined with Soviet refusal to compromise over verification provisions, the decision to pursue a limited agreement was finalized.

Even after Kennedy's dramatic speech given at American University and Averell Harriman's arrival in Moscow as a presidential envoy, it was not clear to participants that a limited agreement was feasible.[41] Successful negotiations required extremely tight control of negotiating instructions and reporting within the U.S. government, no consultation with members of Congress, and strong arguments by Secretary of Defense Robert McNamara and President Kennedy to the Joint Chiefs of Staff. After the agreement was initialed, the White House took great pains to gain congressional support. Interest groups were formed to fight for ratification of the agreement, and bipartisan senatorial opposition was expected.

The dynamics of the ratification fight were fundamentally changed when several public opinion polls showed that support for the treaty varied from the level of consensus to near-unanimity. It then became clear to senators who in the past had opposed President Kennedy on various foreign policy, weapons procurement, and arms control initiatives that there was no political benefit to fighting a life-and-death battle to stop ratification of the Limited Test Ban Treaty.[42]

Public opinion can also have a negative impact at the ratification stage. At the beginning of the policy process, the senior SALT decision-making community ignored both the political forces opposed to East-West arms control that were building outside of the bureaucracy and the warnings of public opinion experts about the limited mass support for specific arms control agreements. As a result, decision makers did not take public opinion into account during the negotiations. By the time preparations for ratification were made inside the administration, the effort was doomed. The public education strategy used by the State Department was based on the top-down, elite-led model of attitude change. It assumed that sending hundreds of executive branch officials around the country to give speeches to local "opinion leaders" would mobilize public support for SALT. The failed strategy also assumed that President Jimmy Carter and several cabinet secretaries could take a high public profile after the treaty was concluded, give numerous speeches for several months, and thereby win over the public to support SALT. The lack of strong public support was a central factor in the failure to ratify SALT II.

Public opinion's indirect role in foreign policy decision making can best be illustrated by two examples, one that occurred at the negotiating stage and one at the implementation stage. As noted earlier in this chapter, in 1946 the public was strongly in favor of U.S. negotiations with the Soviet Union about international control of atomic energy. A group of eminent scientists and diplomats had initiated the Acheson-Lilienthal plan, which was an idealistic effort to negotiate with the Soviet Union in good faith. They were opposed to the appointment of Bernard Baruch as the senior U.S.

negotiator because they understood that his strategy was to present the Soviet Union with a proposal the Soviets would find impossible to support.[43] Bernard Baruch's close watching of opinion polls and systematic monitoring of the media told him that strong public support for verifying any agreement (at the level of preponderant to nearly unanimous) would allow him to introduce draconian verification provisions into his proposal, stimulate Soviet rejection, but still sustain American public support for a virtually nonnegotiable diplomatic position. In this example, public opinion played an important indirect role in determining the process of the negotiations.

In a second example, forty years later, public opinion had a major indirect role during the implementation of the Antiballistic Missile (ABM) Treaty. The ABM treaty was generally on track to be implemented until Ronald Reagan proposed the Strategic Defense Initiative (SDI). Contrary to the impressions of arms control policy specialists and the opinion and policy elite, President Reagan's initiative was presented to an already very sympathetic mass audience. Reagan did not change public attitudes, but he isolated the dimension of the issue for which he would receive the strongest level of public support. By communicating his desire to build a perfect defense against nuclear weapons in understandable language, he mobilized political support for several administration policy positions: defeating the nuclear freeze movement, strengthening his bargaining position vis-à-vis the Soviet Union, "creating" a major strategic program out of a disconnected set of research projects, and distracting opponents' attention so that major components of his offensive nuclear buildup could pass Congress.[44] Ronald Reagan, supported by a professional team of survey researchers and communications experts, understood the opinion-policy relationship and used it to its fullest. His arms control opponents' ignorance of and disdain for public opinion made it impossible for them to develop a political strategy that was capable of challenging the president even though their criticisms of SDI were strong on technical and strategic grounds.

Decision Makers' Understanding of Mass Opinion

The third element of the model of public opinion influence relates to the degree to which competing elites are aware of the multidimensional nature of public opinion. One surprising finding of this research is the consistently sharp contrast between the two sides involved in a policy fight over arms control in terms of their knowledge of public opinion. Bernard Baruch had a far more sophisticated understanding of the public mood than his arms control opponents, and he used this understanding to frame the main issues in the policy and public debate in ways that strengthened his position. His close monitoring of public opinion, content analysis of the media, and timely reaction to the political activities of the advocates of arms control effectively ruled out any change in the U.S. negotiating position.[45]

In the case of the Limited Test Ban Treaty, political efforts by supporters in the 1950s failed due to ignorance about public opinion. Neither Adlai Stevenson, during the 1956 presidential campaign, nor antinuclear interest groups in the late 1950s could build support for a test ban agreement even by focusing on the controversy over radioactive fallout. By misunderstanding public opinion and pressuring the Eisenhower and Kennedy administrations to initiate or continue a test *moratorium* for which there was little public support, arms control advocates undermined their own positions. By helping to obscure strong public support for a bilateral test ban *agreement*, they were indirectly responsible for producing the circumstances that forced President Kennedy to resume testing in the atmosphere and then to conclude a limited, rather than a comprehensive, test ban treaty. President Kennedy learned of strong public support for a bilateral test ban treaty in 1963 only after he made the key decision to negotiate a limited agreement. By that time, his political options were severely constrained.

In a competitive political system, it is possible for public opinion to have a limited impact on policy if both sides of a debate are poorly informed of the true nature of mass attitudes. This situation occurred during the 1969 ABM debate. Limited grass-roots activity around the country but concentrated television news coverage and interest group pressure successfully created the (false) impression that "the public" was against deployment of the ABM.[46] Against the backdrop of the war in Vietnam and increasingly critical public attitudes toward military spending, it is understandable that many congressional leaders, executive branch officials, and political activists incorrectly perceived public opinion as opposed to the ABM. In this case, an incorrect understanding of public opinion helped limit deployment of a full ABM system, but it left the door open for Ronald Reagan's SDI initiative, discussed above.

During the negotiation and ratification of SALT II in the 1970s, advocates of the agreement failed to understand either the nature of public opinion or the relationship between public opinion and the policy process. At first glance, the conclusion that public opinion played an extremely limited role concerning the negotiation of SALT II seems surprising. First, during the time of its negotiation and unsuccessful ratification, more survey research was conducted on this arms control subject than had been obtained about any other arms control or nuclear issue in the atomic age.[47] Unlike earlier arms control efforts, virtually all of this polling was made public, and much of it was extremely sophisticated in methodology and question wording. Thus, at relatively low cost in terms of staff time, policy makers and participants in this arms control debate could have been relatively well informed on public attitudes about strategic arms control. Second, the negotiating process for SALT II was under the firm bureaucratic and intellectual control of highly trained professional legal and policy experts in the executive branch who, unlike Bernard Baruch, were strongly in favor of reaching a serious agreement with the Soviets. Third, by the time SALT II was negotiated, an office in the public affairs bureau of the State Department regularly monitored public opinion and reported its findings to senior decision makers.[48] Fourth, during the Carter administration, presidential pollster Pat Caddell conducted his own surveys, which included questions on SALT. President Carter not only read these surveys but also believed them to be accurate. In light of all these factors, why did public opinion fail to affect decisions made at the negotiation stage?

The answer lies in the perceptions of elite policy makers about the nature and importance of public opinion. Virtually all the participants understood public opinion in simplistic terms and considered it relevant only during ratification. As a result, limited staff resources were devoted to tracking, understanding, and integrating public opinion into the decision-making process.

A failure to understand public opinion occurred at the beginning of the Carter administration when a dispute over interpreting survey data shaped policy makers' understanding of the nature of public opinion toward SALT II. By reviewing one type of survey question, researchers such as Pat Caddell, Louis Harris, and Dan Yankelovich believed that strong public support for arms control negotiations, recorded since 1972, meant that the public would support the SALT II agreement. By reviewing another type of survey question, other public opinion specialists such as Bernard Roshco and Al Richmand (in the State Department) and Bud Roper (president of the Roper Organization) explicitly rejected this interpretation and pointed to polls showing that only a plurality favored SALT II. They believed that support for arms control negotiations and support for the SALT II agreement were two distinct dimensions of public opinion.

Rarely are these abstract debates answered conclusively. In this case, however, one survey organization asked both types of questions on several surveys. One question asked about public support for negotiating strategic arms control in general (i.e., the format used by Pat Caddell, Louis Harris, and Dan Yankelovich). A second question asked specifically about the SALT II agreement. The first type of question was asked from January 1978 through September 1979. Both questions were asked five times throughout 1979.[49] The results showed that generic support for arms control negotiations remained stable at the level of 70 percent throughout the SALT II era. Support for the SALT II agreement, however, was much lower and declined over time. This survey research demonstrated that judgments concerning public opinion made by mid-level public opinion specialists in the State Department, not nationally known experts like Pat Caddell, Louis Harris, and Dan Yankelovich, were correct. The public generally supports arms control negotiations at the level of preponderance, but they do not necessarily support a specific arms control agreement at the same level. Thus, positive interpretation of early polling data showing support for arms control, which was strengthened by Pat Caddell's interpretation, reinforced a proclivity inside the expert SALT II decision-making community to focus on negotiating details and to take public opinion seriously only much later when the agreement was near completion.

At a more abstract level, policy makers in the Carter administration were utilizing the traditional elitist paradigm of public opinion. At the beginning of the administration, many believed that public opinion did not and should not play an important role in the process of negotiating strategic arms limitation with the Soviet Union. According to this perspective, good policy would be made by negotiating a "solid" agreement that did not incorporate ambiguous language, which was the problem with key provisions of SALT I. These officials believed that ambiguous language was responsible for the political controversy over Soviet verification, which in turn undermined public support for arms control.

By relying on the elitist paradigm, SALT II decision makers made several strategic errors. Since administration officials were unaware that mass attitudes are not greatly influenced by attitudes of the opinion and policy elite and that only popular presidents (over 50 percent in job approval rating) can lead public opinion to a limited extent (i.e., by 10–15 percent), there was no discussion concerning the optimal time when to conclude an agreement. As a result, the negotiations proceeded at their own pace. At the time SALT II was signed, Jimmy Carter was well below the 50-percent job approval threshold, so he could not effectively mobilize mass support for SALT II.

The political communication strategy adopted for selling SALT seems to have been developed independently of relevant findings from survey research. Many of the assertions made about public opinion and attitude change in the key planning memorandum written by presidential aide Jerry Rafshoon to Jimmy Carter were inaccurate[50]. In addition, the content of speeches by senior officials such as SALT negotiator Paul Warnke and Secretary of State Cyrus Vance were filled with arms control jargon and concepts and thus made sense only to the small group of people who can be considered part of the opinion and policy elite. For all of these reasons, the political communication strategy for SALT II was destined to influence a very small group in the population. Evidence from survey research shows that approximately 6 percent of the public heard the detail-dominated message and supported the logic of SALT II.

A final example that illustrates the importance of knowing the true nature of public opinion comes from decision making associated with the 1983 Grenada invasion. A typical interpretation of the event has been written by Samuel Kernell, who did not know that early polling results obtained by Richard Wirthlin revealed widespread opposition to the invasion. These results did not change after Ronald Reagan's speech one day after the invasion. Polling results shot up, however, a few days later when the returning students were seen on local and national television news kissing the ground and explaining their belief that the rescue was necessary. This example demonstrates the importance of having accurate data on public opinion.

Reagan political opponents who expressed initial disagreement with the invasion were effectively silenced by Reagan's speech. Had they been aware that the public was equally unsure of the wisdom of the invasion, they could have continued their public opposition.

This example also shows the danger of scholarship that is based almost entirely on a limited number of polls and secondary source documents and includes no pretest survey data before a key event or presidential speech. Scholars who have made these mistakes have exaggerated the importance of opinion leadership and failed to understand the role played by public opinion in the policy process.

Political Communications Compatible with Pre-Existing Attitudes

The fourth and final element of the model that helps to explain the conditions under which public opinion affects policy relates to the integration of public opinion with a political communication strategy. The best example comes from the Eisenhower administration. Contrary to his image, Eisenhower was quite well versed on political issues, public opinion data, and the strengths and weaknesses of survey research.[51] Eisenhower was informed about public attitudes toward a possible test ban agreement between the United States and the Soviet Union before the Democratic presidential candidate Adlai Stevenson raised it as an issue in the 1956 election. Eisenhower was aware that there was public support for a bilateral test ban agreement with strict verification provisions, but there was opposition to the United States' unilaterally stopping testing. Eisenhower thus used Adlai Stevenson's own rhetoric, which called for unilateral U.S. action, to stop testing to kill the Democrats test ban proposal. Stevenson not only failed to track public opinion but was also his own worst enemy in proclaiming that the United States should unilaterally stop nuclear testing.

CONCLUSION

Over the last fifty years public opinion has played an important, but not adequately recognized, role in U.S. foreign policy decision making. If public opinion is sufficiently strong, it can make decisions virtually automatic. Consensus-level mass attitudes can have more impact on decisions than powerful bureaucratic players. However, the public's impact on decision making is not automatic, and the linkages are quite complex. The new emerging paradigm outlined in this chapter specifies the conditions under which public opinion has an impact on policy formation and implementation. In the final analysis, in the competitive American political system, the group that takes public opinion seriously, is well informed about its complex nature, and integrates public attitudes into the decision-making process is more likely to win than is a group that is still operating within the intellectual parameters of the old elitist paradigm.

Notes

1. This chapter develops a theory that applies exclusively to the United States. While some of the concepts may be appropriate to develop hypotheses regarding the impact public opinion might have on decision making in other Western democracies, this author believes that the structure of the American political system has so many unique features that it is unlikely the theory outlined here will be directly applicable to other countries.

2. For a more extensive review of the literature, see Graham 1989b, Russett and Graham 1989, and Russett 1990.

3. Several approaches have been taken to define the attentive public. Caspary 1970, Cohen 1966, Erskine 1963, Free and Cantril 1967, Graham 1988, Kriesberg 1949, National Science Board (NSB)/National Science Foundation (NSF) 1983, and Rosi 1965 used knowledge-based indicators. Galtung 1969, Rosi 1965, and Smith 1961 used socioeconomic indicators. Cohen 1966, Devine 1970, Hughes 1978, Rosenau 1974, and Rosi 1965 used behavioral indicators. Devine 1970, Genco 1984, and NSB/NSF 1983 used exposure indicators. Caspary 1970, Cohen 1966, Deutsch and Merritt 1965, Key 1961, Genco 1984, Mandelbaum and Schneider 1979, NSB/NSF 1983, Rosenau 1961, and Rosi 1965 used interest indicators.

4. A panel study involves interviewing the same group of people more than once to determine changes in their attitudes. Most public opinion surveys use a random sample for each survey and thus do not interview the same people over time. This procedure makes it much more difficult to prove conclusively that attitude change has taken place over time.

5. La Feber 1977; May 1964; Paterson 1979.

6. Various authors use different dates to define the period of the cold war attitude consensus. Levering 1978 uses the dates 1946–68; Holsti and Rosenau 1984 do not mention a beginning date but use the mid-1960s as an ending date; Schneider 1985 uses the dates 1948–68.

7. Kohut 1989.

8. This emerging new paradigm was first noted by Graham in his contribution to Russett and Graham 1989 and in Graham 1989b. A panel of senior professors at the 1991 Annual Meeting of the American Political Science Association—which included Ole Holsti, Joseph Nye, Benjamin Page, and Bruce Russett—reached a large degree of consensus on the outline of this new school of thought.

9. Innovative research conducted in the 1960s and 1970s that supports the new emerging paradigm includes Caspary 1970; Gamson and Modigliani 1966; Modigliani 1972; and Mueller 1979 and 1973. Unfortunately, their work did not cause scholars to reexamine fundamental assumptions of the traditional elitist paradigm until quite recently.

10. See Chapter 10 of this volume for an expansion on these research findings. For supporting evidence, see Shapiro and Page 1988; Page and Shapiro 1983, 1992; Jentleson 1992.

11. Benson 1982; Cantril 1948; Caspary 1970; Foster 1983; Graham 1989a; Hinckley 1992; Watts and Free 1978.

12. One reviewer of this chapter commented that the theory outlined here might be time bound because it draws on data associated with a potentially unique period in history—the cold war. This argument would be persuasive if one could demonstrate either that basic American attitudes toward foreign affairs changed after the end of the cold war or that policy makers began acting differently toward public opinion. Survey data reported by Hinckley 1992 provide persuasive evidence that there has been no fundamental change in American attitudes despite the fact that the cold war has been over for several years. Evaluation of the second criticism of the model will have to await scholarship on the Bush and Clinton administrations. However, evaluation of the relationship between public opinion and decision making during Desert Shield and Desert Storm described in Gergen 1992 do not reveal any substantial change in the pattern of decision making discussed in this chapter.

13. Graham 1988 and 1989b.

14. See Aldrich, Sullivan, and Borgida 1989; Popkin 1991.

15. See Fan 1988; Neuman 1986; and Robinson 1976. For a more recent discussion of the media, see Chapter 8 in this volume.

16. For a discussion of this point, see Graham 1989a.
17. Mueller 1973, 1979.
18. Hinckley 1992, pp. 21–22.
19. Graham 1989a.
20. See Wittkopf 1990, chap. 6; Hinckley 1992, chap. 2.
21. See Graham 1989b; Powlick 1991; Beal and Hinckley 1984; Kuznitz 1984; Hinckley 1992; Jacobs 1993.
22. See Bueno De Mesquita, Jackman, and Siverson 1991; Hinckley 1992, p. 4.
23. Drawing on the work of Lawrence Jacobs, Shapiro and Page's contribution to this volume (Chapter 10) argues that public opinion polling has become more institutionalized in the White House since the Kennedy administration. While this may be true, public opinion had a significant impact on policy making under the Roosevelt and Eisenhower administrations, even if the flow of public opinion data to those leaders was less organized and less systematic than during and after the Kennedy administration.
24. Powlick 1991.
25. Bernard Cohen's survey of mid-level foreign policy specialists concentrated on professionals in various agencies, not in the White House. As a result, his description presented an incomplete view of the policy-making process.
26. Franklin Roosevelt conducted approximately a dozen public opinion polls prior to Japan's bombing of Pearl Harbor, and he used these results to develop the concept of lend-lease and move the United States into World War II.
27. The two-level game theory outlined by Robert Putnam 1988 does not include the agenda-setting and implementation stages of the foreign policy process. This omission limits its power and relevance to the full range of foreign policy decisions.
28. Insights on this point came from primary source documents, Beal and Hinckley 1984; Hinckley 1992; and a dinner meeting of presidential pollsters and senior polling specialists sponsored by the Kennedy School of Government in Washington, D.C. The dinner was attended by Albert H. Cantril, Richard Wirthlin, Bud Roper, Robert Teeter, Harry O'Neill, and Pat Caddell.
29. The Carter administration's unwillingness to take public opinion into account during its decision-making regarding the hostages in Iran was described to this author by the senior National Security Council (NSC) staffer who worked on this issue, Dr. Gary Sick. The failure of the Reagan administration to consult public opinion with respect to Iran before trading arms for hostages was described by NSC staffer Dr. Ronald Hinckley.
30. This conclusion is fully documented for nuclear and arms control issues in Graham 1989b.
31. Typically, survey research conducted using a random sample of approximately 1,500 people has a statistical margin of error of 3 percent. However, the "true" margin of error is substantially higher due to question wording, question order, and other effects.
32. More complete discussion and documentation of the findings summarized in this chapter are included in Graham 1989b.
33. Graham 1989a, pp. 7, 82–84. Documenting the policy impact of nearly unanimous opinion is complicated by the fact that most such polling is conducted *after* a dramatic event has occurred. Since the public tends to support a president after he takes a dramatic international initiative (i.e., the "rally round the flag" effect), one might question whether post facto data in the 80 percent range can sustain the argument. Upon reflection, this author has concluded that the Japan example is valid even though the exact figure of 85 percent may have been lower if a poll had been conducted before the bombing. Several factors are involved. First, other data showed strong American public support for massive conventional bombing of Japan and even for using chemical weapons to force Japanese surrender. Second, the empirical evidence upon which the concept of a rally effect is based is the job performance rating of a president, an unusually volatile indicator of public opinion. As documented by Page and Shapiro, attitudes on substantive foreign affairs issues (as opposed to evaluations of president's performance) are rather stable over time. For these reasons, one can posit that opinion on using nuclear weapons against Japan would have been at the nearly unanimous level if a poll had been conducted prior to the event.
34. Sherwin 1977; Bernstein 1976.
35. Graham 1989b, chap. 5.
36. Graham 1989b, chap. 4.

37. Graham 1989b, p. 170.

38. For a review of relevant public opinion data, see Cantril and Strunk 1951. For a discussion of Franklin Roosevelt's use of polling results, see Cantril 1967; Converse 1987.

39. Burns 1970.

40. Graham 1989b, chap. 5.

41. Seaborg and Loeb 1981, pp. 229–53; Terchek 1970, pp. 22–23. The conventional accounts of the Limited Test Ban Treaty are derived from Jacobson and Stein 1966; Lepper 1971.

42. Twenty-three senators consistently voted against the Kennedy administration on related issues. Thus only a dozen additional votes would have been needed to defeat the Limited Test Ban Treaty. See Terchek 1970, pp. 204–6.

43. A quotation from David E. Lilienthal illustrates this point: "He [Oppenheimer] says that they [Baruch and his staff] are enthusiastic about proceeding right away with negotiations and proposals, but have no hope of agreement. They talk about preparing the American people for a refusal by Russia." Lilienthal 1965, p. 43.

44. For a detailed review of the political and arms control implications of SDI, see Nolan 1989.

45. Baruch's private papers are explicit on his knowledge of public opinion, his negotiating strategy, and the fact that neither President Harry Truman nor Secretary of State Dean Acheson understood exactly what he was doing or why he was doing it. See Graham 1989b, p. 107.

46. Scholars have made the same mistake. The original work on agenda setting, by Cobb and Elder 1972, examined the ABM case and made fundamental errors in both fact and judgment as a result of misreading public opinion polls.

47. One hundred twelve surveys were conducted by eleven different organizations from 1972 to 1987 on SALT II and strategic arms control.

48. United States Department of State 1980, 1979a, 1979b, 1979c, 1979d, 1978a, 1978b, 1978c, 1977a, 1977b, 1976.

49. Eleven survey questions that asked about public support for arms control negotiations were asked by NBC/AP from January 1978 until September 1979. During five such surveys conducted from March 1979 through October 1979, an additional question was added that asked explicitly about support for SALT II.

50. Rafshoon, Jerry. "Development of Public Support for SALT II," December 6, 1978. From Carter Presidential Papers, Staff Offices, Assistant for Communications, Subject Files, SALT [5] folder. Jimmy Carter Presidential Library, Atlanta, Georgia.

51. The president received detailed briefings on public opinion from his brother Milton Eisenhower, who was one of the nation's experts on survey research and its relevance to political communications.

References and Other Sources

Aldrich, John H., John L. Sullivan, and Eugene Borgida. "Foreign Affairs and Issue Voting: Do Presidential Candidates 'Waltz before a Blind Audience?' " *American Political Science Review* 83, no. 1 (March 1989): 123–41.

Almond, Gabriel. *American People and Foreign Policy.* New York: Praeger, [1950] 1960.

______. "Public Opinion and National Security Policy," *Public Opinion Quarterly* 22, no. 2 (Summer 1956): 371–378.

Bardes, Barbara A., and Robert W. Oldendick. "The Attitudes of American Elites on Foreign Policy." Paper presented at the Fortieth American Association of Public Opinion Research meeting, McAfee, N.J. May 16–19, 1985.

Barton, Allen H. "Consensus and Conflict among America's Leaders." *Public Opinion Quarterly* 38, no. 4 (Winter 1974–75): 507–30.

Baruch Papers (Atomic Energy) Princeton University Library, Princeton, N.J.: Acheson, Dean; Barnes, James; Lindsay, Frank; Memoranda, Misc. 1946; Notes on Meetings, 1946; Press Reactions, 1946–1947; Truman, Harry.

Beal, Richard S., and Ronald H. Hinckley. "Presidential Decision Making and Opinon Polls." *Annals of the American Academy of Social Science: Polling and the Democratic Consensus,* no. 472 (March 1984): 72–84.

Benson, John M. "The Polls: U.S. Military Intervention." *Public Opinion Quarterly* 46, no. 4 (Winter 1982): 592–98.

Berelson, Bernard R., Paul F. Lazarsfeld, and William N. McPhee. *Voting*. Chicago: University of Chicago Press, 1954.

Bernstein, Barton. J., ed. *The Atomic Bomb: The Critical Issues*. Boston: Little, Brown, 1976.

Bueno de Mesquita, Bruce J., Robert W. Jackman, and Randalph M. Siverson. "Democracy and Foreign Policy: Community and Constraint." *Journal of Conflict Resolution* 35, no. 2 (June 1991): 181–86.

Burns, James MacGregor. *Roosevelt, 1940–1945: The Soldier of Freedom*. New York: Harvarst/HBJ, 1970.

Cantril, Hadley. *The Human Dimension: Experiences in Policy Research*. New Brunswick, N.J.: Rutgers University Press, 1967.

———. "Opinion Trends in World War II: Some Guides to Interpretation." *Public Opinion Quarterly* 12, no. 1 (Spring 1948): 30–44.

Cantril, Hadley, and Mildred B. Strunk, eds. *Public Opinion, 1935–1946*. Princeton, N.J.: Princeton University Press, 1951.

Caspary, William. "The 'Mood' Theory: A Study of Public Opinion and Foreign Policy." *American Political Science Review* 64, no. 2 (June 1970): 536–47.

Cobb, Roger W., and Charles D. Elder. *Participation in American Politics: The Dynamics of Agenda-Building*. Baltimore, Md.: Johns Hopkins University Press, 1972.

Cohen, Bernard C. *The Public's Impact on Foreign Policy*. Boston: Little, Brown, 1973.

———. "The Military Policy Public." *Public Opinion Quarterly* 30, no. 2 (Summer 1966): 200–211.

Converse, Jeane M. *Survey Research in the United States: Roots and Emergence, 1890–1960*. Berkeley: University of California Press, 1987.

Converse, Philip E. "The Nature of Belief Systems in Mass Politics." In D. E. Apter, ed., *Ideology and Discontent*. New York: Free Press, 1964.

Deutsch, Karl W., and Richard L. Merritt. "Effects of Events on National and International Images." In Herbert C. Kelman, ed., *International Behavior*. New York: Holt, Rinehart and Winston, 1965.

Devine, Donald J. *The Attentive Public: Polarchial Democracy*. Chicago: Rand McNally, 1970.

Erskine, Hazel Gaudet. "The Polls: Textbook Knowledge." *Public Opinion Quarterly* 27, no. 1 (Spring 1963): 133–41.

Fan, David P. *Predictions of Public Opinion from the Mass Media Content: Computer Content Analysis and Mathematical Modeling*. New York: Greenwood Press, 1988.

Foster, H. Schuyler. *Activism Replaces Isolationism: U.S. Public Attitudes, 1940–1975*. Washington, D.C.: Foxhall Press, 1983.

Free, Lloyd A., and Hadley Cantril. *The Political Beliefs of Americans*. New Brunswick, N.J.: Rutgers University Press, 1967.

Galtung, Johan. "Foreign Policy Opinion as a Function of Social Position." In James N. Rosenau, ed., *International Politics and Foreign Policy: A Reader in Research and Theory*. New York: Free Press, 1969.

Gamson, William A., and Andre Modigliani. *Untangling the Cold War*. Boston: Little, Brown, 1971.

———. "Knowledge and Foreign Policy Options: Some Models for Consideration." *Public Opinion Quarterly* 30, no. 2 (Summer 1966): 187–199.

Genco, Stephen James, "The Attentive Public and American Foreign Policy." Ph.D. diss., Stanford University, 1984.

Gergen, David. "Domestic Management of the Persian Gulf Crisis: The Unfettered Presidency." In Aspen Strategy Group, *After the Storm: Lessons from the Gulf War* (Lanham, Md.: Madison Books, 1992).

Ginsberg, Benjamin. *The Captive Public: How Mass Opinion Promotes State Power*. New York: Basic Books, 1986.

Graham, Thomas W. *American Public Opinion on NATO, Extended Deterrence and Use of Nuclear Weapons: Future Fission?* (Lanham, Md.: University Press of America, 1989a.

———. "The Politics of Failure: Strategic Nuclear Arms Control, Public Opinion and

Domestic Politics in the United States, 1945–1980." Ph.D. diss., Massachusetts Institute of Technology, 1989b.

______. "The Pattern and Importance of Knowledge in the Nuclear Age." *Journal of Conflict Resolution* 32, no. 2 (June 1988): 319–34.

______. *Public Attitudes toward Active Defense: ABM & Star Wars, 1945–1985*. Cambridge, Mass.: Center for International Studies, MIT, 1986.

Hinckley, Ronald H. *People, Polls, and Policymakers: American Public Opinion and National Security*. New York: Lexington Books, 1992.

Holmes, Jack E. *The Mood/Interest Theory of American Foreign Policy* (University Press of Kentucky, 1985

Holsti, Ole R., and James N. Rosenau. "The Structure of Foreign Policy Attitudes among American Leaders." *Journal of Politics* 52, no. 1 (February 1990): 94–125.

______. *American Leadership in World Affairs: Vietnam and the Breakdown of Consensus*. Boston: Allen and Unwin, 1984.

Hughes, Barry B. *The Domestic Context of American Foreign Policy* San Francisco: W. H. Freeman, 1978.

Jacobs, Lawrence R. *Public Opinion and the Making of American and British Health Policy* (Ithaca, N.Y.: Cornell University Press, 1993.

Jacobson, Harold, and Eric Stein. *Diplomats, Scientists, and Politicians: The United States and the Nuclear Test Ban Negotiations*. Ann Arbor, Mich.: University of Michigan Press, 1966.

Jentleson, Bruce W. "The Pretty Prudent Public: Post Post-Vietnam American Opinion on the Use of Military Force." *International Studies Quarterly* 36, no. 1 (March 1992): 49–74.

Kernell, Samuel. *Going Public: New Strategies of Presidential Leadership*. Washington, D.C.: Congressional Quarterly, 1986.

Key, V. O. *Public Opinion and American Democracy*. New York: Alfred A. Knopf, 1961.

Klingberg, Frank L. *Cyclical Trends in American Foreign Policy Moods: The Unfolding of America's World Role*. Lanham, Md.: University Press of America, 1983.

______. "The Historical Alternations of Moods in American Foreign Policy." *World Politics* 4, no. 2 (January 1952): 239–73.

Kohut, Andrew. "Stability and Change in Opinions about Nuclear Weapons Policy, 1945–1989." In Aspen Strategy Group, *Deep Cuts and the Future of Nuclear Deterrence*. Lanham, Md.: University Press of America, 1989.

Kriesberg, Martin. "Dark Areas of Ignorance." In Lester Markel, ed., *Public Opinion and Foreign Policy*. New York: Harper, 1949.

Kusnitz, Leonard. *Public Opinion and Foreign Policy: America's China Policy, 1949–1979*. Westport, Conn.: Greenwood Press, 1984.

La Feber, Walter. "American Policy-Makers, Public Opinion and the Outbreak of the Cold War, 1945–1950." In Yonosuke Nagai and Akira Iriye, eds., *The Origins of the Cold War in Asia*. Tokyo: University of Tokyo Press, 1977.

Lepper, Mary Milling. *Foreign Policy Formulation: A Case Study of the Nuclear Test Ban Treaty of 1963*. Columbus, Ohio: Charles E. Merrill, 1971.

Levering, Ralph B. *The Public and American Foreign Policy, 1918–1978*. New York: William Morrow, 1978.

Lilienthal, David E. *Journals of David E. Lilienthal: The Atomic Energy Years, 1945–1950*. Vol. 2. New York: Harper and Row, 1965.

Lippmann, Walter. *Public Opinion*. New York: Free Press, [1922] 1965.

Mandelbaum, Michael, and William Schneider. "The New Internationalisms: Public Opinion and American Foreign Policy." In Kenneth A., Oye, Donald Rothchild, and Robert J. Lieber, eds., *Eagle Entangled: U.S. Foreign Policy in A Complex World*. New York: Longman, 1979.

Marttila, John, and Tom Kiley. *A Survey of American Attitudes about Nuclear Weapons and Arms Control*. Boston: Marttila and Kiley, 1985.

May, Ernest R. "An American Tradition in Foreign Policy: The Role of Public Opinion." In William H. Nelson, ed., *Theory and Practice in American Politics*. Chicago: University of Chicago Press, 1964.

Modigliani, Andre. "Hawks and Doves, Isolationism and Political District: An Analysis of Public Opinion on Military Policy." *American Political Science Review* 66, no. 2 (June 1972): 960–78.

Mueller, John E. "American Opinion and the Gulf War: Trends and Historical Comparisons." Paper presented at Political Consequences of War Workshop sponsored by the National Elections Studies, Center for Political Studies, and Brookings Institution, February 11, 1992.

———. "Public Expectations of War during the Cold War." *American Journal of Political Science* 23, no. 2 (May 1979): 301–29.

———. *War, Presidents, and Public Opinion.* New York: John Wiley and Sons, 1973.

National Science Board/National Science Foundation. *Science Indicators 1982.* Washington, D.C.: Government Printing Office, 1983.

Neuman, W. Russell. *The Paradox of Mass Politics: Knowledge and Opinion in the American Electorate.* Cambridge, Mass.: Harvard University Press, 1986.

Nolan, Janne E. *Guardians of the Arsenal: The Politics of Nuclear Strategy.* New York: Basic Books, 1989.

Page, Benjamin I., and Robert Y. Shapiro. *The Rational Public: Fifty Years of Trends in Americans' Policy Preferences.* Chicago: University of Chicago Press, 1992.

———. "Effects of Public Opinion on Policy." *American Political Science Review* 77, no. 1 (March 1983): 175–90.

Page, Benjamin I., Robert Y. Shapiro, and Glenn R. Dempsey. "What Moves Public Opinion?" *American Political Science Review* 81, no. 1 (March 1987): 23–43.

Paterson, Thomas G. "Presidential Foreign Policy, Public Opinion, and Congress: The Truman Years." *Diplomatic History* 3, no. 1 (Winter 1979): 1–18.

Popkin, Samuel L. *The Reasoning Voter: Communication and Persuasion in Presidential Campaigns.* Chicago: University of Chicago Press, 1991.

Powlick, Philip J. "The Attitudinal Bases for Responsiveness to Public Opinion among American Foreign Policy Officials." *Journal of Conflict Resolution* 35, no. 4 (December 1991): 611–41.

Putnam, Robert D. "Diplomacy and Domestic Politics: The Logic of Two-Level Games." *International Organization* 42, no. 3 (Summer 1988): 427–60.

Rafshoon, Jerry. "Development of Public Support for SALT II." December 6, 1978, Carter Presidential Papers, Atlanta, Georgia, Staff Offices, Assistant for Communications, Subject Files, SALT [5] folder.

Robinson, John P. "Interpersonal Influence in Election Campaigns: Two Step-Flow Hypothesis." *Public Opinion Quarterly* 40, no. 3 (Fall 1976): 304–19.

Rosenau, James N. *Citizenship between Elections.* New York: Free Press, 1974.

———. *Public Opinion and Foreign Policy: An Operational Formulation.* New York: Random House, 1961.

Rosenau, James N., and Ole R. Holsti. "U.S. Leadership in a Shrinking World: The Breakdown of Consensus and the Emergence of Conflicting Belief Systems." *World Politics* 35, no. 3 (April 1983): 368–91.

Rosi, Eugene J. "Elite Political Communication: Five Washington Columnists on Nuclear Weapons Testing, 1954–1958." *Social Research* 34, no. 4 (Winter 1967): 703–27.

———. "Mass and Attentive Opinion on Nuclear Weapons Tests and Fallout, 1954–1963." *Public Opinion Quarterly,* 29, no. 2 (September 1965): 280–97.

Russett, Bruce. *Controlling the Sword: The Democratic Governance of National Security.* Cambridge, Mass.: Harvard University Press, 1990.

Russett, Bruce, and Thomas W. Graham. "Public Opinion and National Security Policy: Relationships and Impacts." In Manus Midlarsky, ed., *Handbook of War Studies.* Winchester, Mass.: Unwin Hyman, 1989.

Russett, Bruce, and Elizabeth C. Hanson. *Interest and Ideology: The Foreign Policy Beliefs of American Businessmen.* San Francisco: W. H. Freemen, 1975.

Schneider, William. " 'Rambo' and Reality: Having it Both Ways." In Kenneth Oye, Robert Lieber, and Donald Rothchild, eds., *Eagle Resurgent? The Reagan Era in American Foreign Policy.* Boston: Little, Brown, 1986.

———. "Peace and Strength: American Public Opinion on National Security." In Gregory Flynn and Hans Rattinger, eds., *The Public and Atlantic Defense.* Totowa, N.J.: Rowman and Allanheld, 1985.

Seaborg, Glenn T., and Benjamin S. Loeb. *Stemming the Tide: Arms Control in the Johnson Years.* Lexington, Mass.: Lexington Books, 1987.

———. *Kennedy, Khrushchev and the Test Ban.* Berkeley: University of California Press, 1981.

Shapiro, Robert Y., and Benjamin I. Page. "Foreign Policy and the Rational Public." *Journal of Conflict Resolution* 32, no. 2 (June 1988): 211–47.

Sherwin, Martin. *A World Destroyed: The Atomic Bomb and the Grand Alliance*. New York: Vintage, 1977.

Smith, Paul A., "Opinions, Publics, and World Affairs in the United States." *Western Political Quarterly* 14, no. 3 (September 1961): 698–714.

Social Science Research Council, Committee on Social and Economic Aspects of Atomic Energy. *Public Reaction to the Atomic Bomb and World Affairs: A Nation-Wide Survey of Attitudes and Information*. Ithaca, N.Y.: Cornell University, 1947.

Terchek, Ronald T. *The Making of the Test Ban Treaty*. The Hague: Martinus Nijhoff, 1970.

United States Department of State. "Polish Crisis Portends Drop in SALT Support." Briefing Memorandum from Public Affairs Bureau to Mr. Newsom, December 11, 1980.

______. "Public Slow to Decide on SALT: Opinions Divided Sharply along Party Lines." Memorandum from Public Affairs to Mr. Nimetz, November 23, 1979a.

______. "Two Polls Show Declining Support for SALT." Memorandum from Public Affairs to Mr. Nimitz, August 8, 1979b.

______. "Public Opinion on SALT: Still in for Formative Stage." Memorandum from Public Affairs to General Signious and Mr. Nimetz, June 1, 1979c.

______. "A Review of the SALT Poll Released by the Committee on the Present Danger." Memorandum from Public Affairs to General Signious and Mr. Nimetz, April 6, 1979d.

______. "Roper vs. NBC: Two SALT Polls are Poles Apart." Memorandum from Public Affairs to General Signious and Mr. Nimetz, December 19, 1978a.

______. "Public Attitudes toward SALT: A Summary of the Poll Data." Memorandum from Public Affairs to Mr. Warnke, Mr. Nimitz, and Ambassador Shulman, September 19, 1978b.

______. "Relationship between Attitudes toward SALT Negotiations and Attitudes toward SALT." Memorandum from Public Affairs to Ms. Morton, March 1, 1978c.

______. "Summary of Caddell Poll Findings." Memorandum from Hodding Carter III to the Secretary, April 22, 1977a.

______. "Public Opinion and Impending Negotiations." Memorandum from Public Affairs to Mr. Lake, February 14, 1977b.

______. "Public Attitudes toward Negotiating with the Soviet Union." Memorandum from Public Affairs to the Secretary, October 27, 1976.

______. *A Report on the International Control of Atomic Energy, Prepared for the Secretary of State's Committee on Atomic Energy by a Board of Consultants*. Publication 2498 Washington, D.C.: Government Printing Office, 1946.

Watts, William, and Lloyd A. Free. *State of a Nation III*. Lexington, Mass.: Lexington Books, 1978.

Wittkopf, Eugene R. *Faces of Internationalism: Public Opinion and American Foreign Policy*. Durham, N.C.: Duke University Press, 1990.

10 / Foreign Policy and Public Opinion

ROBERT Y. SHAPIRO
BENJAMIN I. PAGE

INTRODUCTION

George F. Kennan expressed great concern about the "erratic and subjective nature of public reaction" in foreign affairs, that is, the widely held view that public opinion is volatile, ill informed, and inadequate to deal with the important and complex issues of foreign policy. It would seem to follow from this view that, given the tremendous and complicated changes that have occurred recently in world politics—including a stunning war in the Middle East, the restructuring of Germany and eastern Europe, and (most striking and unexpected of all) the collapse of the former Soviet Union—the public would be in worse shape than ever, since it faces a transformed world without simplifying cold war, East versus West principles to guide it. But we disagree with this pessimistic depiction of the American public. It is not supported by the evidence—neither by recent trends in public opinion nor by trends going back through five or six decades.

This chapter argues that such a cynical view of public opinion has never been correct, at least not during the fifty or sixty years for which systematic survey data are available. True, the American public must now wrestle with the implications of tremendous changes on the world scene. But there is no cause for alarm; the public is quite capable of the task, so long as its leaders and the nation's media of mass communications do their part. That the American public has this capacity is fortunate, because (as we will see in the second section below) a growing body of evidence indicates that public opinion has quite a substantial impact on the making of U.S. foreign policy.

THE RATIONAL PUBLIC

For several years we have been analyzing a large body of data, drawn from many thousands of questions about Americans' policy preferences that were asked in hundreds of different opinion surveys between the 1930s and 1991. We have paid particular attention to *trends* in *collective* public opinion, that is, to changes in the percentages of Americans supporting one policy or another, as indicated by their answers to questions that were repeated in identical form. Many of our results are reported in our recently published book, *The Rational Public*.[1]

As the title of the book indicates, our main theme is that the American public, collectively, is "rational," in a certain specific sense of that term. Despite the well-known instability and changeability of individuals' survey responses, for example, collective public opinion is real, consisting of something more than "nonattitudes" (i.e., people knowing little or nothing and giving random responses to survey interviewers); it is accurately measurable by survey questions; and it is highly stable. Despite the general absence of elaborate political belief systems among most Americans (at least the absence of the sort of widely shared, uniform belief systems that elites are used to and that would show up easily in survey research), collective public opinion is highly differentiated, patterned, coherent, consistent, and reflective of values that endure over long periods of time and seem to be deeply held. Despite the well-documented paucity of political information among most individual citizens, collective public opinion responds to new information and to objective changes in the world. It responds in ways that are regular, predictable, and indeed generally sensible, given the values that citizens hold and the information that is made available to them.

There are several reasons why collective public opinion is much more "rational" (in this sense) than it is often thought to be. Some reasons involve statistics and measurement. When we add up the responses of many individuals, for example, random measurement errors and erratic responses by particular individuals tend to cancel each other out, leaving more solid, accurate, stable measures of collective public opinion. Similarly, when we examine trends over time in collective responses to identically repeated survey questions, we are not distracted by shifting responses to differently worded questions. (Most such "question-wording effects," we believe, actually result from sensible public reactions to questions with different meanings.) Other reasons have to do with what we call "collective deliberation." New information and new ideas are produced and disseminated through society in ways that allow even citizens who lack specific political information, but who receive cues from trusted opinion leaders, to take all available information into account in adjusting their opinions.

Be that as it may, these generalizations about the rational public apply to opinions concerning foreign policy as well as domestic issues. To the extent that there exist some differences between the two, the differences do not reflect at all badly on public opinion. And increasing world interdependence, together with the heightened salience of foreign economic and environmental rather than military issues, probably means that the differences are decreasing.

Before turning to the question of what effects public opinion has on government policy, we will illustrate some of these points about the nature of public opinion, taking account of new data.

Differentiation and Coherence in Collective Public Opinion

At any particular moment in time, public opinion about foreign policy tends to be *differentiated,* that is, the American public makes sharp distinctions among policies, favoring some and opposing others. Moreover, these distinctions tend to be coherent and consistent with each other: they fall into regular patterns that make sense and that fit with an overall system of values.

Differentiation and distinctions in opinion are most obvious when a series of different questions are asked with exactly the same format, so that the meaning varies in clearly identifiable ways. Year after year during the cold war, for example, the Roper Organization asked about "selling arms and weapons" to a list of different countries. And year after year, the proportions of Americans thinking the United States should sell arms varied markedly from one country to another. A high proportion (about 70 percent during the 1980s, excluding "don't know" responses, as is our usual convention) always approved of selling arms to England, a significantly lower proportion (about 50 percent) approved of selling them to West Germany, still lower (about 40 percent) to Japan, Greece, South Korea, and Turkey, and much lower still (less than 10 percent) to Iran. (See Figure 10-1.)

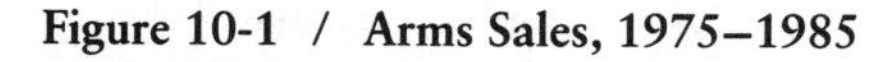

Figure 10-1 / Arms Sales, 1975–1985

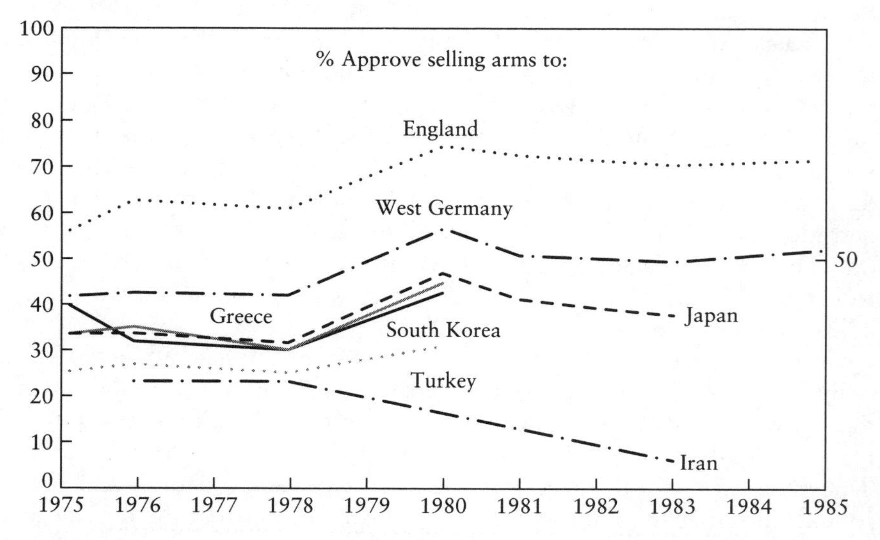

Question (Roper): "You may have differing opinions about selling arms and weapons to certain specific countries. Here is a list of some different countries. Would you go down the list, and for each one tell me whether you think the United States should or should not sell them arms? . . . England, . . . West Germany, . . . Turkey, . . . Greece, . . . South Korea, . . . Japan, . . . Iran?" Includes volunteered responses of having mixed feeling about. Survey dates: 8/75, 11/76, 3/78, 8/80, 8/81, 8/83, 1985.

Our point is not that this ranking of countries necessarily reflects what we would consider an ideal foreign policy, but rather that the public makes distinctions which are coherent. These distinctions obviously reflect varying degrees of closeness in military, economic, and cultural relations between the United States and other countries. Moreover, as the parallel lines in the graph indicate, these same distinctions endured rather steadily over time, with the understandable exception of the drop in relative support for arms sales to Iran after the overthrow of the shah and the taking of American hostages in Teheran.

During the same period the public made similar distinctions among various hypothetical uses of military force. Year after year, from the mid-1970s to the mid-1980s, large majorities (in the 70–80 percent range) told Gallup that the United States should come to the defense of "major European allies" with military force if necessary "if any of them are attacked by the Soviet Union." In each survey about 10 percent fewer said a similar thing about Japan. And Roper Organization questions about "the use of U.S. troops" elicited fairly high levels of support (in the 50–60 percent range) "if the Soviet Union invaded western Europe," or the Soviets invaded West Berlin, or Cuban troops were involved in a takeover of a Central American country, but much lower support (closer to 30 percent) for defending Yugoslavia or Poland from a hypothetical Soviet invasion.

Similar kinds of distinctions among policies can be found throughout our data. In 1973, for example, the Harris organization found a very high (73 percent) level of support for military involvement and the use of U.S. troops if there were danger of a communist takeover of neighboring Canada and nearly as much (66 percent and 60 percent, respectively) in the cases of the Panama Canal zone and England, but markedly less in the event of a communist threat to western Europe, Australia, or West Berlin (46–49 percent), and still less support for military involvement in the event of danger of a communist takeover of Brazil (40 percent), Japan or Israel (37 percent), Taiwan (33 percent), Greece (32 percent), South Korea (30 percent), Thailand (28 percent), or India (27 percent). There was somewhat lower support for the use of military force in such hypothetical circumstances in 1973 than there had been in 1969 and 1970, but the ranking of countries remained quite similar—save for a sharp post-Vietnam drop in enthusiasm for resisting communism in Asia.

Such clear distinctions concern different kinds of programs as well as different countries or regions. Take the case of aid to El Salvador to combat leftist guerrillas during the early 1980s. Questions by Roper and by ABC/*Washington Post* revealed virtually the same distinctions in a number of surveys: about half the public favored keeping "military advisors" in El Salvador to "help train" government troops; about 10 percent fewer favored giving increased economic aid; fewer than that favored using U.S. air and naval force to stop Cuban supply ships and planes; and still fewer favored giving increased military aid. Only about 20–30 percent wanted to "send U.S. troops" even if this were "the only way to prevent the government of El Salvador from being overthrown by the leftist guerillas." (See Figure 10-2.)

A strong aversion to using U.S. troops and a preference for negotiated settlements, arms control, and cooperative relations run through decades of public opinion data. The American public is willing to fight when it perceives a clear threat to U.S. interests but is very reluctant to do so unless there is no alternative.

Figure 10-2 / Aid to El Salvador, 1982–1985

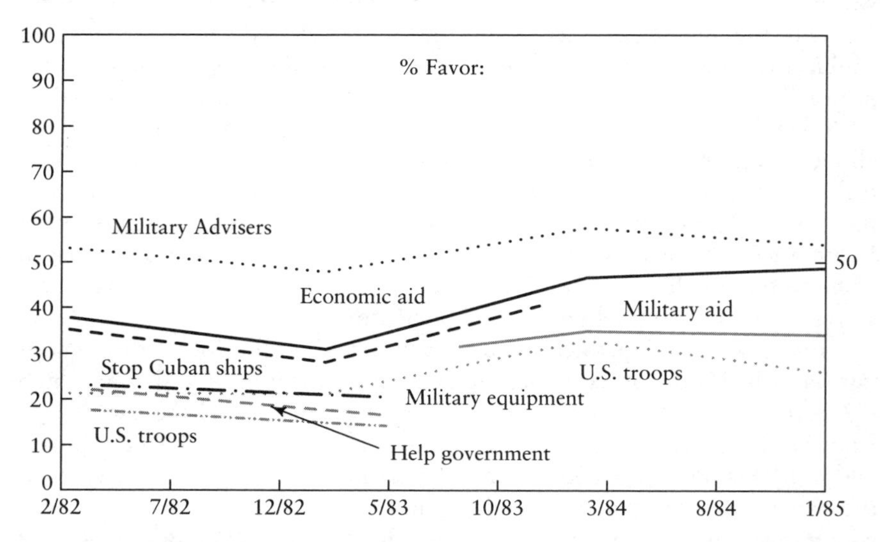

Question (ABC/*Washington Post*): "Would you approve or disapprove of the United States sending troops to fight in El Salvador?" Survey dates: 3/82, 5/83. *Question* (Roper): ". . . Here is a list of possible steps the U.S. could take to protect its interests in El Salvador and other parts of Central America and the Caribbean. [Card shown respondent.] For each one, would you tell me if you favor it strongly, generally favor it, generally oppose it, or oppose it strongly? First, . . . keep U.S. military advisors in El Salvador to help train the El Salvador government troops in their fight against leftist guerrillas there." Survey dates: 2/82, 2/83, 9/83, 2/84, 1/85. *Question* (ABC/*Washington Post*): "The Reagan administration wants to send an increased amount of military equipment and weapons to the government of El Salvador. Do you approve or disapprove of that?" Survey dates: 3/82, 5/83. *Question* (ABC/*Washington Post*): "Which side should the United States help: the government in El Salvador, the rebels in El Salvador, or should the United States stay out of the situation?" Survey dates: 3/82, 5/83. *Question* (Roper): (Same as previous Roper question on El Salvador) ". . . send increased military aid to El Salvador to help its government fight against the leftist guerrillas?" Survey dates: 9/83, 2/84, 1/85. *Question* (Roper): (Same) ". . . use U.S. air and naval forces to stop Cuban ships and planes carrying military supplies to the leftist guerrillas in El Salvador?" Survey dates: 2/82, 2/83, 2/84. *Question* (Roper): (Same) ". . . send increased economic aid to El Salvador to help its government cope with the damage caused by the conflict with the leftist guerrillas?" Survey dates: 2/82, 2/83, 2/84, 1/85. *Question* (Roper): (Same) ". . . send U.S. troops to El Salvador if this is the only way to prevent the government of El Salvador from being overthrown by the leftist guerrillas?" Survey dates: 2/82, 2/83, 2/84, 1/85.

Many other cases of clear distinctions in the public mind could be added. Before World War II in 1937, for example, a 59 percent majority said that (if forced to choose) they would rather live under the kind of government in (Nazi) Germany than under the kind in (communist) Russia; yet, presumably for strategic reasons, 82 percent in 1938 said they would rather see Russia than Germany win a hypothetical war between the two. During World War II, there was always 15–20 percent more support for negotiating peace with the German army (assuming the army had overthrown Adolf Hitler) than there was for discussing peace terms with Hitler himself. In repeated surveys during the 1950s and 1960s, when few Americans said Communist China "should" be admitted as a member of the United Nations, many more (often 30 percent more) still said the United States should "go along" with the U.N. decision if a majority of members decided to admit China. In later years, large majorities (over 80 percent) of the public said they wanted "friendlier relations" with the People's Republic of China and approved of President Richard Nixon's visit to that country; somewhat fewer favored official recognition; only about half wanted to "strengthen" U.S. ties with China (when asked amid a list of other countries); but only 10–20 percent wanted to "end" the U.S. "defensive alliance" with Taiwan. Thomas W. Graham describes in Chapter 9 the case of public opinion toward the use of nuclear weapons by the United States.

Again, our point is not that the public necessarily favors policies that are correct or optimal, but rather that the public makes clear and coherent distinctions about which policies it favors and which it opposes.

This capacity to make distinctions is also evident in the transformed world of the 1990s. The patterns that have occurred in the past in the case of foreign aid are clearly evident in the new era of foreign policy. A September 1991 NBC News/*Wall Street Journal* survey, for example, reported substantial support for economic aid (". . . help that country with economic aid?") to the Soviet Union (64 percent) and Poland (65 percent); in contrast, support for such aid to Israel and Egypt was 15–20 percent lower (49 percent and 44 percent, respectively).

The Stability of Opinion

Collective public opinion concerning foreign policy is not only differentiated and structured; it is generally stable.

Many years ago William Caspary threw a great deal of doubt on the "mood theory" of public opinion (the notion that the public lurches into interventionist moods and then "snaps back" to isolationism) by showing that a high, stable proportion of the public thought the United States should take an "active part" in world affairs, with little change during the tumultuous years from the early 1940s to the middle 1950s.[2] Our data on this question, including slight corrections of Caspary's, are shown in Figure 10-3.

Figure 10-3 / Opinion Stability: Foreign Policy Activism, 1942–1956

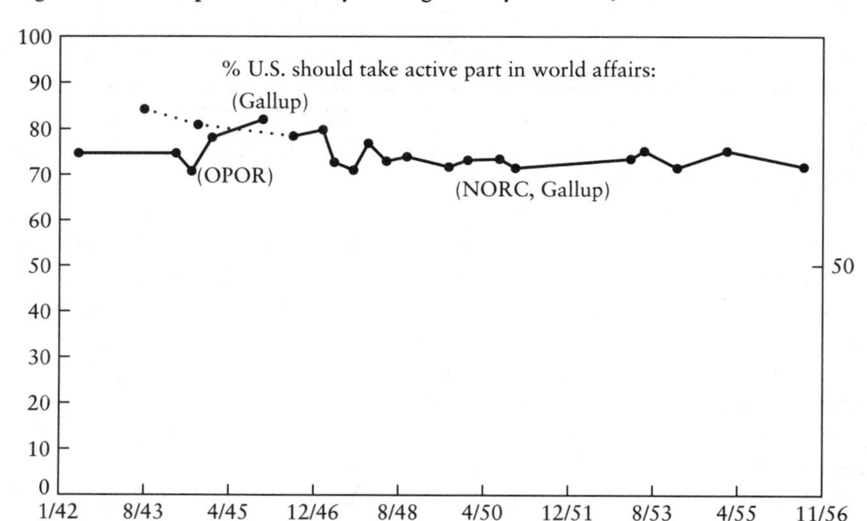

Question (Gallup, NORC): "Do you think it will be best for the future of this country if we take an active part in world affairs or if we stay out of world affairs?" Survey dates: (Gallup) 10/45, 2/46, 11/46, 8/47, 11/50; (NORC) 3/47, 6/47, 3/48, 6/48, 9/49, 1/50, 12/50, 10/52, 2/53, 9/53, 4/54, 3/55, 11/56. *Question* (Gallup): Reported as the above (*Public Opinion Quarterly*, "The Quarterly Polls"). Survey dates: 3/43, 5/44; *Question* (OPOR): "Which of these two things do you think the United States should do when the war is over—stay out of world affairs (as much as we can) or take an active part in world affairs?" Survey dates: 1/42, 5/43, 11/43, 4/44, 6/44, 3/45.

Our study of many hundreds of responses to survey questions on other foreign policy issues has also generally revealed a substantial degree of opinion stability. Considering 215 survey questions about foreign and defense policy that were repeated at least once, for example, there was no significant opinion change at all (that is, no change in collective responses of 6 percent or more), on 51 percent of them. When significant changes do occur, they tend to be quite small. Forty-three percent of the significant opinion changes on foreign and defense issues involved only 6–9 percentage points of change; only 12 percent of them fell in the 20–29 percent range, and only a tiny 3 percent involved changes of 30 percentage points or more.[3]

To be sure, opinions changed even less frequently on domestic issues, where 63 percent of our items showed no significant change at all. Moreover, foreign policy opinions tend to change more abruptly than domestic opinions, in response to sudden international events. More than half (58 percent) of our foreign policy opinion changes, as opposed to only 27 percent of domestic changes, occurred at a rate of 10 percentage points per year. (Most of these abrupt foreign policy opinion changes involved rather small percentage changes over very brief time intervals.) As discussed further below, however, abrupt opinion changes in response to changing world events are not necessarily erratic or irrational. Indeed, a public that failed to take new realities into account could hardly be called rational.

In any case, stability is the rule for foreign as well as domestic issues. When opinion changes do occur, many do so quite gradually. This conclusion should be evident from the figures already presented. It is apparent in many other examples as well.

One striking example involves the M-X multiple-warhead, intercontinental ballistic missile.[4] The questions about production and deployment of the M-X that came up in the early and middle 1980s are just the sort of complicated, technical issues about which we would expect ordinary citizens to have "nonattitudes." Indeed, at the beginning of 1983, when policy debates were fairly far advanced, Roper found that 55 percent of Americans admitted they knew "very little" about the M-X; only 7 percent claimed they knew "a lot." Repeated measurements of individuals' stated opinions (if such repeated measurements existed) would probably show a lot of switching back and forth in more or less random fashion. Yet such randomness canceled out in the whole population: *collective* public opinion about the M-X was quite stable. In May–June 1985, for example, Harris found that 59 percent of the public thought that Congress, which favored cutting back on the missile, was "more right" than President Ronald Reagan, who opposed cutbacks; a July survey came up with an identical 59 percent agreement with Congress. (See Table 10-1.) A year earlier, about the same results occurred in response to a different Harris question that asked whether people favored or opposed various bills that had been passed by the House of Representatives, with Speaker Thomas P. ("Tip") O'Neill (D-Mass.) taking a leadership position: 64 percent of those with opinions said they favored delaying spending of funds for the future development of the M-X until April 1985.

Opinion stability is even more obvious in the American public's attitude toward foreign aid in general, which—whether we like it or not—has long been highly unpopular. In nearly three dozen surveys between 1971 and 1991 by Roper and the National Opinion Research Center's General Social Survey (NORC-GSS), only about 5 percent of Americans have said "we" were spending "too little" on foreign aid, while a steady—and much larger—70–80 percent said we were spending "too much." In order to put this response into the context of Americans' overall spending priorities, and to illustrate once again the public's capacity to distinguish among policies (in addition to its preferences toward which countries should receive different types of assistance), we have graphed the tiny but steady proportion saying that "too little" is spent on foreign aid along with the much higher "too little" proportion on fighting crime, the high and rising proportion on education (roughly matched by support for spending on the environment and medical care, not displayed), the medium support for highways and mass transportation, and the low support for spending on the space program. (See Figure 10-4.)

Table 10-1 / Collective Opinion Stability on the M-X Missile Issue

Question: "Recently, President Reagan has had some serious disagreements with Congress. Now who do you think was more right—Reagan or Congress—in their differences over . . . cutting back on the M-X missile, which is favored by Congress and opposed by Reagan?"

	May–June 1985	*July 1985*
Reagan	41%	41%
Congress	59	59

Source: Harris surveys. "Not sure" responses (omitted in this table for clarity about the balance of opinion among those with opinions) were 7 percent in the first survey and 6 percent in the second. "Both" or "neither" responses were less than 0.5 percent each time.

There are a great many other examples of opinion stability and gradual change in addition to those shown in the above figures. The recent uproar over trade and protectionism, for example, can be placed in perspective. The American public has long expressed ambivalence about barriers to trade, favoring the concept of "free trade" (and low-priced imports) in general but also advocating tariffs or quotas to protect vulnerable U.S. industries and workers, especially during economic hard times.

In 1946, for example, 70 percent told NORC that they opposed "letting goods come into this country which would sell for less than our goods," and 71 percent remained opposed in 1952. On the other hand, throughout the 1950s, Gallup polls found majorities or pluralities (among those "informed" about tariffs) favoring lower rather than higher tariffs: 56 percent for lower tariffs in 1953, for example (with only 23 percent for higher tariffs). This proportion favoring higher tariffs rose by 13 percent by late 1961, with most of the rise occurring by 1959, perhaps in response to increased competition from rebuilt European and Japanese industry and to the sluggishness of the U.S. economy after the 1958 recession. Through the 1970s and 1980s large majorities favored some measure of protectionism: a Roper poll, for example, found 70 percent in late 1972 saying the United States should "place restrictions" on imports. This proportion held steady in 1975 and 1977 and then rose modestly to 76 percent in the difficult year 1979. Likewise, 1973 and 1978 Harris surveys found 63 percent and 65 percent, respectively, favoring more restrictions.

Harris and other surveys have, however, found substantial public support for free trade when the question-wording framed the issue by stressing

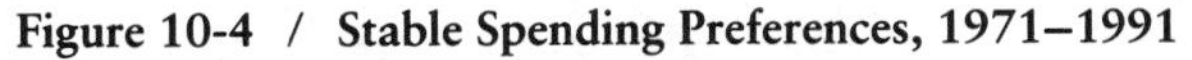

Figure 10-4 / Stable Spending Preferences, 1971–1991

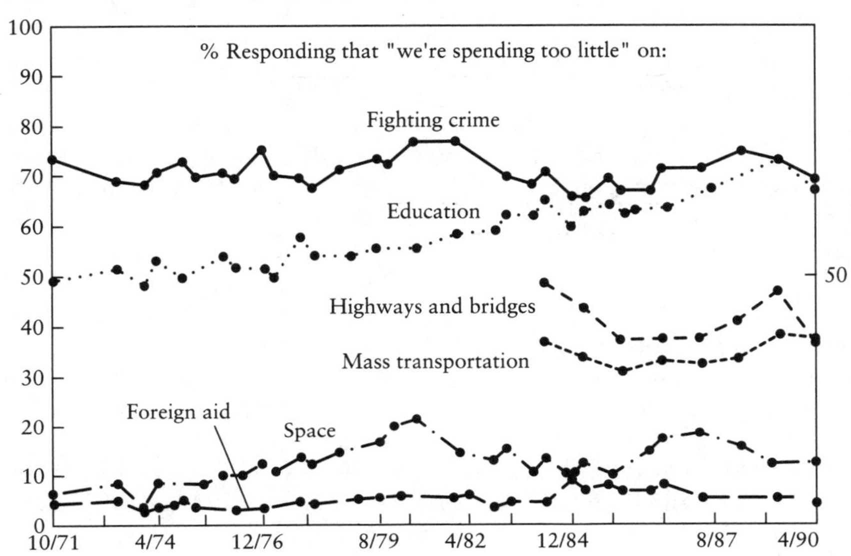

Question (Roper, NORC-[GSS]): "We are faced with many problems in this country, none of which can be solved easily or inexpensively. I'm going to name some of these problems, and for each one I'd like you to tell me whether you think we're spending too much money on it, too little money, or about the right amount, . . . Are we spending too much, too little, or about the right amount on . . . Halting the rising crime rate? . . . Improving the nation's education system? . . . Mass transportation? . . . Highways and bridges? . . . Space exploration program? . . . Foreign aid?" Survey dates (Roper): 10/71, 12/73, 12/74, 12/75, 12/76, 12/77, 12/78, 12/79, 12/80, 12/81, 12/82, 12/83, 12/84, 12/85, 12/86; (NORC-GSS): 3/73, 3/74, 3/75, 3/76, 3/77, 3/78, 3/80, 3/82, 3/83, 3/84, 3/85, 3/86, 3/87, 3/88, 3/89, 3/90, 3/91.

reciprocity and the potential for U.S. exports. When given the choice between restricting Japanese and European imports or a system that encourages more U.S. exports and foreign imports, the latter was preferred by 74 percent in February 1983, 72 percent in April 1983, and 76 percent in September 1985. But the public has overall been defensive when it comes to domestic industries, with steady support from 1988 to 1991 in CBS and *New York Times* surveys for trade restrictions to prevent harm to domestic industries from foreign competition: 62 percent in 1988, 60 percent in 1989, 57 percent in 1990, and rising (as might be expected) to 71 percent during the recession that followed in 1991.

Other cases of noteworthy opinion stability include the essentially high and stable support for U.S. involvement in world affairs from World War II to the present and support for the continued significant presence of U.S. troops in Europe. In the case of U.S. policy toward South Africa, 1990 and 1991 Gallup polls found that support for economic sanctions against South Africa remained steady and solid, at better than the 70 percent level.

Responsiveness of Opinion to New Information and New Realities

We do not mean to imply that collective public opinion never changes suddenly—quite the contrary. When the world changes, the American public tends to change its opinions accordingly and sometimes in a much sharper fashion than the foreign trade and other examples cited above. Collective public opinion generally responds to new information and to objective changes in ways that are regular, predictable, and generally sensible—given the values that citizens hold and the information presented to them.

Some of the most striking changes over the last fifteen years have occurred in public attitudes toward U.S.-Soviet relations. Figure 10-5 summarizes this important transformation, taking the country a long way from the peak of a new cold war and accusations about "the evil empire" during the early 1980s.[5] Other data show striking increases in support for economic aid to the former Soviet enemy, as tracked by several NBC News/*Wall Street Journal* surveys and CBS/*New York Times* surveys from 1989 to 1991. CBS/*New York Times* polls, for example, reported a nearly 30 percent change (from 17 percent to 46 percent) in support for the use of U.S. "government funds" to help the Soviets to buy American grain.

Many such major changes, including some abrupt and sharp ones, have occurred in opinions toward foreign policy in the past, more so than toward domestic issues for which events and changes in real world conditions tend to change less suddenly. One of the most dramatic changes, shown in Figure 10-6, involved the impact of the Tet offensive during the Vietnam War in January 1968. The public's response was at first belligerent, but between February and early April, as many leaders and commentators interpreted the offensive as a disaster, casting doubt on the U.S. prospects for victory, 22 percent fewer Americans described themselves as "hawks" (versus "doves") about the war, and public opinion continued to move in that direction. Clearly, the mass media conveyed the kinds of reactions to Tet that made this major reversal in public opinion toward the war explicable and predictable.

This case involving the Tet offensive and others during the Vietnam War, along with several other important foreign policy issues—including the controversy over the gulf of Tonkin, the alleged Kennedy-era "missile gap," and the more recent "window of vulnerability" in national security—raise critical questions about the possibilities for, and actual occurrence of, manipulation and deception of public opinion. We do not argue that opinion changes are always based on factually correct information and interpretations; rather, we conclude that the public reacts sensibly to the information it is given.[6] For deficiencies in political processes involving public opinion, one is wiser to look to the role of political leaders, "experts," and important figures in the mass media on whom the public relies for information, insight,

Figure 10-5 / Changing Views toward the Soviet Union

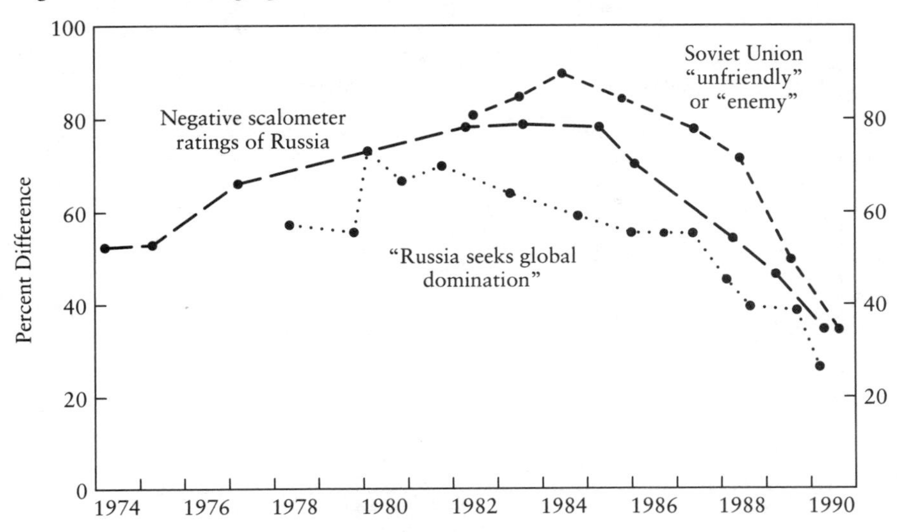

Source: Alvin Richman, "The Polls—Poll Trends: Changing American Attitudes toward the Soviet Union,"*Public Opinion Quarterly* 55 (1991): 135–48. Robert Y. Shapiro, as editor of *POQ*'s "Poll Trends," and his research assistant, Sara Offenhartz, originally created this figure for Richman's article, and it is reproduced here. *Question* (Roper): "In your opinion, which of the following best describes Russia's primary objective in world affairs? A. Russia seeks only to protect itself against the possibility of attack by other countries. B. Russia seeks to compete with the U.S. for more influence in different parts of the world. C. Russia seeks global domination, but not at the expense of starting a major war. D. Russia seeks global domination and will risk a major war to achieve that domination if it can't be achieved by other means." Survey dates: 6/78, 10–11/79, 2/80, 11/80, 9/81, 3/83, 9/84, 9/85, 3/86, 4–5/87, 1/88, 7/88, 8/89, 2/90. *Question* (Roper): "I'd like to have your impression about the overall position that some countries have taken toward the U.S. [Respondent shown list of countries.] Would you read down the list, and for each country, tell me if you believe the country has acted as a close ally of the U.S., has acted as a friend but not a close ally, has been more neutral toward the U.S., has been mainly unfriendly toward the U.S., but not an enemy, or has acted as an enemy of the U.S.? . . . The Soviet Union." Survey dates: 6/82, 6/83, 5–6/84, 12/85, 5/87, 5/88, 7/89, 7/90. *Question* (NORC/GSS—Scalometer ratings of Russia): "You will notice that the boxes on this card go from the highest position of plus 5 for a country which you like very much to the lowest position of minus 5 for a country which you dislike very much. How far up the scale or how far down the scale would you rate Russia?" Survey dates: 3/74, 3/75, 3/77, 3/82, 3/83, 3/85, 3/86, 3/88, 3/89, 3/90.

and standards of judgment regarding government policy, rather than to any substantial incapacity of ordinary citizens. The role of the media in the foreign policy process is illustrated using the case of the Persian Gulf War and others by W. Lance Bennett in Chapter 8.

A number of other recent cases provide additional examples of the public's reaction to events and new information and how they tended to be interpreted. In the several well-known trends in public support for defense

Figure 10-6 / Abrupt Opinion Changes: The Vietnam War

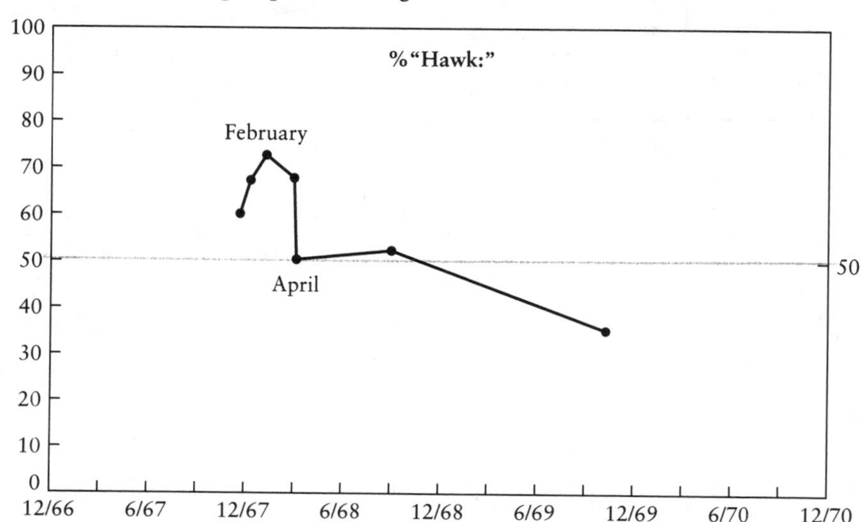

Question (Gallup): "People are called 'hawks' if they want to step up our military effort in Vietnam. They are called 'doves' if they want to reduce our military effort in Vietnam. How would you describe yourself—as a 'hawk' or as a 'dove'?" Survey dates: 12/67, 1/68, 2/1–6/68 and 2/22–27/68 (averaged), 3/68, 4/68, 9/68, 11/69.

spending (one is shown in Figure 10-7), the most salient feature of these data is the sharp increase in support for spending more on defense that occurred as the result of the Iran hostage crisis and the Soviet invasion of Afghanistan. One major survey organization (NORC-GSS) found 60 percent responding in the spring of 1980 that "we" were spending "too little" on defense, a 30 percent increase from the spring of 1978. This sharp rise was followed by an equally sharp decline within the next few years, to about where opinion stood in the early 1990s, except for a short-lived reversal of opinion during the Persian Gulf War. During that war, Gallup (for *Newsweek*) picked up a 15 percent drop in the percentage responding that "we" are spending "too much" on national defense from August 1990 to March 1991 (to 27 percent; NORC-GSS recorded a drop from 44 to 28 percent March 1990 to March 1991); this movement was more than reversed (by 22 percent; to 49 percent) by August 1991. Two additional predictable changes as a result of the war occurred: support for the United States' taking an "active part in world affairs" rose from 71 percent to 75 percent (NORC-GSS; Gallup recorded an increase from 69 percent in 1990 to 76 percent in a September 1991 poll), and support for U.S. membership in the United Nations increased a few points to a high of 90 percent. Not surprisingly, Gallup also found increased "sympathy" for Israel as opposed to the Palestinian Arabs, which rose 12 percent from

October 1990 to March 1991 (to the two-thirds level); although, reflecting the complexity of war and the need for negotiations toward peace in the Middle East, the level of support for "the establishment of an independent Palestinian nation" was 56 percent in October and remained about steady, even rising slightly to 60 percent in March 1991.

EFFECTS OF PUBLIC OPINION ON POLICY

It is fortunate that public opinion has the rational characteristics that we have outlined, because it has substantial effects on policy making.

We take as a given that public opinion acts as a "constraint" on foreign policy. We would agree with V. O. Key and others who have referred to "dikes" or other metaphors in describing the constraining role that public opinion can play. Indeed, this constraint is usually the most that policy makers themselves will concede in addressing the influence of public opinion on foreign policy. It was evident, for example, in the case of U.S. aid to the Contras in Nicaragua, which is described quite thoroughly in a work on that subject,[7] and the same kind of influence of public opinion was raised in the U.S. policy on the Persian Gulf crisis beginning in late 1990. But our conclusion goes beyond this limiting effect of the public: Public opinion has also been able to move government policies in different directions—in ways exceeding simple constraints.

Critics of this view can point to evidence indicating very little, if any, impact of public opinion on foreign policy. The pioneering study by Warren E. Miller and Donald E. Stokes on constituency influence on congressional roll call voting, for example, suggested this lack of impact.[8] This study, however, had serious measurement and methodological problems that we and others, such as Graham in Chapter 9, think underestimated the impact of district preferences. Moreover, it did not examine the actual enactment and implementation of policy. It may also have been time bound or issue specific; one recent and quite sophisticated study has shown that constituency influence was quite important in the very early Reagan defense buildup, during which congressional representatives' votes on necessary defense legislation substantially reflected public opinion in their districts.[9]

Studies comparing mass and elite opinions toward foreign and domestic policies have highlighted significant mass-elite differences as well as similarities in policy preferences, though these studies, too, do not directly examine policies formulated and enacted by policy makers.[10] In contrast, many other studies of connections between opinion and policy have indicated that important effects of public opinion on policy do occur and that the linkage mechanisms connecting opinion and policy have become more apparent. Graham, in Chapter 9, describes these effects and mechanisms further, along with the conditions under which the impact of public opinion on decision making is most likely.

Studies comparing aggregate opinion and government policy have shown a substantial frequency of consistency or "congruence" between public opinion and foreign policies on issues salient enough to appear in national polls. Examining majority opinion concerning proposed changes in policy and subsequent policy, Alan D. Monroe found consistency between opinion and policies in 64 percent of more than two hundred cases, including a striking 92 percent in the case of foreign policy (though 71 percent of his Vietnam War cases and under 50 percent for cases classified under national defense).[11] In our own major study we found a similar 66 percent frequency (based on well over two hundred cases) of congruence between *changes* in opinion and subsequent changes in policy. We found such congruence in 62 percent of foreign and defense policy issues, with 71 percent (again) in the case of Vietnam War issues, 100 percent in the case of World War II issues, about 65 percent in the case of foreign aid and U.S.-Soviet relations, and just over 50 percent in other foreign and defense issues.[12]

Not all of the opinion-policy congruence in these aggregate studies necessarily represented cases of policy responding to opinion, and these studies could not rule out the possibility for the reverse effect of policy on opinion. Nonetheless, and bolstered further by the examination of individual cases, public opinion was apparently an important *proximate* cause of policy change in many cases. The two-thirds figure for the frequency of opinion-policy congruence is one that warrants serious consideration.

Additional statistical evidence—and perhaps the most persuasive—is provided by some of the available time series data. These data, shown in Figure 10-7 and Figure 10-8a and b provide evidence concerning the relationship between public preferences and defense spending and between public opinion and U.S. troop withdrawals from Vietnam.[13] The correlation between the movements of public opinion and policy are visually apparent, and the results from regression analysis substantiate this correlation in a more precise fashion. In the case of defense spending, when the rate of change in defense spending (in constant dollars) is compared statistically for its relationship with opinion, we find that a 10 percent rise in support for defense spending leads to a 2 percent increase in the rate of defense spending.[14]

Similarly, in the case of the month-to-month changes in troop levels and public preferences concerning troop withdrawals, the proportion of responses that the withdrawals were "too slow" was a moderately strong predictor of the actual withdrawal rate one month later: An additional 1 percent saying that withdrawal was too slow was on average followed by a withdrawal of nearly 500 additional troops per month.[15] Repeating a similar analysis with the longer and better-known Vietnam War (was a) "mistake" series revealed that a 1 percent increase in those saying that U.S. involvement was a "mistake" led to a reduction of approximately 930 more troops per

Figure 10-7 / Opinion-Policy Relationship: Defense Spending

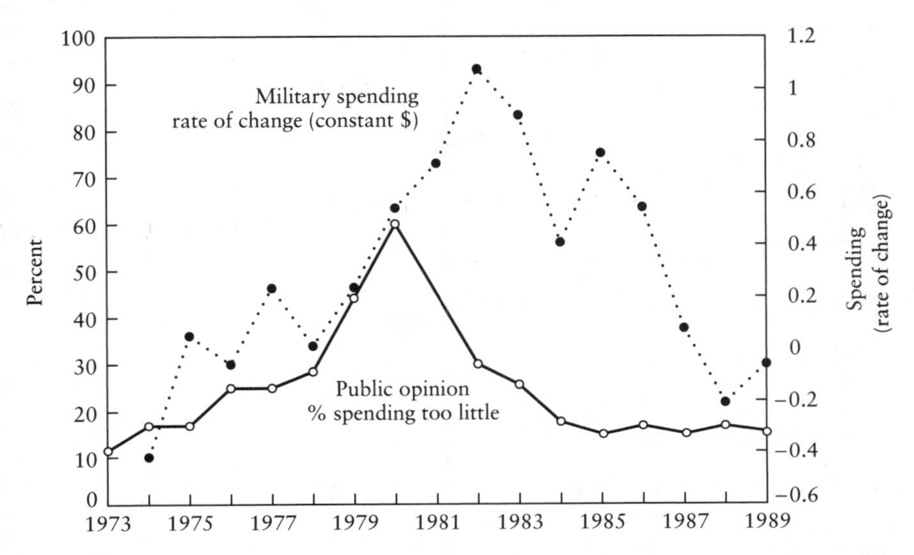

Question (NORC-GSS) (see Fig. 10-4): ". . . The military, armaments and national defense?" Defense spending data were obtained from *The Budget of the United States* and were deflated.

month.[16] This longer opinion time series, offering additional time points and covering a lengthier time period, allowed us to deal with the problem posed by the possible (indeed, likely) reverse causal effect of troop levels on opinion. In the Vietnam case, especially, this reverse effect could not be ruled out well by examining time lags. It was possible, however, to reestimate the relationship between opinion and policy using a more appropriate econometric method.[17] The result was a small reduction in the effect of opinion from 930 troops per month to 740. This reestimation increases our confidence about the direction and strength of the opinion-policy relationship.

For these impressive correlations that we have reported to represent *causal* influences of opinion on policy making—of opinion directly affecting policy makers or otherwise "seeping" into policy deliberations—there have to be linkage processes by which public preferences make their way through the policy-making process. Until relatively recently there has been scant evidence that public opinion has had much opportunity to do this analysis in foreign policy. A prominent study of the State Department by Bernard C. Cohen about thirty years ago provided little evidence that the necessary linkage processes exist.[18] But there is now increasing contrary evidence.

Anything more than a brief summary of this new evidence showing a rather direct influence of public opinion on policy is not necessary because that is the focus of Chapter 9. We think the traditional view of little effect of

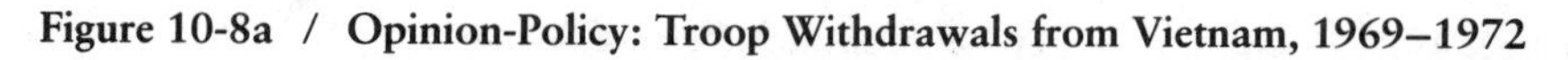

Figure 10-8a / Opinion-Policy: Troop Withdrawals from Vietnam, 1969–1972

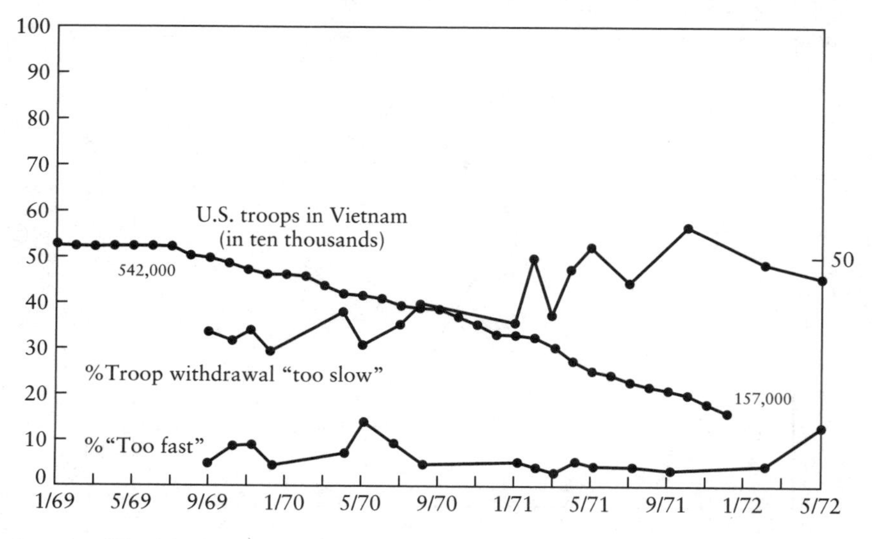

Question (Harris): (Version A) "In general, do you feel the pace at which the president is withdrawing troops from Vietnam is too fast, too slow, or about right?" Survey dates: 10/69, 12/69. (Version B) adds "Nixon." Survey date: 8/70. (Version C) omits "from Vietnam." Survey date: 10/69. (Version D) adds "U.S." Survey date: 4/70. (Version E) puts "slow" first. Survey date: 1/71. (Version G) substitutes "think." Survey dates: 5/70, 7/70. (Version H) adds "Nixon" and "slow" first. Survey dates: 3/71, 4/71. (Version I) omits "in general," puts "slow" first. Survey date: 10/71. (Version J) long prologue on RN plan, substitutes "more rapid rate," "slower rate," "satisfactory." Survey date: 10/71. (Version K) "Do you feel the pace of withdrawal of U.S. troops from . . . ? Survey dates: 2/72, 5/72. Source for U.S. troop data: SS-2 Unclassified Statistics on Southeast Asia, Comptroller, Office of the Secretary of Defense, Table 6. Reported in Raphael Littauer and Norman Uphoff, eds., *The Air War in Indochina* (Boston: Beacon Press, 1972), pp. 265–272, table 6.

opinion on foreign policy has to change. First, one important recent study has shown that officials at the State Department since the time of Cohen's study have apparently been more responsive to, though not necessarily more confident in, public opinion.[19] Further, presidents and their advisers have been the most important foreign policy makers, and it is quite evident by now that the use of public opinion polls and public opinion analyses have followed presidents' election campaigns into the White House—and have, in fact, been institutionalized. This process began in earnest with John F. Kennedy's 1960 campaign (it became well known that Louis Harris was Kennedy's pollster and one of his political advisers), and it has been particularly visible—even strikingly open—during the Johnson, Carter, Reagan, and Bush administrations.[20] Table 9-3 offers one assessment of presidents' understanding and use of polls. Prior to Kennedy, such attention to polling information was kept highly secret.

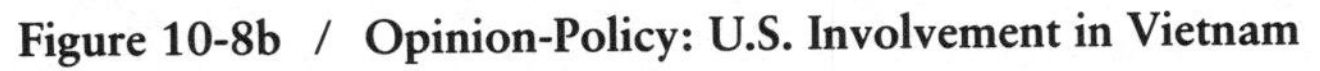

Figure 10-8b / Opinion-Policy: U.S. Involvement in Vietnam

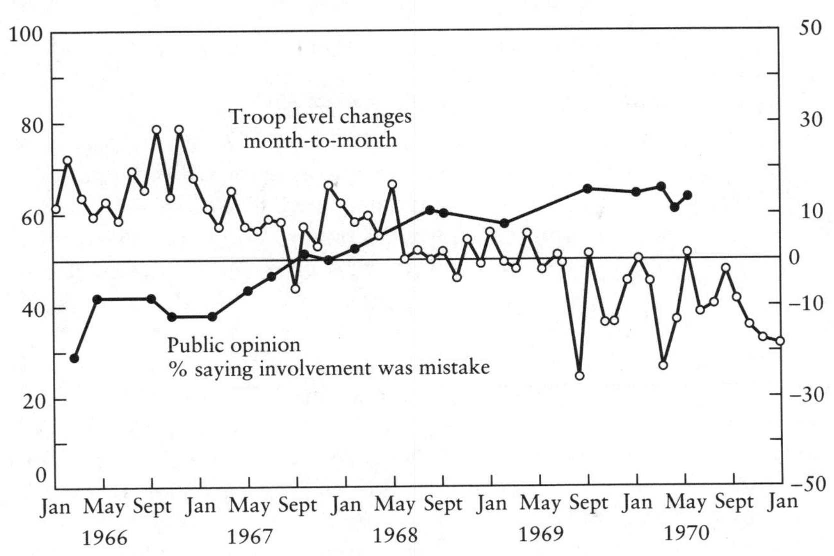

Question (Gallup): "In view of the developments since we entered the fighting in Vietnam, do you think the U.S. made a mistake sending troops to fight in Vietnam?" Survey dates: 8/65, 3/66, 5/66, 9/66, 11/66, 2/67, 5/67, 7/67, 10/67, 12/67, 2/68, 3/68, 4/68, 8/68, 10/68, 2/69, 9/69, 1/70, 3/70, 4/70, 5/70, 1/71, 5/71.

It is far from clear, however, that the effect of opinion on policy is direct—with presidents and their advisers literally reading and acting immediately based upon poll results. The effect is very likely often a more subtle one, which occurs as political leaders attempt to use information about public opinion for the purpose of leading, persuading, or manipulating the public. But, as the research of Lawrence Jacobs, for example, has recently uncovered and described, political leaders' and policy makers' strategies for leading or manipulating public opinion can backfire in subtle or not so subtle ways. According to Jacobs, a "recoil effect" very often occurs as policy makers take public sentiments into account in developing ways to lead or otherwise direct public opinion: the policies that are formulated through this process and later presented to the public have, in fact, been significantly shaped by public opinion.[21]

A number of linkage processes, then, may be at work, and our conclusion is that the effect of public opinion on foreign policy is real. It is a phenomenon that must be taken seriously. The implications of this for the workings of democracy, to be sure, depend upon the performance of our political leaders in interacting with the public and upon the characteristics and qualities of public opinion itself.

CONCLUSION

The apparent effects of public opinion on the nation's foreign and defense policies might seem to be a regrettable state of affairs to many policy analysts and others (including buffs of the American founders) who would prefer to insulate foreign policy from public pressures. We strongly disagree. Based upon the evidence we have described showing the collective rationality of public opinion, we are quite sanguine. There is little or no reason to *fear* the effects of public opinion on foreign policy; people react sensibly to information. If there is a case to be made for a given policy, leaders should make it; win or lose, ideas of democracy suggest that elected leaders should pay some heed to their constituents.

Our evidence, and that of others, indicates that political leaders *do* in fact respond to public opinion a substantial amount of times. If the two-thirds rule of thumb is correct, however, there remain a good many cases in which they do not so respond. And we must remember that some unknown fraction of apparent responsiveness is illusory: It consists of action that corresponds with manipulated or manufactured opinion. For a more democratic foreign policy, we should work for better and more complete information to be given to the public, to ensure that authentic public opinion is formed, and for greater, not less, official responsiveness to it.

Notes

1. Benjamin I. Page and Robert Y. Shapiro, *The Rational Public: Fifty Years of Trends in Americans' Policy Preferences* (Chicago: University of Chicago Press, 1992).
2. William R. Caspary, "The 'Mood Theory': A Study of Public Opinion and Foreign Policy," *American Political Science Review* 64 (1970): 536–47.
3. Page and Shapiro, *Rational Public,* pp. 45–46.
4. For a more detailed treatment of the M-X example, see ibid., pp. 17–23.
5. See Alvin Richman, "The Polls—Poll Trends: Changing American Attitudes toward the Soviet Union," *Public Opinion Quarterly* 55 (1991): 133–48.
6. See Page and Shapiro, *Rational Public,* esp. chap. 8 on the extent to which the public may be given incorrect information and interpretations or otherwise be misled. Chapters 5–6 recount important examples of the manipulation of public opinion in foreign affairs.
7. Richard Sobel, ed., *Public Opinion in U.S. Foreign Policy: The Controversy Over Contra Aid,* Boston: Rowman and Littlefield, 1993.
8. Warren E. Miller and Donald E. Stokes, "Constituency Influence in Congress," *American Political Science Review* 57 (1963): 45–56.
9. Larry M. Bartels, "Constituency Opinion and Congressional Policy Making: The Reagan Defense Buildup," *American Political Science Review* 85 (1991): 457–74.
10. See Russell J. Dalton, *Citizen Politics in Western Democracies* (Chatham, N.J.: Chatham House, 1988), chap. 10; John E. Rielly, ed., *American Public Opinion and U.S. Foreign Policy, 1975* (Chicago: Chicago Council on Foreign Relations, 1975), and editions published in 1979, 1983, 1987, and 1991.
11. Alan D. Monroe, "Consistency between Public Preferences and National Policy Decisions," *American Politics Quarterly* 7 (1979): 3–19. But cf. Joel E. Brooks, "Democratic Frustration in the Anglo-American Polities: A Quantification of Inconsistency between Mass Public Opinion and Public Policy," *Western Political Quarterly* 30 (1985): 250–61.
12. Benjamin I. Page and Robert Y. Shapiro, "Effects of Public Opinion on Policy," *Ameri-*

can Political Science Review 77 (1983): 175–90; Robert Y. Shapiro, "The Dynamics of Public Opinion and Public Policy" (Ph.D. diss., University of Chicago, 1982).

13. These analyses and many additional ones are described in Steve Farkas, Robert Y. Shapiro, and Benjamin I. Page, "The Dynamics of Public Opinion and Policy," paper presented at the Annual Meeting of the American Association for Public Opinion Research, Lancaster, Pa., May 17–20, 1990. See also Page and Shapiro, *Rational Public,* pp. 237–39, on the Vietnam case. On public opinion and defense spending, see also Bruce Russett, *Controlling the Sword: The Democratic Governance of National Security* (Cambridge, Mass.: Harvard University Press, 1990).

14. The rate of change in defense spending is used because of the changing referent in the question wording. In this statistical analysis the rate of change is regressed on opinion (lagged one year); the slope coefficient of .2 (significantly different from zero at the .05 level, r=.56) represents an estimate of the effect of opinion on policy.

15. The slope coefficient is −.488, r=−.43, significant at better than .2.

16. r=.80, significant at the .05 level.

17. Two-stage least squares, using the cumulative logarithm of the rate of American casualties as the necessary exogenous variable.

18. Bernard C. Cohen, *The Public's Impact on Foreign Policy* (Boston: Little, Brown, 1973).

19. Philip J. Powlick, "The Attitudinal Bases for Responsiveness to Public Opinion among American Foreign Policy Officials," *Journal of Conflict Resolution* 35 (1991): 611–41.

20. See Lawrence R. Jacobs and Robert Y. Shapiro, "Democracy, Leadership, and the Private Polls of Presidents Kennedy and Johnson: Beginnings during the Kennedy Campaign," paper presented at the Annual Meeting of the American Political Science Association, Washington, D.C. August 29–September 1, 1991.

21. Lawrence R. Jacobs, "The Recoil Effect: Public Opinion and Policymaking in the U.S. and Britain," *Comparative Politics* 24 (1992): 199–217. See also Thomas W. Graham, "The Politics of Failure: Strategic Nuclear Arms Control, Public Opinion and Domestic Politics in the United States, 1945–1980" (Ph.D. diss. Massachusetts Institute of Technology, 1989).

Part VII / The Comparative Context

11 / Masses and Leaders: Public Opinion, Domestic Structures, and Foreign Policy

THOMAS RISSE-KAPPEN

INTRODUCTION

This book deals primarily with the domestic politics of American foreign policy. However, our understanding of the peculiarities of the American system can be substantially enhanced if a comparative perspective is added. Comparisons with the foreign policy decision-making processes of other liberal democracies help us to sort out which features of the American system are unique to the United States and which can be generalized across countries.

This chapter attempts to provide a model for studying the public impact on foreign policy decisions from a comparative perspective. It takes the revisionist perspective in the research on public opinion and foreign policy as its point of departure.[1] Earlier analyses on the subject assumed that mass public opinion (1) does not know much about foreign policy, (2) does not consider external affairs as important as, for example, economic issues, and (3), as a result, tends to be highly volatile. The revisionists, however, showed that large portions of the mass public (1) regularly follow foreign policy in the media, (2) consider external affairs among the most important issues facing their country, (3) know enough about the world to build informed opinions, and (4) hold rather stable attitudes on core foreign policy issues.

While the public, thus, appears to be quite rational (see Chapter 10), it does not follow that public opinion actually influences foreign policy decisions. Rather than assuming a simple top-down (no influence) or bottom-up (major impact) process, I argue in the following that the public impact on foreign policy is mediated by the *domestic structure* of the country in question. The nature of the political institutions as well as the character of the state-society relationship determine to what degree decision makers have to take public opinion into account. I look at three countries with different domestic structures—the United States as a comparatively *society-dominated* system, France as a *state-dominated* country, and Germany as an example of the *democratic-corporatist* type. To illustrate the argument, the chapter investigates two cases: the reactions of the three countries to (1) the end of the cold war, and (2) the Persian Gulf crisis and war in 1990–91.

THE PUBLIC OPINION: FOREIGN POLICY CONUNDRUM

Bottom Up or Top Down?

Most of the previous scholarship on the public impact on foreign policy decisions in liberal democracies falls into two camps. First, the *bottom-up approach* assumes that mass public opinion has a measurable and distinct impact on foreign policy decisions. Leaders follow masses. Of course, this approach agrees with the liberal pluralist theory of democracy according to which democratic systems are governed by the people for the people.[2] Second, the opposite viewpoint suggests a *top-down process,* according to which popular opinion follows the elites and elite divisions trickle down to mass public opinion. This argument, which represented the conventional wisdom in the literature for decades, is consistent with the *power elite* theory of public policy. Modern political systems are essentially ruled by elite coalitions, and the mass public is excluded from this process except for general elections.[3] It then becomes a question of one's political convictions whether one supports the exclusion of the mass public from the political process ("because the masses do not understand the complexities of foreign policy") or whether one regrets it ("if the public took part in foreign policy decisions, there would be fewer wars").

Both approaches suffer from conceptual problems. First, they treat elites and masses as unitary actors. While certain segments of the public may be manipulated by government propaganda, others may resist efforts to influence them. Moreover, elites are frequently divided among themselves, and the various elite groups often try to convince their constituencies in the general public of their respective viewpoints. When it comes to public opinion, one should at least distinguish between (1) the *mass public,* (2) the *attentive public* with a more-than-average interest in politics, and (3) various *issue publics* that can be mobilized on specific problems. With regard to mass public opinion, one needs to differentiate further between attitudes on questions that people regard as highly significant for them (high issue salience) and those of lower importance. One should also bear in mind that it makes a big difference whether there is mass consensus or majority agreement on a particular issue or only a plurality of people hold a specific attitude (see Chapter 9).

Second, rather than viewing the interaction between public opinion and foreign policy as simple top-down or bottom-up processes, the general public might influence policy decisions at various stages of the process. It can directly affect decisions by influencing the policy goals and means of decision makers or by narrowing their range of choices. More important, public opinion can also influence coalition-building processes among elites and lead to realignments among interest groups, political parties, and bureaucratic actors within governments. These indirect policy effects are difficult to trace and are, therefore, easily overlooked.

Finally, it is an erroneous assumption that public opinion and elites interact with each other and affect policy decisions in the same way across different countries. There is no reason to expect that the Japanese public has the same impact on the foreign policy of their country as, say, Americans have on presidential decisions. More often than not, mass public opinion in several countries shares attitudes on specific foreign policy issues, while the choices made by their respective governments are remarkably different.[4]

I suggest in the following that the interaction between elites and the public as well as the differences in policy impact across countries can be better understood if domestic structures are considered as intervening between public opinion and policy outcomes.

Domestic Structure As an Intervening Variable

The domestic institutions of each country are distinctive. The political institutions in Japan work differently from the American ones, and both are different from the Canadian political system. On the other hand, countries are not completely unique, and their political and societal characteristics can be grouped under similar categories. Domestic structure approaches allow for sufficient differentiation among political systems to account for variations in policy outcomes but also provide a limited number of categories to enable meaningful comparisons. Originally developed in the field of comparative foreign economic policy, the concept has generated empirical research across issue areas to explain variations in state responses to similar international pressures, constraints, and opportunities.[5] Domestic structure approaches should be well suited to account for the variation in the impact of public opinion on foreign policy decisions.

The notion of *domestic structures* refers to the institutional characteristics of the state, to societal structures, and to the policy networks linking state and society. Domestic structures encompass the organizational apparatus of political and societal institutions, their routines, the decision-making rules and procedures as incorporated in law and custom, and, finally, the norms of the political culture prescribing appropriate behavior. Earlier expressions of the domestic structure concept focused, above all, on organizational features of state and society, especially on the degree of their centralization. The "new institutionalism" adds to those expressions the logic of communicative action, duties, social obligations, and appropriate behavior. The normative and cultural context of state-society relations is not fully grasped if one focuses only on organizational and legal characteristics of political and social institutions. For example, the Japanese decision-making norm of reciprocal consent, the German understanding of social partnership, and the American notion of liberal pluralism are only partly embodied in explicit regulations but nevertheless constitute powerful cultural norms that define appropriate methods for decision making in the political system.

Three tiers of domestic structures can be distinguished.[6] First, the structure of the *political institutions*—the state—can be analyzed in terms of its centralization or fragmentation. To what extent is executive power centralized in the hands of a small group of decision makers? Can the national government control the legislative process? How far does the domestic power of the national government reach into the governance of regions and local communities (central or federal structures)? Does the political culture emphasize the state as a benign institution taking care of its citizens or as a threat to individual liberties?

Second, the *structure of demand-formation in the society* can be examined with regard to the internal polarization in terms of ideological and/or class cleavages, the degree to which societal demands can be mobilized, and the organizational strength and centralization of interest groups and societal coalitions. How heterogeneous is the society in terms of ideological and/or class cleavages? How well organized are social coalitions and interest groups in their ability to express grievances and mobilize people for their demands?

Third, the coalition-building processes in the *policy networks* linking state and society and the norms regulating these processes have to be investigated. How strong are intermediate organizations such as political parties channeling societal demands into the political process? Does the political culture emphasize consensual decision making or distributive bargaining and dissent?

These components can be regarded as a three-dimensional space. The three axes are formed by (1) the state structure (centralization v. fragmentation as the two ends of a continuum), (2) the societal structure (weak v. strong degree of social mobilization and/or power of societal organizations), and (3) the policy networks (consensual v. conflictual as the two ends). This three-dimensional space allows one to locate the domestic structures of specific countries. However, most empirical domestic structures can probably be aligned along a continuum that ranges from state-dominated to society-dominated. One can then distinguish three ideal types linked to specific propositions regarding the policy impact of public opinion.

Society-dominated domestic structures are to be expected in countries with comparatively strong societal organizations, a high degree of societal mobilization in interest groups, but comparatively decentralized and fragmented political institutions as well as weak intermediary organizations in the policy networks. Society-domination of the political process will be strengthened by a political culture emphasizing individualism, pluralistic norms, and distributive bargaining.

Among the highly industrialized democracies, the United States is frequently cited as an example of a society-dominated domestic structure. It has a comparatively decentralized foreign policy-making structure, even in the realm of security policy.[7] The built-in tensions within the executive branch among the Pentagon, the State Department, and the National Security Council lead to continuous infighting over foreign and defense policies.

Moreover, Congress has more authority over the conduct of U.S. foreign policy than most other Western parliaments due to (1) the weakness of the American party system, which severely limits executive power over congressional decisions, and (2) institutional provisions such as the two-thirds majority requirement for the ratification of international treaties. On the other hand, the ability of societal actors to mobilize support for their demands and to organize themselves seems to be comparatively well developed. One can mention as examples the significant power of the American defense industry and the importance of countervailing public interest groups lobbying for arms control and disarmament.[8]

Concerning the policy networks, constantly shifting coalitions among societal actors and political elites are fairly common in the American foreign and security policy-making process. The openness of the political system provides society with a comparatively easy access to the decision-making process. Accordingly, even tight networks such as the so-called military-industrial complex, linking business, military, and political interests, have had only a limited impact on foreign policy decisions beyond the weapons procurement process and demands for a certain level of defense spending.[9]

With regard to society-dominated domestic structures such as the United States, the following assumptions seem to be plausible about the impact of public opinion on foreign policy decisions. First, given the openness of the political system, one can hypothesize that societal demands, including the various components of public opinion, should have no trouble reaching decision makers in Congress and the administration. At the same time, one would assume that political leaders closely monitor public attitudes and patterns of opinion formation. Second, however, the fragmentation and the polarization of the political process frequently lead to the formation of countervailing coalitions whenever important choices are at stake. As a result, one would assume that the short-term impact of public opinion on foreign policy decisions should be greater in the United States than in other countries with more significant state roles in the domestic structure. By the same token, though, the long-term influence is expected to be limited, given the lack of strong intermediary organizations in the policy networks that could otherwise serve to institutionalize majority attitudes in public opinion.

State-dominated domestic structures, on the other hand, encompass comparatively centralized political institutions with strong national executives able to manipulate the political process. The level of societal organization is usually rather weak, whether because of strong cross-cutting ideological or class cleavages or—in the case of authoritarian systems—because of the oppressive character of the state. The political culture of such systems often emphasizes the state as caretaker of the needs of its citizens. Many former communist systems with centrally planned economies and various authoritarian third world states seem to fit this description.

Among the Western democracies, France probably comes closest to the state-dominated type.[10] The Fifth Republic institutionalized a centralized political system that all presidents since Charles de Gaulle have reinforced. The power of the French bureaucracy adds to the strength of the executive. This general feature of the French political system is particularly relevant for foreign and defense policy making, the *domaine réservée* of the president. Furthermore, the French parliament plays an almost negligible role in foreign policy. In sum, the centralization of the French decision-making apparatus seems to be even greater in foreign policy than in other issue areas. Even when Socialist President François Mitterrand was forced into "cohabitation" with a conservative government from 1986 to 1988, he continued to control foreign and defense policy.

This is not to suggest that public opinion is completely irrelevant in state-dominated domestic structures such as France. But the process by which public attitudes influence foreign policy decisions is expected to be different from the society-dominated case. First, there are few institutionalized access points for societal demands to reach the political system. As a result, mass public opinion should be able to affect policy decisions only if it reaches the top decision makers, that is, in France, the president. Second, however, once this hurdle is overcome and top decision makers are prepared to listen to public opinion, the policy impact may be profound. If not, it should be almost irrelevant. In other words, the public impact on foreign policy in state-dominated domestic structures depends almost entirely on the degree to which the national leaders are prepared to take its views into account.

Finally, *corporatist* domestic structures are likely in cases in which strong and comparatively centralized political institutions are faced with equally strong societal and/or powerful intermediary organizations, for example, political parties.[11] Political and societal actors are engaged in continuous bargaining processes in a give-and-take environment. Consensus-oriented decision norms frequently prevail, resulting in decision-making processes geared toward political compromises. The domestic structures of many of the smaller European states come close to this model.

In the Federal Republic of Germany, executive control over foreign and defense policy is generally stronger than in the United States, and the role of parliament has been fairly limited.[12] This weakness, however, does not result simply in strengthening executive power, as it does in France. Rather, the political parties constrain both the legislative *and* the executive. Partly as a result of peculiar election procedures,[13] the party system is smaller and less polarized than in France. Coalition governments require constant consensus building on major foreign policy decisions among the ruling parties. The "party democracy" permeates the government bureaucracy insofar as internal divisions usually occur with regard to the party affiliation of the respective minister rather than along traditional bureaucratic roles.

German society, on the other hand, seems to be less fragmented than both the United States and France in terms of ideological and class cleavages. Moreover, the country enjoys comparatively strong social organizations with a high level of participation. The three most important and centralized social organizations—business, trade unions, and churches—never hesitate to speak out on foreign and security issues, usually on the "dovish" side of the debate. Moreover, new social movements, encompassing, above all, environmentalists and peace groups, have increasingly become a well-established part of the German political life.[14]

The party system forms the most important link between society and the political system. The two major parties—the Christian Democratic Union (CDU) and the Social Democratic party (SPD)—are essentially catch-all organizations and integrate rather divergent societal demands. Sometimes the inner-party divisions are stronger than the cleavages between the parties, as was the case with the CDU and *Ostpolitik* during the 1970s and with the SPD and nuclear deterrence during the 1980s. Additionally, institutional arrangements as well as the political culture emphasize consensus building and the mutually beneficial settlement of diverging societal interests (*Interessenausgleich*).

In democratic-corporatist domestic structures such as Germany, one would expect public opinion to play a somewhat more limited short-term role in the decision-making processes than in the American case, since the political system is less permeable than in society-dominated structures. On the other hand, since corporatist structures tend to have strong intermediary organizations in the policy networks, public demands are likely to be adopted by, for example, political parties and then channeled into the political process. As a result, the impact of public opinion is expected to last longer, because corporatist structures are usually geared toward achieving long-term institutionalized consensus on policies.

To evaluate these various propositions, I will now look at two cases. I examine public attitudes and their policy impact in the United States, France, and Germany with regard to (1) the end of the cold war, and (2) the Persian Gulf crisis and war of 1990–91.

PUBLIC OPINION AND FOREIGN POLICY: TWO CASE STUDIES

The End of the Cold War

The "end of the cold war" means different things to different people. One could refer to the Gorbachev revolution in Soviet foreign policy, to the conclusion of major arms control treaties during the late 1980s, to the opening of the Berlin wall and the revolutions in eastern Europe in the fall of 1989, or to German unification in 1990. In the following, I use the term "end of the cold war" as the point in time when the "Soviet threat," which dominated Western foreign and defense policy outlooks throughout the post-1945 era, ceased to exist. Such a definition allows us to measure both mass public opinion and the foreign policies of the states under consideration here. When did public opinion no longer perceive a Soviet threat? When did Western policy makers change their rhetoric from an emphasis on the "evil empire" (Ronald Reagan) to a focus on partnership and even on alliances with the former Soviet Union? When did they match words with deeds—from the conclusion of arms control agreements to pledges for economic assistance and a new emphasis on multilateral institutions that included their former opponents?

As we will see in the following, the "end of the cold war" occurred at different points in time in the three countries considered here.

The United States: Cautious Response Among the three countries under consideration, the American people were the most reluctant to declare the cold war over. Threat perceptions and negative feelings toward the Soviet Union had reached the level of mass public consensus during the early 1980s, when more than 75 percent of Americans held unfriendly attitudes toward Moscow. Since then, threat perceptions gradually decreased. During the first three years of Gorbachev's tenure, however, a majority of U.S. public opinion was still deeply suspicious of the Soviet Union. For example, when the pacesetting treaty banning medium-range missiles (the Intermediate Nuclear Forces Treaty, or INF) was concluded in 1987, signifying Gorbachev's acceptance of Reagan's "zero option," more than 50 percent of Americans were still convinced that "Russia seeks global domination."[15]

The cold war ended in the eyes of the American public some time between late 1988 and 1990. Between May and November 1989, for example, the percentage of those who thought that the Soviet Union was trying to dominate the world decreased from 50 percent to 34 percent. At about the same time, more and more Americans expressed either favorable or at least neutral feelings toward the Soviet Union. Moreover, the U.S. public liked Mikhail Gorbachev much more than the Soviet Union in general (47 percent as opposed to 30 percent in November 1989).[16] One can conclude from these and other data that American public opinion declared the cold war over not so much as a result of Gorbachev's foreign policy initiatives but as a consequence of the democratic revolutions in eastern Europe and of the fall of the Berlin wall. This interpretation makes sense if one thinks about the significance that the Sovietization of eastern Europe during the late 1940s had for the creation of a cold war consensus in the United States. Moreover, one should remember the importance of Berlin as the symbol of freedom in the midst of communist oppression.

How did the change in threat perceptions translate into attitudes on policies toward the Soviet Union? First, it should be noted that the public distrust of the Soviet Union during the early 1980s did not result simply in demands for confrontational policies. The majority support for increased defense spending and against the ratification of arms control treaties that contributed to Ronald Reagan's election quickly evaporated during the early

1980s. Beginning in 1982, more Americans favored decreased defense expenditures than increased military budgets. There was also an overwhelming support for the resumption of arms control negotiations.

Toward the end of the 1980s, more and more Americans were inclined to support reforms in the country of the former enemy. Between 1989 and 1991, the percentage of those favoring government funds to assist Soviet purchases of American grain increased from 17 percent to 46 percent. In November 1989, 57 percent supported extension of most-favored-nation (MFN) privileges to the Soviet Union. At the same time, however, Americans were almost equally divided over whether or not the government should encourage U.S. business to invest in the Soviet Union.[17] In interpreting these data, however, one should keep in mind the general tendency of Americans to oppose both foreign aid and an active government role in the private economy.

American foreign policy toward the Soviet Union during the 1980s was roughly in line with the trends in mass public opinion. George Bush's cautious but supportive approach to Gorbachev and, later, Boris Yeltsin, as well as his promotion of German unification within the North Atlantic Treaty Organization (NATO), paralleled closely these tendencies in the American public. It is hard to say, though, who followed whom on such a level of generalities. Nevertheless, there are three instances during the 1980s in which American foreign policy seems to have been directly influenced by trends in public opinion.

First, during the early 1980s, Reagan's military buildup and cold war rhetoric quickly eroded the public support for his defense policy. Public support for arms control and the long-standing opposition against the (first) use of nuclear weapons provided issue publics such as the nuclear freeze movement with an opportunity to affect the political debate. The nuclear freeze movement represented genuine grass-roots efforts that, at first, were not well connected with the liberal arms control establishment in Washington. But the peace groups created a powerful issue public determined to reverse the arms race. In 1983, the House of Representatives adopted a freeze resolution, and Congress in general became increasingly active in arms control issues. Responding to these coalition-building processes in the policy network and to allied pressures, the power balance in the Reagan administration between hard-liners in the Pentagon and more pragmatic conservatives in the State Department slowly shifted in favor of the latter.[18] By 1984 and throughout 1985, that is, *before* Gorbachev began his peace initiatives, the Reagan administration had considerably softened its rhetoric and adopted a more compromising stance on arms control. While the peace movement did not achieve its immediate objective of a nuclear freeze, it successfully used mass public opinion and allied pressures to move the Reagan administration back on the arms control track.

The second instance occurred in the spring of 1989. When the Bush administration came into power, government officials conducted a comprehensive review of Reagan's policies and apparently concluded that a more cautious approach should be adopted toward Moscow. By that time, however, the Soviet threat was eroding in the American public. When Bush first outlined his foreign policy of "status quo plus," the public reaction was rather negative. The administration quickly adjusted to the public mood, and the president assured the public and the allies that the United States was firmly supportive of Gorbachev and the politics of *perestroika*. Throughout the remainder of 1989 and 1990, U.S. foreign policy tried to reassure the Soviet Union that the West would not exploit its turmoil and that in eastern Europe to promote U.S. interests.

A final case in which American foreign policy seems to have been influenced by trends in public opinion concerns the issue of economic assistance for democratizing Russia. While the United States spent trillions of dollars to fight the cold war, it showed reluctance to invest even moderate amounts of money to assist eastern Europe and the successor states of the Soviet Union, particularly if compared with the amounts spent by Germany and other European states. In conjunction with the domestic economic crisis of 1990–92, there is not much public support for expensive foreign aid programs. The Democrats also fell victim to the constraints of public attitudes in this area.[19]

While public opinion did not simply determine U.S. policies toward the Soviet Union during the 1980s, it left a discernible mark on the decision-making process. The U.S. domestic structure, as a comparatively open system, allowed societal actors to mobilize support and to affect the balance of forces within the policy network. On the other hand, the fragmentation and decentralization of the political system worked against the stabilization and institutionalization of policies. The long-term public impact was, therefore, more limited, mainly because the domestic structure does not provide institutional support for a lasting foreign policy consensus, except on a fairly general level.

France: Erratic Response French public opinion toward the Soviet Union was less hostile than American public opinion during the early 1980s.[20] However, later in the decade it showed trends similar to those in the United States. By about 1988–89, the Soviet threat evaporated in the French eyes, too. In 1991 only 17 percent of the French considered the Soviet Union a serious military threat to France. Nevertheless, the French people were as reluctant to show favorable feelings toward the former Soviet Union as were the Americans. Even as late as May 1991 only 36 percent of the French held a favorable opinion of the Soviet Union, while 53 percent still did not like it. Only after the coup attempt in August 1991 did a majority of the French public express favorable feelings toward the former Soviet Union. Similar to American public opinion, the French liked Mikhail Gorbachev far more than the former Soviet Union.

In contrast to the American public, however, considerably more French people supported economic assistance to the former Soviet Union. In September 1991, 83 percent agreed that their country should give economic aid to the Soviet Union, and 55 percent supported the view that large-scale Western aid would strengthen democracy in the former Soviet Union.

French foreign policy toward the former Soviet Union during the 1980s did not seem to be too concerned about public opinion. In the early 1980s, for example, the newly elected Socialist President Mitterrand adopted a somewhat more hostile approach toward the Soviet Union than had his predecessor. While this change, which marked a departure from the independent policy of détente begun under President Charles de Gaulle, did not reflect trends in mass public opinion, it was in line with parts of the issue public and the political elites who had become very hostile toward Moscow. Mitterrand's policy was also in part motivated by the domestic desire to isolate and to marginalize the French Communist party.[21]

When Gorbachev came into power and initiated the turnaround in Soviet foreign policy, the French public and their president were more closely in touch. Mitterrand joined in the emerging European consensus that support for Gorbachev's policies was in the Western interest. Mitterrand rarely took the lead on the issues, however, and left this role primarily to the German government (see below). Toward the late 1980s and in contrast to the Bush administration, French policy strongly advocated economic assistance for the Soviet Union and the new democracies in eastern Europe. The new European Bank for Reconstruction and Development, for example, originated from a French initiative.

There was one instance, however, when French policy and French public opinion were out of touch. Almost two-thirds of the French public endorsed German unification. As to the future German role in Europe, 72 percent of the French did not believe that unified Germany would pose a military threat to their country, while 62 percent thought that it would present an economic threat. Moreover, in May 1991, a majority of French public opinion advocated a more active German role in world politics, a proportion that seems to indicate that the French were not overly concerned about growing German power. Forty years of German-French friendship had apparently paid off.[22]

In sharp contrast to mass public opinion, the initial reaction of the French foreign policy elite to German unification represented an attempt to pursue traditional power politics. In December 1989, for example, Mitterrand visited the Soviet Union and apparently tried to convince Gorbachev to cooperate with Paris in order to slow down the process of German unification. Shortly afterward, Mitterrand paid a visit to the communist East German regime, which was rapidly disintegrating at the time. These and other initiatives created the impression in Europe that the French government was trying to prevent German unification. It took Bonn about six months to reassure a nervous French foreign policy elite.[23] To the extent that Mitterrand's policy had anything to do with domestic politics, it responded to concerns among the French political elites, not among public opinion.

In sum, the public opinion–foreign policy conundrum in the French case does not offer a clear-cut answer to the question of who follows whom. But the somewhat contradictory evidence seems to confirm the above-stated assumptions about the role of public opinion in state-dominated domestic structures. If it is up to the president to determine the degree to which public opinion is taken into account, a clear pattern of public impact on foreign policy decisions is not to be expected. Moreover, as in the case of German unification, French policy seemed to be more concerned about elite than about public opinion.

The Federal Republic of Germany: Enthusiastic Response For the Germans, the cold war was over before it was actually over, that is, before the Berlin wall came tumbling down in November 1989. Already in 1986–87, three years before American public opinion turned around, a majority of (West) Germans did not perceive a Soviet threat any longer. One could even argue that the turnaround in Soviet foreign policy initiated by Mikhail Gorbachev accelerated a trend in German public opinion that had already been under way. In December 1987, 73 percent of the West Germans agreed that the Soviet Union was becoming more trustworthy as a result of the reforms begun under Gorbachev. In 1991, 81 percent liked Mikhail Gorbachev, while 65 percent held a favorable opinion of the Soviet Union.[24] The German enthusiasm for Gorbachev, which some in the West termed "Gorbimania," is adequately reflected in these numbers.

Moreover and in contrast to the American public, German public opinion was "dovish" throughout the 1980s. While Americans briefly turned "hawkish" during the Afghanistan crisis in 1980–81, Germans continued to support cooperative relations with eastern Europe and the Soviet Union (*Ostpolitik*), while arms control efforts enjoyed near consensus. In 1987 more than 50 percent of Germans supported unilateral Western arms reduction to encourage Soviet disarmament. Four years later, there was near consensus that the West should provide economic assistance to the Soviet Union.

What was the policy impact of these trends in public opinion? At first glance, there is a striking congruence between the public support for détente and arms control and (West) German policy toward the Soviet Union throughout the 1980s. During the early 1980s when the superpower relationship was quickly deteriorating, the Bonn government under the chancellorship of the Social Democrat Helmut Schmidt worked hard on "damage limitation" to preserve *Ostpolitik* and the relations with eastern Europe and the Soviet Union. Schmidt's conservative successor Helmut Kohl continued the path of détente and arms control despite the fact that his party, the Christian Democrats, had bitterly opposed *Ostpolitik* throughout the 1970s. When Gorbachev came into power and initiated the revolution in Soviet foreign policy, the German government became the first among the major Western powers to embrace the reforms wholeheartedly. "Genscherism"—named after Bonn's longtime Foreign Minister Hans-Dietrich Genscher—became the catchword for such a supportive approach, which was widely criticized in the West as going too far.[25]

The congruence between German public opinion and Bonn's policy toward the East alone does not answer the question of who followed whom. But a closer look at the public opinion–policy relationship in Germany seems to confirm the proposition that the German domestic structure of democratic corporatism mediates the public influence on the decision-making process.

First, it is hard to explain why the conservative Christian Democrats continued to pursue détente during the early 1980s if one does not take trends in public opinion into account, in particular among the conservative constituency. After all, the CDU had opposed détente throughout the 1970s, and Chancellor Kohl could have followed the lead of his ideological ally, Ronald Reagan, by bringing German policy toward the East more closely in line with the confrontational American stance at the time. Only public opinion can explain why the center-right Bonn government continued to support arms control and détente. Public attitudes were crucial in affecting the coalition-building process within the party system and in bringing about a German consensus on détente. Moreover, this consensus became institutionalized and deeply ingrained in the policy networks of the party system, thus allowing for a highly active and consistent *Ostpolitik*.

Second, the German peace debate of the early 1980s also crucially influenced the coalition-building processes within the party system and prepared the ground for the positive German response to the Gorbachev revolution later in the decade. While the peace movements and other issue publics failed in their immediate goals to prevent the deployment of new medium-range nuclear missiles in West Germany, their long-term impact was profound. Support for vigorous arms control efforts increased in all parties, particularly within the conservative CDU. "Common security" and "security partnership" with the East, favored by the center-left, the churches, and the trade unions and supported by general public opinion, emerged as the new security consensus in the country. Moreover, "disarmers" entered the party system and not only affected coalition-building processes within the Social Democratic party but also raised their voice through a new party, the Greens.

In sum, the coalitions within the policy network gradually moved to the left, embracing "common security" and energetic arms control efforts in order to overcome the East-West division. In contrast to France, the West German domestic structure was open enough for societal influences to allow for these changes in the first place. Unlike the United States, the nature of a strong party system in the policy network allowed for the institutionalization of the societal consensus. When Gorbachev came into power and profoundly changed Soviet foreign policy, the Germans were well prepared. Gorbachev's initiatives fit right into their security consensus.

The Persian Gulf Crisis and War, 1990–1991

The United States: Presidential Leadership At first glance, the Persian Gulf crisis and war do not seem to substantiate the proposition that the society-dominated American domestic structure provides public opinion with considerable influence on U.S. foreign policy. Opinion poll data taken before and after President George Bush decided to send American troops to the Persian Gulf as well as before and after the hostilities started in January 1991 indicate that the American public followed rather than affected the president's choices.[26] Prior to the Iraqi invasion of Kuwait, there was no majority support in public opinion for the commitment of U.S. troops to such a contingency. Fifty-six percent of Americans opposed the idea of sending troops to defend Kuwait on August 4, 1990, three days before the president decided to do just that. Immediately after the decision was taken, however, two-thirds to three-fourths of the American public approved Bush's action.

Five months later, a similar turnaround occurred. In mid-November, only 21 percent approved of starting military action against Iraq, while 71 percent favored waiting to see what Iraq was up to. On January 15, 1991, one day before the war started, Americans were about equally split on whether the United States should start military action against Iraq if Saddam Hussein did not withdraw from Kuwait or should continue to see if the economic sanctions worked. Only three days later, 75 percent of the American public endorsed military actions, while support for continued sanctions had dropped to 20 percent.

These dramatic shifts in public opinion cannot be explained by low issue salience. Americans knew what they were judging. They paid greater attention to the Persian Gulf crisis than to any other more recent international event, including the fall of the Berlin wall. The opinion change can only be accounted for by the well-known "rally round the flag" effect, according to which the American public enthusiastically supports the president in wartime, at least initially. Here, masses clearly followed leaders.[27]

What about the society-dominated U.S. domestic structure? A closer look at the structure of public attitudes on the issues involved reveals that President Bush, who provided strong leadership during the crisis and the subsequent war, nevertheless acted within the constraints of and with an eye on mass public opinion. Two examples serve to illustrate this point.

First, there is the legacy of Vietnam in American public opinion. As one would predict with regard to the attitudes of those who ultimately have to pay the costs of war, public support for military action is closely correlated with expectations about the level of casualties and about the time length of a war. Throughout the Persian Gulf crisis and war, the American people supported military action all the more, the less U.S. casualties they expected and the less time they anticipated the war to last. Enduring American support for the presidential decisions, thus, required military planning that concentrated on avoiding American losses of life as much as possible and ensuring a quick victory for the coalition forces. While military reasoning as such might have allowed for different force postures, the enormous force buildup in the Persian Gulf, the large-scale air war against Iraq prior to the ground offensive, and the massive land attack against the remaining Iraqi forces can hardly be explained without taking the constraints posed by public opinion into consideration.[28]

Second, President Bush worked hard not only to ensure congressional support for his decisions but also to assemble a broad international coalition against Iraq. He secured twelve United Nations resolutions that approved economic sanctions and later sent an ultimatum to Saddam Hussein after which military intervention to expel Iraqi forces from Kuwait was permitted. These efforts not only legitimized U.S. decisions after they had been taken and prevented negative international repercussions of American military actions; they were also crucial for gaining the support of the American public. Public opinion almost unanimously demanded congressional approval. More important, by gathering a large international coalition against Iraq and winning U.N. approval, the president was able to cobble together a *domestic* coalition comprising three of the four clusters of American public opinion on foreign policy that Eugene Wittkopf has identified: "internationalists" who support both aggressive and cooperative foreign policies; "hard-liners" who endorse the use of force; "accommodationists" who favor nonmilitary policies and cooperation with allies and the United Nations in general. Only "isolationists" à la Pat Buchanan on the right and pacifists on the left remained outside this coalition.[29] In other words, assembling an international coalition was crucial to gathering a domestic "winning coalition" in public opinion and in Congress.[30]

France: The Desire to Be Independent The French are usually accused of being the most anti-American nation in western Europe and of frequently insisting on conducting an independent foreign policy.[31] Not so during the Persian Gulf crisis! From the outset, French public opinion supported the coalition effort. More than 80 percent thought that French national interests were affected by the crisis. Seventy-eight percent supported the American decision to send troops to the Persian Gulf, and a majority thought that the U.S. response was appropriate rather than simply an expression of a desire to become the "policeman of the world" (the questioning tried to tap directly into potential anti-American sentiments). Almost two-thirds of the French approved the use of military force to enforce U.N. sanctions against Iraq. Moreover, there was near consensus in French public opinion throughout the crisis and the war that France's contribution to the coalition forces was appropriate. Two-thirds endorsed the view that French forces should remain in the Persian Gulf as long as necessary rather than be brought home as soon as possible. Only in December 1990 were the French (like the Americans) split on whether to continue with economic sanctions or to take military action. In January 1991, however, a majority endorsed the view that efforts to find a peaceful solution to the crisis had been adequate, and 70 percent agreed that the United States had done all it could to avoid the use of force. In sum, not only was French public opinion extremely supportive of the American decisions during the crisis; it also endorsed military action stronger and more consistently than the American people.[32]

If there was French apprehension about U.S. policies during the crisis and the war, it was an elite phenomenon and did not reflect public attitudes. In fact, there was considerable elite opposition to the French participation in the coalition effort. Ironically, the most outspoken critic of the war was the French defense minister, Jean-Pierre Chevènement, who also used to represent the left wing of the Socialist party. Chevènement not only expressed general reluctance to support the president's decision to commit French troops to the coalition forces; he also insisted that "should there be a war and should we take part, it must be France's war" (i.e., rather than an American war).[33] His opposition against the planned ground offensive finally led to his resignation two weeks after the air war had begun.

President Mitterrand, on the other hand, supported the American-led effort to drive Iraq out of Kuwait early on and pledged roughly ten thousand French troops to the coalition forces. But on at least two occasions, he tried to conduct an independent French foreign policy. First, during a speech at the United Nations in late September 1990, he proposed new Middle East peace talks if Iraq agreed to withdraw its forces from Kuwait. The United States and the British were strongly opposed to such an explicit linkage that came very close to ideas promoted by Saddam Hussein himself. Second, in January 1991 the French government conducted a last-ditch effort to convince the Iraqis to retreat from Kuwait just days before the air offensive started.

While the French support for the coalition effort was strongly endorsed by public opinion, it is, of course, difficult to discern who followed whom. However, the various attempts by President Mitterrand to play an independent role during the crisis could not have been motivated by public criticism, because public opinion strongly supported the U.S. leadership. If these moves were at all motivated by domestic concerns, they were inspired by the need to keep the left wing of the French political elite in Mitterrand's own party in line. The case, thus, shows a pattern similar to that observed with respect to the French reaction to German unification. As one would predict in countries with a state-dominated domestic structure, political leaders in Paris were at least as much concerned about elite support as about public endorsement.

The Federal Republic of Germany: "Taxation without Representation" German public opinion widely shared the attitudes of the Americans and the French during the crisis.[34] In August 1990, for example, overwhelming majorities approved the American decision to send forces to the Persian Gulf swiftly without waiting for a multinational response. At the time, more Germans (52 percent) than even Americans supported the use of force to enforce the U.N. embargo against Iraq. A shift similar to that in American public opinion also occurred in Germany. In December 73 percent of the Germans preferred to continue economic sanctions before military action should be taken. In January 1991, however, almost two-thirds endorsed the use of force against Iraq. While German public opinion was slightly less supportive of the use of force than the French, it did not endorse the peace movement activities against the war when hostilities started.

On the other hand, there was a clear limit to German support for the war effort throughout the crisis. At no time did public opinion endorse the participation of German troops in the coalition effort. Various polls between October and January showed that at least two-thirds of the public opposed the deployment of *Bundeswehr* forces in the Persian Gulf. German public opinion was also slightly more reluctant than French public opinion to endorse specific military goals of the war that would go beyond the U.N. mandate. Finally, in January 1991 the Germans were almost equally split on whether or not the efforts to seek a peaceful solution to the crisis had been adequate. However, while German public opinion rejected the use of German troops in the Persian Gulf, it overwhelmingly supported financial contributions to the war effort. A majority also approved of the decision to send *Bundeswehr* aircraft to Turkey which was intended to support a NATO ally in case of an Iraqi attack.

The ambivalence of German public opinion toward the Persian Gulf War—general support for the coalition effort but opposition to German military participation—was not surprising; it reflected a long-standing German ambiguity toward military activities outside the NATO area. While the overwhelming majority of German public and elite opinion favored cooperative internationalism and strongly endorsed the United Nations, there was a domestic consensus that the German constitution did not allow the use of troops outside the NATO area, that is, outside the Tropic of Cancer. This consensus was institutionalized in the party system and prevailed during various "out-of-area" crises of the 1980s when the United States demanded a stronger involvement of Germany in the Middle East. Only *after* the Persian Gulf War was this consensus gradually challenged in the domestic debate.[35]

The policy of the Kohl government during the Persian Gulf crisis and the war reflected the domestic attitudes. When the Persian Gulf crisis unfolded, Bonn was preoccupied with the problems of German unification, which took place in the fall of 1990. The Bonn government endorsed the coalition effort and the U.N. resolutions but refrained from actively influencing American and U.N. policies. Chancellor Kohl made it clear from the outset that Germany could not contribute troops to the coalition effort. Even though the right wing of his party would have supported such a move, the Bonn government could not have committed troops to the Persian Gulf War without risking a serious domestic and constitutional crisis. Instead, Kohl favorably reacted to American pressure with regard to burden sharing and pledged more than $10 billion to the war effort. Moreover, Germany contributed large amounts of military equipment, partly taken from the former East German army, and served as a logistical platform from which United States and British forces were transported to the Gulf.[36] But at no point during the crisis and the war did the German government use its substantial financial contribution to increase its leverage over U.S. actions or to influence U.N. decisions. Rather, the Germans used their deutsche marks to buy themselves out of a political responsibility for the coalition effort.

CONCLUSIONS: SUMMARY AND ALTERNATIVE EXPLANATIONS

I have argued in this chapter that domestic structures mediate the impact of public opinion on the foreign policy of states. Similar issue salience and similar public attitudes do not necessarily lead to similar policy influence across countries. Rather, the differences in impact can be accounted for by the variation in domestic structures. I have illustrated this argument by two case studies involving countries with society-dominated, state-dominated, and corporatist domestic structures. The society-dominated United States provides domestic actors and public opinion with multiple access points into the political system. In the case of American policy toward the former Soviet Union, there were various instances in which the choices of the president were either severely constrained by public opinion or in which the public moved policy in the desired direction. During the Persian Gulf crisis and war, in which the president exerted strong leadership, George Bush's choices were nevertheless heavily influenced by the need to ensure public support.

In France, on the other hand, which resembles the state-dominated type of domestic structure, decision makers need not worry too much about mass public opinion. To the extent that domestic politics mattered at all, the sometimes erratic moves of French foreign policy with regard to both the end of the cold war and the Persian Gulf crisis have to be accounted for by the president's desire to maintain the support of elite groups more than mass public opinion.

Finally, the Federal Republic of Germany comes closest to the corporatist type of domestic structure, in which the policy networks with strong intermediate organizations linking state and society play a significant role in the decision-making process. Concerning both the end of the cold war and the Persian Gulf crisis, German public opinion did not seem to have influenced governmental decisions as directly as in the U.S. case. However, German public attitudes were mediated in both cases by the party system, leading to an institutionalized policy consensus that encompassed both public and elite opinion. While the public's influence on policies was more indirect, it nevertheless left a clear mark on governmental decisions.

A comparative foreign policy analysis that emphasizes domestic political processes does not only have to show that its propositions make empirical sense. It also has to deal with alternative explanations focusing on international structures. Realist approaches, which claim that foreign policy is mainly determined by a country's position in the international distribution of power, are among the most prominent of such accounts.[37] Why bother about domestic politics if reference to international structures can explain outcomes more concisely?

The problem with most realist accounts is that they are at best underdetermining when it comes to explaining specific foreign policy choices.[38] Too much gets lost on the way from international power structures to foreign policy decisions.

In the case of the United States, for example, realists could argue that the cautious but increasingly favorable response to the changes in the former Soviet Union can be explained by the nuclear deterrence system that restrained the actions of both sides. Other realists, however, claim that challengers in fierce hegemonic rivalries such as the cold war more often than not exploit the weaknesses of their opponents (which the United States obviously refrained from doing).[39] Since the Soviet Union was equally constrained by the nuclear deterrence system, such an alternative course of action entailed only moderate risks for the United States. In other words, while there is a realist account of the cautious American approach to the end of the cold war, other realists would have expected a riskier U.S. behavior that took advantage of its former opponent's weakness. Thus, reference to international structure alone does not tell us which of the two realist options the U.S. administration chose.

Concerning the American response to the Iraqi invasion of Kuwait, a realist account would probably make two points. First, a shift in the power balance in the Middle East had to be prevented given the oil resources in the region. Second, the Persian Gulf crisis and war presented the United States with the opportunity to establish itself as the world hegemon after its power rival of the cold war had given up the competition. This explanation can account for the president's initial decision to deploy American forces. But the specifics, such as the nature of the military response and the gathering of an international coalition, including U.N. support, are beyond the scope of the argument. Why would the world's only remaining superpower need to entangle itself in an international alliance against Iraq instead of going alone? Both the huge military buildup and the multilateral instead of unilateral approach can only be explained by the need to ensure domestic support.

French foreign policy toward the former Soviet Union and during the Persian Gulf crisis seems to be more easily accounted for by a realist explanation. The French concern with German unification, for example, can be analyzed in terms of an effort to prevent a change in the regional power balance in Europe and to preserve the French position vis-à-vis Germany. Similar concerns about a great power role might have motivated both the French participation in the coalition effort during the Persian Gulf War and Mitterrand's various attempts at independent policies. However, this analysis does not contradict the account presented here. Leaders of countries with state-dominated domestic structures need not worry too much about domestic politics but can concentrate on the constraints and opportunities presented by international structures.

German foreign policy in the two cases discussed above seems to present the greatest anomaly for an explanation focusing on international structure. The stubborn continuity of (West) German *Ostpolitik* and the country's enthusiastic response to the Gorbachev revolution seem to be at least counterintuitive if one assumes that national security is the number one priority in foreign policy. The German security situation at the frontline of the Iron Curtain, with Berlin even situated inside the opposing bloc, was most precarious throughout the cold war and ultimately depended on the American nuclear guarantee. Why, then, did Bonn risk a major conflict with the United States during the early Reagan years, and, when Gorbachev entered the scene, why did Bonn not pursue a more cautious approach by letting others take the lead in the West?

Of course, one could argue that it was precisely West Germany's precarious geopolitical situation at the East-West borderline that led the country to pursue accommodationist policies toward the Soviet Union. Why not try to ensure one's security through détente and arms control if the country would be the first battleground if the cold war turned hot? But then again, the precariousness of the geopolitical location of Germany allows for at least two policy responses. If we want to understand why the Bonn government opted for détente instead of simply following U.S. policies, domestic politics has to enter the picture.

With the cold war over and the country unified, a realist account would also have expected a different German response to the Persian Gulf crisis. Similar to France, the Bonn government should have played out the country's newly won sovereignty and established itself as one of the great powers in world politics once its dependence on the United States for security had ceased to constrain its actions. Rather, the German government not only did not commit German troops to the coalition effort but even paid the American bills without insisting on policy influence ("taxation *without* representation"). It is hard to reconcile this action with a realist explanation according to which states are supposed to increase their relative power position in the world.

In sum, alternative explanations that emphasize the significance of international power structures in determining a state's foreign policy do not seem to present better or more plausible accounts for the cases discussed here. Propositions about international structural constraints on state behavior either allow for various policy responses (in the American and German cases) or do not contradict a domestic structure explanation (in the French case). As a result, we have to look at domestic politics if we want to understand how states react to the constraints and opportunities presented by the international system. The domestic structure concept presented in this chapter offers such an approach. While domestic politics explanations for foreign policy behavior are often accused of being fuzzy and confusing, domestic structure accounts seem to be concise enough to allow for testable propositions. At the same time, they contain enough richness to accommodate the empirical variety among countries from a comparative perspective.

Notes

A first draft of this paper was presented at the Fifth Thomas P. O'Neill Symposium in American Politics, Boston College, April 3–4, 1992. I thank the conference participants, particularly David Deese and Andrew Moravcsik, for their valuable comments. I am also grateful for suggestions by Bruce Russett and an anonymous reviewer. The chapter builds upon and expands an argument first presented in Thomas Risse-Kappen, "Public Opinion, Domestic Structure, and Foreign Policy in Liberal Democracies," *World Politics* 43, no. 4 (July 1991): 479–512.

1. See Chapters 9 and 10 in this volume. See also Ronald Hinckley, *People, Polls, and Policymakers: American Public Opinion and National Security* (New York: Lexington Books, 1992); Benjamin I. Page and Robert Y. Shapiro, *The Rational Public: Fifty Years of Trends in Americans' Policy Preferences* (Chicago: University of Chicago Press, 1992); Bruce Russett, *Controlling the Sword: The Democratic Governance of National Security* (Cambridge, Mass.: Harvard University Press, 1990); Eugene Wittkopf, *Faces of Internationalism: Public Opinion and American Foreign Policy* (Durham N.C.: Duke University Press, 1990). For an excellent review of the revisionist literature see Ole R. Holsti, "Public Opinion and Foreign Policy: Challenges to the Almond-Lippmann Consensus," *International Studies Quarterly* 36, no. 4 (December 1992): pp. 439–466.

2. See, for example, Robert Dahl, *Who Governs? Democracy and Power in an American City* (New Haven: Yale University Press, 1961). For an analysis of the public opinion–foreign policy relationship confirming this argument, see Benjamin I. Page and Robert Y. Shapiro, "Effects of Public Opinion on Policy," *American Political Science Review* 77, no. 1 (1983): 175–90.

3. For the *power elite* theory, see C. Wright Mills, *The Power Elite* (New York: Oxford University Press, 1956). For studies arguing this view in the public opinion literature, see, for example, Noam Chomsky and Edward Herman, *Manufacturing Consent* (New York: Pantheon, 1988); Benjamin Ginsberg, *The Captive Public: How Mass Opinion Promotes State Power* (New York: Basic Books, 1986); James N. Rosenau, *Public Opinion and Foreign Policy: An Operational Formulation* (New York: Random House, 1961).

4. Compare, for example, mass attitudes toward nuclear weapons and arms control in the United States, France, and Germany. These attitudes have been remarkably similar throughout the cold war, while the policies of the three countries differed considerably. For data on U.S., French, and German public opinion on nuclear weapons, see Richard Eichenberg, *Public Opinion and National Security in Western Europe* (Ithaca N.Y.: Cornell University Press, 1989); Thomas W. Graham, *American Public Opinion on NATO, Extended Deterrence and Use of Nuclear Weapons* (Cambridge, Mass.: Center for Science and International Affairs, Harvard University, 1989); Russett, *Controlling the Sword,* chap. 4.

5. See, for example, Peter Katzenstein, ed., *Between Power and Plenty* (Madison: Univer-

sity of Wisconsin Press, 1978); Peter Katzenstein, *Small States in World Markets* (Ithaca, N.Y.: Cornell University Press, 1984); Peter Gourevitch, *Politics in Hard Times* (Ithaca, N.Y.: Cornell University Press, 1986). See also Michael Barnett, "High Politics Is Low Politics: The Domestic and Systemic Sources of Israeli Security Policy, 1967–1977," *World Politics* 42, no. 4 (1990): 529–62; Matthew Evangelista, *Arms and Innovation* (Ithaca, N.Y.: Cornell University Press, 1988); Michael Evangelista, "Domestic Structure and International Change," in Michael Doyle and G. John Ikenberry, eds., *New Thinking in International Relations,* forthcoming; G. John Ikenberry, *Reasons of State* (Ithaca, N.Y.: Cornell University Press, 1988); G. John Ikenberry et al., eds., *The State and American Foreign Policy* (Ithaca, N.Y.: Cornell University Press, 1988); Risse-Kappen, "Public Opinion, Domestic Structure, and Foreign Policy."

6. The following builds upon Risse-Kappen, "Public Opinion, Domestic Structure, and Foreign Policy," p. 486. See also Peter Katzenstein, "Introduction" and "Conclusion," in Katzenstein, ed., *Between Power and Plenty,* pp. 3–22, 295–336; Evangelista, "Domestic Structure and International Change"; Gourevitch, *Politics in Hard Times;* Ikenberry, "Conclusion: An Institutional Approach to American foreign economic policy," in Ikenberry et al., eds., *State and American Foreign Policy,* pp. 219–243.

7. For details on the American domestic structure, see Chapters 2–6 in this volume. See also Dan Caldwell, *The Dynamics of Domestic Politics and Arms Control: The SALT II Treaty Ratification Debate* (Columbia, S.C.: University of South Carolina Press, 1991); James M. Lindsay, *Congress and Nuclear Weapons* (Baltimore, Md.: Johns Hopkins University Press, 1991); David Mayhew, *Divided We Govern* (New Haven: Yale University Press, 1991).

8. On the U.S. armament industry see, for example, Gordon Adams, *The Politics of Defense Contracting: The Iron Triangle* (New York: Council on Economic Priorities, 1984); Fen O. Hampson, *Un-guided Missiles: How America Buys Its Weapons* (New York: W. W. Norton, 1989). On the U.S. peace movements, see David S. Meyer, *A Winter of Discontent: The Nuclear Freeze and American Politics* (New York: Praeger, 1990); Frances B. McCrea and Gerald E. Markle, *Minutes to Midnight: Nuclear Weapons Protest in America* (Newbury Park, Calif.: Sage, 1989).

9. This is the conclusion of various studies. See, for example, Aaron Friedberg, "Why Didn't the United States Become a Garrison State?" *International Security* 16, no. 4 (Spring 1992): 109–42; Hampson, *Un-guided Missiles;* Gert Krell, *Rüstungsdynamik und Rüstungskontrolle: Die gesellschaftlichen Auseinandersetzungen um SALT in den USA, 1969–1975* (Frankfurt/M.: Haag and Herchen, 1976); Stephen Rosen, ed., *Testing the Theory of the Military Industrial Complex* (Lexington, Mass.: Heath, 1973).

10. For the following, see Robert Aldrich and John Connell, eds., *France in World Politics* (London: Routledge, 1989); Douglas Ashford, *Policy and Politics in France: Living with Uncertainty* (Philadelphia: Temple University Press, 1982); Paul Godt, ed., *Policy-Making in France: From De Gaulle to Mitterrand* (London: Pinter, 1989); Alfred Grosser, *Affaires Extérieures: La Politique de la France, 1944–1984* (Paris: Flammarion, 1984).

11. This definition is broader than the original concept of *corporatism,* which primarily referred to the interaction patterns between labor, business, and state agencies. See, for example, Philippe Schmitter and Gerhard Lehmbruch, eds., *Trends towards Corporatist Intermediation* (London: Sage, 1979); Peter Katzenstein, *Corporatism and Change* (Ithaca, N.Y.: Cornell University Press, 1984).

12. On the German foreign policy-making process, see Helga Haftendorn, ed., *Verwaltete Außenpolitik* (Köln: Wissenschaft and Politik, 1978); Barry Blechman et al., *The Silent Partner: West Germany and Arms Control* (Cambridge, Mass.: Ballinger, 1988). On the German domestic structure in general, see Peter Katzenstein, *Policy and Politics in West Germany* (Philadelphia: Temple University Press, 1989).

13. These election procedures include the proportionate vote and the provision that a party has to gain at least 5 percent of the votes nationwide in order to be represented in parliament.

14. See, for example, Rob Burns and Wilfried van der Wills, eds., *Protest and Democracy in West Germany: Extra-Parliamentary Opposition and the Democratic Agenda* (New York: St. Martin's Press, 1989); Russell Dalton and Manfred Küchler, eds., *Challenging the Political Order: New Social and Political Movements in Western Democracies* (New York: Oxford University Press, 1990); Thomas Risse-Kappen, *Die Krise der Sicherheitspolitik: Neuorientierungen und Entscheidungsprozesse im politischen System der Bundesrepublik Deutschland, 1977–1984* (Mainz-München: Grünewald-Kaiser, 1988); Thomas Rochon, *Mobilizing for*

Peace: The Antinuclear Movements in Western Europe (Princeton, N.J.: Princeton University Press, 1988).

15. For data see, for example, figure 10-5 in this volume. See also Risse-Kappen, "Public Opinion, Domestic Structure, and Foreign Policy," pp. 494–95.

16. For data, see CBS News/*New York Times* poll, "The Malta Summit," New York, 12/2/89; "Americans Much Warmer toward Soviets, Poll Finds," *New York Times,* December 3, 1989.

17. For data, see Chapter 10 in this volume; CBS News/*New York Times* poll, "The Malta Summit"; Risse-Kappen, "Public Opinion, Domestic Structure, and Foreign Policy," pp. 496–97.

18. For details, see Strobe Talbot, *The Master of the Game: Paul Nitze and the Nuclear Arms Race* (New York: Knopf, 1988). The best study on the nuclear freeze movement is Meyer, *Winter of Discontent.*

19. Cf. the fate of the initiative by Sam Nunn (D-Ga.) and Les Aspin (D-Wis.) to spend $1 billion of the Pentagon budget to assist the dismantling of former Soviet nuclear weapons. In the end, only $400 million was allocated and much less was spent. On the Bush administration's foreign policy toward the former Soviet Union, see Michael Beschloss and Strobe Talbott, *At the Highest Levels The Inside Story of the End of the Cold War* (Boston: Little, Brown, 1993).

20. For the following, see data in Eichenberg, *Public Opinion and National Security;* Michel Girard, "L'opinion publique et la politique extérieure," *Pouvoirs,* no. 51 (1989); United States Information Agency, "May 1991 Security Survey: German Data Tables," November 4, 1991; USIA, "Post Soviet Coup Telephone Survey," September 1991.

21. For details, see Julius W. Friend, *Seven Years in France* (Boulder, Colo.: Westview Press, 1989).

22. Data in Spiegel-Spezial, *Das Profil der Deutschen* (Hamburg, January 1991), p. 27; USIA, "May 1991 Security Survey."

23. A very good source of French-German relations in 1989–90, when German unification was negotiated, is Horst Teltschik, *329 Tage* (Berlin: Siedler, 1991). Teltschik was Chancellor Helmut Kohl's foreign policy adviser at the time. See also Ronald Tiersky, "France in the New Europe," *Foreign Affairs* 71, no. 2, (Spring 1992): 131–46.

24. See the data in Hans Rattinger, "The INF Agreement and Public Opinion in West Germany," in David Dewitt and Hans Rattinger, eds., *East-West Arms Control* (London: Routledge, 1992), pp. 165–90; Thomas Risse-Kappen, "Anti-Nuclear and Pro-Détente? The Evolution of the West German Security Debate," in Don Munton and Hans Rattinger, eds., *Debating National Security: The Public Dimension* (New York: Lang Publishers, 1991), pp. 269–99; Risse-Kappen, "Public Opinion, Domestic Structure, and Foreign Policy," pp. 494–95; USIA, "May 1991 Security Survey." The following data are also taken from these sources.

25. On German *Ostpolitik,* see Helga Haftendorn, *Security and Détente: West German Foreign Policy, 1955–1983* (New York: Praeger, 1985); Wolfram Hanrieder, *Germany, America, Europe* (New Haven: Yale University Press, 1989); Clay Clemens, *Reluctant Realists: The Christian Democrats and West German Ostpolitik* (Durham, N.C.: Duke University Press, 1989); Risse-Kappen, *Krise der Sicherheitspolitik.*

26. The following opinion polls are taken from Hinckley, *People, Polls, and Policymakers,* pp. 109–11; "Americans More Wary of Gulf Policy, Poll Finds," *New York Times,* November 20, 1990; "Americans Don't Expect Short War," *New York Times,* January 15, 1991, p. A-11; "America's Success in the Gulf Healed Several Wounds at Home," *New York Times,* March 4, 1991, p. A-11.

27. On the "rally round the flag" effect, see John Mueller, *War, Presidents, and Public Opinion* (New York: Wiley, 1973). For a critical discussion, see Russett, *Controlling the Sword,* pp. 34–51.

28. The military recommendations to the president by the Joint Chiefs of Staff and by Gen. Norman Schwarzkopf were to a large extent influenced by their concern about a loss of support from the American people in case of high casualties and a drawn-out war of attrition, as was experienced in Vietnam. See the account in Bob Woodward, *The Commanders* (New York: Simon and Schuster, 1991). See also Lawrence Freedman and Efraim Karsh, *The Gulf Conflict, 1990–1991* (Princeton, N.J.: Princeton University Press, 1993). On opinion poll data with regard to this point, see Hinckley, *People, Polls, and Policymakers,* pp. 118–19; "Americans Don't Expect Short War."

29. I owe this point to Bruce Russett. See his "The Gulf War as Empowering the United Nations," in Edward Greenberg et al., eds., *War and Its Consequences: Lessons from the Persian Gulf Conflict* (New York: Harper Collins, 1994). On the four clusters see Eugene Wittkopf, *Faces of Internationalism.*

30. Thus President Bush's efforts closely resemble what Robert Putnam has termed "two-level games," emphasizing the interaction of international and domestic coalition-building attempts. See Putnam, "Diplomacy and Domestic politics: The Logic of Two-Level Games," *International Organization* 42 (1988): 427–60.

31. For research assistance on this part, I thank Richard Tanksley.

32. The opinion poll data reported above are taken from various USIA-commissioned surveys (courtesy of Ronald Hinckley). See also his "World Public Opinion and the Persian Gulf Crisis," paper presented at the Annual Meeting of the American Association for Public Opinion Research, Phoenix, May 16–19, 1991.

33. Jean-Pierre Chevènement, quoted in "Mitterrand Takes a Strong Stand," *New Leader,* September 3, 1990, p. 11. On French policy during the Persian Gulf crisis and war, see also "The Center Holds—For Now," *Time,* September 3, 1990, "The Gaul of It," *Economist,* November 17, 1990, p. 15; "French Policy Upsets Friend and Foe, at Home and Abroad," *New York Times,* January 24, 1991, p. A-10; "French Minister's Stand on War Draws Criticism," *Washington Post,* January 25, 1991, p. A-30; "French Defense Chief Quits, Opposing Allied War Goals," *New York Times,* January 30, 1991, p. A-11; "France Tries to Reconcile Role in Gulf War with History of Strong Arab Ties," *Christian Science Monitor,* February 1, 1991, p. 5.

34. The following opinion poll data are taken from USIA-commissioned telephone surveys (courtesy of Ronald Hinckley). See also the data in *Der Spiegel,* January 28, 1991.

35. There is an additional ambiguity when it comes to German policies in the Middle East. On the one hand, there is the special German relationship with Israel resulting from history and the Holocaust. On the other hand, Germany as a "trading state" has strong economic ties with the Arab world. As a result, Bonn has always been reluctant to pursue an active Middle Eastern policy. For details on German policies in the area, see Shahram Chubin, ed., *Germany and the Middle East: Patterns and Prospects* (London: Pinter, 1992).

36. For details, see Isabelle Grunberg, "The 'Two Congresses': Raising Funds abroad and at Home for the Persian Gulf War," paper presented at the Pan-European Conference on International Relations, Heidelberg, Germany, September 16–20, 1992; Dorothee Heisenberg, "A Comparison of the U.S. Attitude towards Japan and Germany during the Persian Gulf War," Yale University, February 25, 1992, unpublished; Helmut Hubel, *Der zweite Golfkrieg in der internationalen Politik,* Arbeitspapiere zur internationalen Politik (Bonn: Europa Union, 1991); Chubin, ed., *Germany and the Middle East.*

37. See, for example, Kenneth Waltz, *Theory of International Politics* (Reading, Mass.: Addison-Wesley, 1979); Joseph Grieco, *Cooperation among Nations* (Ithaca, N.Y.: Cornell University Press, 1990).

38. On this point, see in particular Robert Keohane, "Theory of World Politics: Structural Realism and Beyond," in Robert Keohane, ed., *Neorealism and Its Critics* (New York: Columbia University Press, 1986), pp. 158–203.

39. See, for example, Robert Gilpin, *War and Change in World Politics* (Cambridge: Cambridge University Press, 1981).

12 / Making American Foreign Policy in the 1990s

DAVID A. DEESE

In the new global environment of the 1990s the challenges to American foreign policy will originate at home as often as overseas. With the end of the Soviet Union and the cold war and the heightening of multiple, diverse global threats, the U.S. government must articulate a new overarching rationale for American foreign policy. And as the purposes change, so will the nature of the people, ideas, and institutions that make foreign policy in the United States. The effectiveness of U.S. leadership in world affairs under these new conditions will depend heavily upon Americans' ability to understand and respond to "the new politics of American foreign policy."

Yet the vast majority of books and analysts of American foreign affairs ignore how policy is actually made. They study the "issues" but largely overlook the people, ideas, and institutions that make U.S. foreign policy. Those who do analyze how policy is made most often follow the conventional, "top-down" view, which focuses heavily on the president as the dominant player. Public opinion in this view is volatile, unpredictable, inconsistent, and generally not very important to American foreign policy. The media and Congress are viewed as hindrances that must be managed or even manipulated on the way to making effective foreign policy.[1] The prevailing wisdom also tends to emphasize the differences between the domestic and foreign realms, the "high-politics" nature of foreign policy, and the need to reestablish "bipartisanship" in foreign affairs. Officials and analysts of this school are likely to be particularly concerned about any erosion of presidential or American power that may be brought about by the new global environment of the 1990s.

This volume offers a new direction. It presents American foreign policy as a continuous but dynamic process from the American founding, but it emphasizes change in the aftermath of two defining wars in the twentieth century: World War II and the cold war, with particular attention to the period since the Vietnam War. It argues that the initiative for change, even for leadership in U.S. foreign policy, can begin in various quarters. It can be external, such as a response to foreign decisions, leaders, events, and movements, as argued in Chapter 1. As other chapters have demonstrated, however, foreign policy change is increasingly a presidential response to American public opinion, mass media, interest groups, political movements, Congress, and of course the executive agencies and departments. The "new politics" argument synthesizes elements of the "top-down" and "outside (international)-in" perspectives, but it is unique in its focus on the "bottom-up" character of American foreign policy making. It emphasizes not only the increasing overlap and intermeshing of domestic and foreign policy issues but also the ways in which international coordination, coalitions, and institutions are required to accomplish domestic as well as foreign objectives. But the principal argument of this chapter is that new issues, more actors, and a wider range of interests do not necessarily create a less effective, "politics as usual" foreign policy process. On the contrary, the American foreign policy process could be strengthened and improved by a more consensus-oriented process that is driven by "deliberative politics."

U.S. FOREIGN POLICY LEADERSHIP AND THE NEW GLOBAL ENVIRONMENT

More than ever, the post–cold war environment demands artful foreign policy initiatives and follow-through by presidents and top congressional leaders who are willing to educate, negotiate, and build consensus. With basic change in American politics and the end of the clear and present international danger posed over the past three-quarters of a century, William Schneider argues in the Introduction that "it is reasonable to expect Americans to have less interest in world affairs." In fact, as Schneider explains, the growth of "politics as usual" in U.S. foreign affairs could "turn off" the American public and undermine basic support for American global leadership in the 1990s. Without insightful and creative foreign policy leadership, it will be exceedingly difficult for top officials to sustain American support for important efforts, such as funding the rapidly escalating number of United Nations Security Council peacekeeping and peacemaking operations worldwide, generating financial support for reform in Russia and the surrounding republics, strengthening the General Agreement on Tariffs and Trade, and enhancing trade and cooperation in the western hemisphere through the North American Free Trade Agreement.

In a world that "is far less manageable than it was fifty years ago,"[2] creative leadership for American foreign policy becomes all the more necessary. The president, members of Congress, and independent experts will have to devote considerable energy to devising and targeting political strategies to create and sustain American support for the highest priority policies. It is crucial that the president in particular enunciate a coherent set of goals for American foreign policy in the 1990s, in order to avoid policies which are driven by a series of hasty responses to immediate crises. The rising number of new states, escalating assertiveness by ethnic and religious movements, and the complexity of "mixed interests" and conflicting principles, means a decreasing U.S. ability to exclusively control global markets and institutions, foreign governments and peoples. The bombing of the World

Trade Center in 1993 by immigrants associated with Islamic fundamentalism in Egypt highlights the permeability of borders and the vulnerability of the United States to transnational groups and violence. Americans could become frustrated at what foreign policies cannot accomplish in the complex international and domestic environments of the 1990s and skeptical of attempts at creative solutions.

At home most foreign policy issues will confront a more divided, highly decentralized, and more partisan policy process, which can produce policies with multiple and conflicting objectives. The support of allies and international mechanisms, such as the U.N. Charter and Security Council, the International Monetary Fund, and the World Bank will be central not just to effective foreign policy but also to selling policies in the United States. For example, President George Bush's intense and effective efforts at mobilizing broad international support for the use of force against Iraq in 1990–91 were even more important to the domestic political mission of achieving congressional and public support than they were essential to the success of military operations on the battlefield. As Robert Scigliano argues, Congress and the president were both using the moral and political suasion of the U.N. Charter and Security Council support, as well as the agreement of traditional allies and nations in the Middle East region, to bolster their positions relative to each other. Congressional leaders argued that Congress would not support war against Iraq if the United States did not act in concert with the United Nations and U.S. allies, while the president relied heavily upon U.N. and allied support to justify to members of Congress, independent experts, and the public his intention to force Iraq from Kuwait by all necessary means. Similarly, at a December 1990 hearing on the advisability of offensive military action against Iraq, Senator Sam Nunn (D-Ga.), the leading authority and political force in the Senate on defense matters, pointed to the effectiveness of the U.N.-imposed economic sanctions against Iraq to question and influence President Bush's decisions on the use of force against Iraq.

In another case, the need to mobilize international financial aid to support economic, financial, and political reform in Russia, a critical U.S. foreign policy goal, simply cannot be accomplished without major assistance from international institutions. "At a time when most industrial nations are plagued by anemic growth and budget deficits, the Clinton Administration and Group of Seven Governments are looking increasingly to the International Monetary Fund and World Bank to help Russia."[3] In light of the generally strong U.S. public aversion to foreign aid, as explained by Bert A. Rockman, and the need to reduce the U.S. federal deficit, among other urgent domestic priorities, the only way to help Russia manage its extremely difficult and important challenges is to rely heavily on these institutions. A large majority of the twenty-eight billion dollar aid package approved at an emergency meeting of the Group of Seven industrial nations in April of 1993 was provided by the International Monetary Fund and the World Bank. And almost all of the loans, stabilization funds, and loan guarantees were to be administered by these institutions, with some support as well from the European Bank for Reconstruction and Development.[4]

Thus it is urgent that groups and publics understand that central U.S. goals at home as well as abroad simply cannot be achieved without effective coordination of foreign policies across countries and international institutions. American opinion leaders and publics must thoroughly deliberate the use of international coalitions and institutions to focus, conserve, and leverage scarce U.S. foreign policy resources under these new international conditions. In a fundamental sense, state sovereignty and American power or influence in particular, are not at odds with strengthened international institutions. By the 1990s, for most issues American power and resources may be reinforced, not undermined, by involving the United Nations, the International Monetary Fund, or the World Bank in the foreign policy decision-making and execution process. Rather than the deeply ingrained prevailing view that states must decide to, or not to, cede power to an international organization, in what game theorists call a zero-sum relationship, even the most powerful nations increasingly will find that they can multiply their influence at home and abroad and strengthen the basic foundations of their "sovereignty" over the longer term through international coalitions and institutions. In the context of the European Community this new approach has been termed "pooled sovereignty."[5] For most other nations it can be seen as an opportunity to "enhance" their sovereignty. What new global and domestic political environments have taken away in influence from the state, international coordination may in part replace.

The challenge is clear. At a time of overwhelming domestic demands in the United States and major industrial nations, the end of both the cold war and its clear threat to the United States, and a complex and confusing new international environment, it is entirely natural that Americans might turn inward and be skeptical of what can be accomplished by U.S. foreign policy. International issues and regions are less stable and predictable in the 1990s, as are the politics of making American foreign policy. At the very time that foreign policy leadership at home and abroad is most needed, it may also be the most difficult to accomplish.

"POLITICS AS USUAL" OR "DELIBERATIVE POLITICS"?

Given the complexity of the new international environment and the "new politics" of American foreign policy, is it likely that the American public will dismiss foreign policy as "politics as usual," just as it has domestic policy, and even reject strong U.S. international leadership? Despite the conventional wisdom that presidents must control the process of making U.S. foreign policy, it is difficult to determine whether the foreign policy process of the 1980s or 1990s is less effective than that of the 1950s or 1960s. In fact, the changes that characterize the "new politics" of foreign policy point to the possibility of a stronger, more effective, and more democratic policy process. A longer, more deliberative discussion and debating process does not necessarily indicate weakness. Argument, study, and contention, if focused and structured by a combination of presidential, congressional, and independent expert participation, are more likely to lead to consensus than to paralysis, to consistency than to instability, to moral choices than to expediency.

Because of the decentralized nature of U.S. political institutions for making foreign policy, including a relatively strong role for Congress, interest groups, and society generally, at times the system appears to be overloaded or to break down under the strain of multiple, conflicting demands from all directions at once. As Thomas Risse-Kappen explains, interest groups generally have easy access to the foreign policy institutions in the United States, but the public policy system lacks a coherent, consistent means to connect the society and the institutions. At times, therefore, it is difficult to move from the active debate or conflict stage to consensus building and agreement. With the declining role of U.S. political parties by the 1970s and 1980s, interests such as the nuclear freeze movement of the early 1980s are strongly represented for a relatively brief time but cannot sustain or "institutionalize" their position as they could within the strong political parties in Germany. What David W. Rohde terms "conditional party government" in Congress fulfills this role in part, but it does appear that the United States would be well served by more stable representation of ideas and interests over time than by brief, sometimes overwhelming bursts of interest group activity and influence.

In the "new politics" deliberation can be a crucial means to develop the consent and knowledge base of the community. "To moderate partial views and yet learn from them, to insure meritorious outcomes, and to serve the goal of civic education, government officials need to foster a continuing conversation about public policy and the values and ideals upon which such decisions depend."[6] In an era when the foreign policy process will more closely resemble that of domestic policy, congressional and executive branch leaders, both elected and appointed, will be called upon to provide constructive leadership through "deliberative politics." Deliberation demands that officials and opinion leaders help to make available more information for consideration and discussion, with the aim of arming public opinion and interest groups with a fuller, richer view of the foreign policy alternatives on a given issue. In light of the findings of Thomas W. Graham and Robert Y. Shapiro and Benjamin I. Page that many Americans have the ability to understand and evaluate foreign policy issues, leaders should be engaging the public that follows foreign policy to the greatest possible extent. Leaders should focus on establishing a process that clearly maps out not only the problems and complexities of reaching effective policies but also the range of competing values and preferences among the public and engaged groups. The goal is ultimately to encourage people to consider the conflicts between their deeper values and their practical policy preferences on an issue, to communicate their choices to leaders, and to accept policies that differ from their preferences. Furthermore, public opinion sometimes expresses competing preferences, for example, support for President Yeltsin but opposition to foreign aid that could be essential to his political survival. Therefore, it is particularly important for the public to engage in this process of contrasting values and preferences, and of understanding the real resource constraints confronting leaders.

The "new politics" also means that it is likely to take more time to deal effectively with the new mix of foreign issues in the 1990s, to incorporate the new actors, and to shape competing values and interests. Public opinion leaders need to be more patient with the deliberative process, and to weigh carefully the question of when the time is right to push for closure in reaching decisions. Legislators should exercise the substantial ability of Congress to represent ideas and values held by the general public when they are in conflict with policies advocated by special interests. The media can resist being manipulated by government officials and more often bring independent expert review to bear on policies under deliberation. In most cases the longer process will turn out to be time well spent, in light of the long-term nature of the foreign policy challenges now confronting the United States. In some respects the new environment both requires and lends itself to a greater degree of deliberation, involving a greater number of participants in the process. On this dimension, the U.S. foreign policy process could and should become more like that in Germany, Sweden, or Japan. The outcome could be a process more representative of basic American values and beliefs. In general, a longer debate, involving more press coverage and more public input and participation, will result in policies that are more thoroughly tested and supported, more durable, and therefore more likely to succeed.

DISAGREEMENT, DEBATE, AND THE MEDIA

A crucial driving force in determining how the American foreign policy process works for any particular issue and point in time is the degree of disagreement voiced by executive branch officials, members of Congress, and the opinion leaders. Throughout this volume the authors, and W. Lance Bennett in particular, point to the extent of executive branch and executive-congressional disagreement as critical in determining the extent of media coverage, public debate and participation, and overall openness of the foreign policy process on any given issue. In general, when officials do not express disagreement in public or with members of the press, there is considerably less press coverage and congressional and public debate. When they do disagree, the controversy may engage congressional attention and opposition, which in turn almost guarantees significant media coverage and the availability of much more information to the public and interest groups. Clearly, in almost every case the extent of the media debate is central to determining the type of foreign policy process for a given issue.

In the new global environment, live television coverage of international events and new communications technologies allowing rapid information transmission introduce news to attentive publics and government officials simultaneously. With segments of the public, interest groups, politicians, and foreign policy officials all reacting to media coverage of the same international events in the same day, there are naturally multiple, complex influences across all the actors in the foreign policy process. With trade, environmental, and regulatory issues increasingly in the forefront of the news and media programming, it is more difficult to forge strong agreement across government agencies and departments, and the policy stages, from agenda setting, to negotiation, to political approval and execution, are more open and less time urgent. The "rally round the flag" effect whereby presidential popularity usually rises in times of crisis (unless the public blames political elites for creating the crisis) may occur less often and less strongly in the "new politics," in part because there may be fewer opportunities for executive branch officials to manipulate the information. Furthermore, members of Congress, the public, and interest groups are more likely to monitor and oppose decisions with which they disagree, such as those of the Reagan administration on Central America.

Under these circumstances, Donald L. Hafner's argument for strong presidential leadership of the foreign policy bureaucracy takes on particular importance. If the president is consistently attentive and engaged in leading the bureaucracy on a policy issue over time, it increases the chances for deliberation and consensus building and reduces the likelihood of misunderstandings, unresolved differences, and public disagreements within the White House or executive departments and agencies, and between the president and executive branch officials. Presidential participation helps to minimize the strong inclinations of both executive departments to compete with each other and the White House staff to focus excessively on winning favorable public opinion ratings for the president. Without active presidential involvement, for example, the deep divides of the Bush administration over foreign economic policy will also haunt the Clinton administration. Strong presidential leadership of the bureaucracy, whether focused through an interagency committee, the National Economic Council, the National Security Council, or leadership by a particular department, also lends itself more readily to the early involvement of members of Congress and interest groups. Particularly in an era when, as I. M. Destler explains, many of the executive departments and agencies involved in foreign policy have domestic issues as their primary responsibility, increased consensus within the executive branch and the involvement of Congress may reduce the active opposition of Congress at the later stages of approving and executing policies. Even when congressional or interest group representatives oppose a particular initiative, it is generally preferable to have the reasons for opposition or disagreement voiced early, when it is relatively easy to make adjustments and build consensus on a policy.

As this volume has amply demonstrated, American foreign policy was more political in the 1970s and 1980s than during the 1950s and 1960s. Indeed, despite the new period of "one-party" government, with Democratic control of the White House and Congress beginning in 1993, there are good reasons to be concerned about possible changing priorities and attitudes among elite and broad public opinion in the United States. Even with the long recession of the early 1990s behind, many Americans and legislators may still want to keep the focus in the mid-1990s on deficit reduction, job creation, and domestic policy issues such as health care and education. Yet at a minimum, American foreign policy opinion leaders must convince the members of Congress and the various publics that the nation's security, prosperity, and international competitiveness are tightly interwoven with its global trade, monetary, financial, diplomatic, and military relations. Enlightened leadership by the president, and attempts at public education by members of Congress, the media, some interest groups, and opinion leaders outside government will be influential to public attitudes and priorities.

THE AMERICAN PUBLIC, OPINION LEADERS, AND PRESIDENTS

The preceding chapters raise important questions concerning the prevailing understanding of presidential leadership and the role of public opinion in American foreign policy. Very much contrary to the conventional perspective, this volume finds that "public opinion regarding foreign policy is not inconsistent or incoherent. Rather it appears to be stable and clear about a number of fundamental issues relating to the U.S. position in international affairs. The public displays rational, responsible, sensible, and even sophisticated opinion about national security matters."[7] As Thomas W. Graham and Robert Y. Shapiro and Benjamin I. Page have forcefully argued, public opinion plays a significant, sometimes central role in U.S. foreign policy. Its influence on policy makers and policy may be either direct or indirect, but it always plays some role in the minds of elected decision makers and their political appointees. Its influence can occur early in the policy process when issues first are being moved on and off the agenda or when policy options are being formulated and negotiated within the executive and/or legislative branches, as well as in the coalitions of experts analyzed by John T. Tierney. The influence of public opinion also occurs later in the process as policies are formalized and approved, for example in congressional authorization or appropriations committees, and, finally, as they are executed by the multiple federal government agencies and departments in the economic and security complexes analyzed by I. M. Destler.

The direct influence of public opinion can be quite subtle and difficult to measure, such as a decision by officials not to pursue a desired policy as a result of known and anticipated public opposition, as, for example, major U.S. military intervention against Nicaragua under President Ronald Reagan. As Thomas W. Graham explains, presidents since Franklin Roosevelt (and even earlier in some cases) have commissioned their own private public opinion polls as a means to gauge whether policy options are likely to be politically feasible. Private polls have also been relied upon to determine how specific components of public opinion might be used to mold public attitudes in a particular way, such as when President Jimmy Carter undertook a major, nationwide public relations effort to gain passage of the Panama Canal treaties. Further, as overlapping foreign and domestic issues come to dominate the agenda, we can increasingly expect leaders to routinely assess public opinion in foreign countries and to make use of this information in formulating and executing U.S. foreign policy, as, for example, in the case of the Reagan administration's tracking and responding to the peace movement in western Europe in the early 1980s.

Presidents, other top officials, and leaders in Congress certainly make multiple uses of opinion data. In some cases data may help them to avoid, or to reformulate policies that are expected to raise strong opposition among publics at home and in other countries, interest groups, rank-and-file members of Congress, or the bureaucracy. The indirect influence of public opinion occurs, for example, when top officials derive from opinion data the information they need to undermine the influence of determined interest groups or other actors in the foreign policy process who oppose their proposals. Opinion poll results can also be publicized in order to rally additional public support for a favored policy. Additional types of indirect effects occur when the media, interest groups, or independent experts use opinion data to highlight an issue in order to support or oppose a policy at any particular point in the process. No matter what the uses of the poll data, it has become crucial for leaders to gauge the overall level of public opinion support and opposition. As Thomas W. Graham argues, when public opinion is nearly unanimously in favor of a particular course of action, it is very likely to predominate, and whenever it is 60 percent or more in agreement it is likely to play a significant role in shaping policy.

In the 1990s presidents and top congressional leaders must choose carefully their highest priority foreign policy issues and strategically design, target, and time their political strategies for domestic as well as foreign audiences. Even strong, relatively popular presidents will be humbled by the relatively small shifts in public opinion that they can expect on foreign policy issues that they target with public education and persuasion campaigns. Given the inherent limitations on the extent to which any president can change public opinion and the rapid advances in the ability to collect, analyze, and apply opinion data, there will likely be many temptations for leaders to do what is politically expedient: follow opinion rather than educate and lead it. While it is important in democratic societies for public preferences to be translated into policy, the danger is that the quality of foreign policy will be undermined when it is fashioned in a way that reflects "the lowest common denominator." If top U.S. foreign policy leaders, particularly the president, consistently follow the politically expedient path over the longer term, it will make it more likely that the public will find foreign policy to be little more than "politics as usual." In U.S. trade relations, for example, the politically expedient path is generally to assume that foreign producers are in the wrong and to grant the protection, such as antidumping or countervailing duties, requested by U.S. commercial or financial interests. Under these circumstances there are usually heavy and continuing costs to the consumer, who is underrepresented in Washington, D.C., and to the national economy as a whole. Given these characteristics of the "new politics," if leaders are to be effective they must not only forge compromises with the political interests that are well represented in Washington, D.C. but also insist on doing what is right for the underrepresented and the broader public interest.

When a foreign policy is unpopular, constructive leadership from the president and cabinet officials is particularly important. Excessive presidential persistence, stubbornness, or overly strong personal commitments are destructive to the purposes of American foreign policy. Sometimes top officials encounter strong opposition to an initiative from public opinion, Congress, or their own associates, but decide to persist on their course. In these cases officials must be alert to the possibility that the policy has become counterproductive, and the checks inherent in the American political system must be free to operate, such as congressional oversight and media presentations of independent expert opinions. On issues such as human rights in China beginning in 1989, nuclear proliferation and arms transfers in the third world, or diplomatic relations with authoritarian regimes such as that of Ferdinand Marcos in the Philippines in 1986, Congress has shown that it can constructively act to protect fundamental American values and interests. In those cases where a foreign policy has met strong public opposition for several months, the only tolerable reason for persisting is a clear, strategic objective that can be clearly and fully articulated to the foreign policy bureaucracy, Congress, involved groups, and the media.

It is perhaps an appropriate point in history for Americans to remind themselves of Thomas Jefferson's basic belief that "the objectives of foreign policy were but a means to the end of protecting and promoting the goals of domestic society, that is, the individual's freedom and society's well-being."[8] It is an important reminder that the foreign policy process must adapt to fundamental changes in American society and political institutions, just as the ideas represented in American foreign policies must be in accord with basic American values and beliefs. Indeed, as Senator J. William Fulbright argued, "If America has a service to perform in the world . . . it is in large part the service of its own example."[9] If the nation cannot renew the fundamental domestic sources of its own strength, as well as meet the basic needs of its own people, it is destined to be less legitimate, influential, and effective in world affairs.

Yet in the global environment of the 1990s, it is equally important to recognize that Americans, including Thomas Jefferson both as secretary of state and as president, have always defined themselves as distinctively dedicated to certain principles, both at home and abroad. To be true to its own principles, and even to defend its own interests and values, the nation cannot turn inward and ignore the demand for strong U.S. global leadership during a time of rapid and uncontrollable changes worldwide. The greatest challenge may be developing a new fundamental U.S. foreign policy strategy that avoids both the "arrogance of power" and the temptation to shirk global responsibility. The nation must beware of, in Fulbright's terms, "excessive preoccupation" with either foreign or domestic affairs. The solution is to provide at home the best possible example of democracy, deliberation, and social justice, while sharing global leadership responsibilities with traditional allies and international institutions, and promoting regional and global cooperation on the crucial issues of the 1990s: the overwhelming stresses on third world societies created by rapid population growth, the explosion of ethnicity and religion as political forces, extreme forms of nationalism and the creation of new states, serious environmental degradation, and the global spread of arms and advanced weapons systems.

Notes

1. An important exception with respect to the role of Congress is Thomas E. Mann, ed., *A Question of Balance: The President, the Congress, and Foreign Policy* (Washington, D.C.: Brookings Institution, 1990). Although most of the volume focuses on issues, Mann's opening chapter thoughtfully analyzes the paths to constructive executive-legislative relations in the making of American foreign policy.

2. Roy C. Macridis, "The United States in a New World," in Roy C. Macridis, ed., *Foreign Policy in World Politics,* 8th edition (New York: Prentice Hall, 1992), p. 411.

3. Steven Greenhouse, "I.M.F. Easing Rules for Moscow Aid," *New York Times,* April 10, 1993, p. 5.

4. David E. Sanger, "7 Nations Pledge $28 Billion Fund to Assist Russia," *New York Times,* April 16, 1993, p. 1.

5. See Robert O. Keohane and Stanley Hoffmann, eds., *The New European Community* (Boulder, Colorado: Westview, 1991), pp. 13 and 30.

6. Marc K. Landy, Marc J. Roberts, and Stephen R. Thomas, *The Environmental Protection Agency: Asking the Wrong Questions* (New York: Oxford University Press, 1990), pp. 13–14.

7. Ronald H. Hinckley, *People, Polls, and Policymakers: American Public Opinion and National Security* (New York: Lexington Books, 1992), p. 139.

8. Robert W. Tucker and David C. Hendrickson, "Thomas Jefferson and American Foreign Policy," *Foreign Affairs* (v. 69, Spring, 1990), p. 139.

9. J. William Fulbright, *The Arrogance of Power* (New York: Random House, 1966), p. 21; see also pp. 257–58. Thomas Jefferson believed that: "A just and solid republican government maintained here, will be a standing monument and example for the aim and imitation of the people in other countries." See Merrill D. Peterson, ed. *Thomas Jefferson: Writings* (New York: The Library of America, 1984), pp. 1084–85.

Index

About the Contributors

DAVID A. DEESE is associate professor of policital science and director of the International Studies Program at Boston College. He is also senior research associate at Harvard University and senior associate at Cambridge Energy Research Associates. Over the past thirteen years he has also taught the foundation courses in American foreign policy and international politics at the Harvard University Extension School. Professor Deese co-edited *Energy and Security* and *Nuclear Non-proliferation.* He has published numerous articles and contributed chapters on international political economy and American foreign policy. He has served as expert adviser to several scholarly journals, nonprofit organizations, international organizations, foreign governments, corporations, and federal and state agencies in the United States, as well as the U.S. Congress.

W. LANCE BENNETT, professor of political science at the University of Washington, is the author of numerous books and articles on public opinion, the news media, and political communication. His recent works include *News: The Politics of Illusion* and *The Governing Crisis: Media, Money, and Marketing in American Elections* (St. Martin's Press, 1992). He is also a member of the Foreign Policy Studies Committee of the Social Science Research Council and was chair of the council's workshop series on "The Media and Foreign Policy."

I. M. (MAC) DESTLER is professor at the School of Public Affairs, University of Maryland; director of the Center for International and Security Studies at Maryland (CISSM); and visiting fellow, Institute for International Economics. His books on U.S. policymaking include *Presidents, Bureaucrats, and Foreign Policy* and *American Trade Politics.* He was a member of the Commission on Government Renewal, which presented recommendations to President Bill Clinton in 1992.

THOMAS W. GRAHAM is a senior program adviser for international security at the Rockefeller Foundation. Dr. Graham, a member of the Council on Foreign Relations and the International Institute for Strategic Studies, was a foreign affairs officer at the U.S. Arms Control and Disarmament Agency. His publications on public opinion include *American Public Opinion on NATO* and articles in *Public Opinion Quarterly* and *The Journal of Conflict Resolution.*

DONALD L. HAFNER is professor of political science at Boston College. During the Carter administration, he served in the U.S. Arms Control and Disarmament Agency as an adviser to the Strategic Arms Limitation Talks and the Anti-Satellite Arms Control initiative. Professor Hafner has written widely on arms control and American foreign policy issues in such journals as *Daedalus, International Security,* and *Survival.*

BENJAMIN I. PAGE is the Gordon Scott Fulcher Professor of Decision Making at Northwestern University. His books include *Choices and Echoes in Presidential Elections, Who Gets What from Government, The American Presidency* (with Mark Petracca), and most recently, *The Rational Public: Fifty Years of Trends in Americans' Policy Preferences* (with Robert Shapiro). Since 1973, Dr. Page has worked on the Chicago Council on Foreign Relations studies of American public opinion and U.S. foreign policy.

THOMAS RISSE-KAPPEN is professor of international politics at the University of Konstanz, Germany. Previously, he taught at the University of Wyoming, Yale University, and Cornell University. Dr. Risse-Kappen is the author of *Cooperation among Democracies: Alliance Norms, Transnational Relations and the European Influence on U.S. Foreign Policy* and various publications on public opinion and foreign policy in liberal democracies.

BERT A. ROCKMAN is University Professor of Political Science and research professor in the University Center for International Studies at the University of Pittsburgh. He is also a senior fellow at the Brookings Institution. Most recently, he has co-edited and contributed to *Do Institutions Matter? Government Capabilities in the United States and Abroad* and *Researching the Presidency: Vital Questions, New Approaches.*

DAVID W. ROHDE holds the Manning Dauer Eminent Scholar Chair in political science at the University of Florida. Previously he taught for over twenty years and served as department chair at Michigan State University. From 1988 to 1990, Dr. Rohde served as editor of the *American Journal of Political Science.* His most recent book is *Parties and Leaders in the Postreform House.*

WILLIAM SCHNEIDER has been the Speaker Thomas P. O'Neill, Jr., Visiting Professor of American Politics in the department of political science at Boston College since 1990. He is also a resident fellow at the American Enterprise Institute in Washington, D.C., and political analyst for Cable News Network. Dr. Schneider is co-author of *The Confidence Gap: Business, Labor and Government in the Public Mind* and a contributing editor to *The Los Angeles Times, National Journal,* and *The Atlantic,* where his articles on American politics and public opinion appear regularly.

ROBERT SCIGLIANO is professor of American Politics at Boston College. His books include *South Vietnam: Nation under Stress* and *The Supreme Court and the Presidency.* He has written essays on warmaking under the Constitution, including "The War Powers Resolution and the War Powers" (in Bessete & Tulis, eds., *The Presidency in the Constitutional Order*). He is now coauthoring a book on American liberal democracy.

ROBERT Y. SHAPIRO is associate professor of political science and international and public affairs at Columbia University. He is the author of several articles on public opinion, foreign policy, and American politics. His most recent book is *The Rational Public: Fifty Years of Trends in Americans' Policy Preferences* (with Benjamin I. Page). He is also coediting the forthcoming volume, *Research in Micropolitics: New Directions in Political Psychology.*

JOHN T. TIERNEY is associate professor of political science at Boston College. Coauthor of *Organized Interests and American Democracy,* he is also the author of two other books and many articles on varied aspects of American politics, including government organization and administration, health care politics, American foreign policy, and the political linkages between members of Congress and interest groups.